C0-ABO-847

American Technology: Are We Falling Behind?

Edited by Melinda Maidens

Facts On File
460 Park Ave. So., New York, N.Y. 10016

American Technology: Are We Falling Behind?

Published by Facts On File, Inc.
460 Park Ave. So., New York, N.Y. 10016

Library of Congress Cataloging in Publication Data

Maidens, Melinda.
American technology.

Includes index.
1. Technology—United States. I. Title.
T21.M34 1982 338.973 82-15603
ISBN 0-87196-677-8

International Standard Book Number: 0-87196-677-8
Library of Congress Catalog Card Number: 82-15603
9 8 7 6 5 4 3 2 1
PRINTED IN THE UNITED STATES OF AMERICA

Contents

Preface

The light bulb, the automobile, the airplane, the record player, the telephone, the camera, the elevator, the television—these items are only a fraction of the necessities of modern life that came from America. The same pioneering spirit that drove Americans westward produced our technological achievements. Inventiveness is considered the hallmark of our national character.

Today, that innovative spirit seems to have vanished. We look elsewhere for our cameras, cars, televisions, even some of our computer equipment. "America is backing away from technology," declared one participant at a 1978 National Academy of Sciences forum. Another speaker blamed the "failure of our political system to understand the really revolutionary, positive role of technology." Technology, he added, "is the one single factor that has been the major contributor to the advances in the American standard of living since the middle of the last century." Despite its contribution, technology has come under suspicion in recent years. This mistrust is expressed in part by government regulations, many of them directed to commendable ends such as controlling pollution. Nevertheless, the emphasis on regulation has had a negative cumulative effect on innovation. Each advance, whether in medicine, biological sciences or energy, is scrutinized for all possible side effects. The majority of technological advances are thus made to suffer for the sins of a few.

If there is a failure of American technology, it is a failure reflected in all aspects of American society: political, industrial and educational. It is significant that worry over our industrial and technological inferiority has emerged at the same time as concern over serious deficiencies in our educational system. A generation of children has completed high school and college with considerable problems in reading, writing and reasoning.

The editorials in this book address all of these problems, from the automobile crisis to the "space race." They were selected from newspapers in every state, and they illustrate many aspects of the technology problem at the national and local level. Although most editors agree that the problem exists, few agree on what to do about it. Their recommendations form the basis of a lively discussion.

The problem of American technology must be solved soon, for it has long-term implications. In the words of Sen. Harrison Schmitt, who explored the moon in 1972 with the last Apollo mission: "It is through technology that this country will survive. . . . If we don't have a technology base from which we can deal from strength with the rest of the world on whatever issue happens to come up internationally, then we won't survive."

August, 1982 — Melinda Maidens

Part I: Industry

How serious is the challenge of Japan? The question might seem simple-minded, considering that the American auto industry is in a shambles, American steel is limping and even American high-technology industries are under pressure from foreign competition. First glances are deceptive, however. In spite of Japan's trade surplus there are many areas in which it does not hold the lead. Technology comprises a wide variety of fields, from mechanics to biology, and the U.S. is ahead in quite a few. The pressure of foreign competition is increasing, though, and not only from Japan but also from Western Europe.

Japan's strength is in manufacturing, that is, turning out better-quality consumer products for less money, especially electronics: televisions and stereos, video and camera equipment. It began in 1955, when Sony introduced the world to the transistor radio. By 1978, Japan exported more technology than it imported, and its world trade surplus reached a staggering $25 billion. The ingredients of Japan's success have been analyzed endlessly: management-labor harmony, emphasis on long-term planning, high labor productivity, keen salesmanship and a high rate of industrial investment in research and development. Government helps by advising industry of the growth areas of the future and subsidizing research in those fields.

Manufacturing aside, Japan is turning more and more to inventing instead of merely making practical use of other countries' ideas. The Japanese have invested heavily in developing computers, and while the U.S. leads in the field, there is suspicion that Japan's determination to get ahead soon will bear fruit. Japan is also pursuing advances in pharmaceuticals and in the new fields of genetic engineering and producing biological compounds.

The challenge of Japan is significant less for understanding Japanese skills than for its effect on the American psyche. America has grown accustomed to leading the world in innovation. The shock of a serious challenge to that lead has forced America to examine some basic national assumptions. Do Americans respect competition only as long as they are the winners? Experts agree that American industry has only itself to blame for many of its problems. Industry, they say, is more interested in quarterly profits than in long-term developments; industry and labor have not yet learned to cooperate, and industry has become reluctant to take risks, an essential part of innovation. Of course, an ailing economy does not help these problems, and government regulations often get in the way. Industry is not entirely at fault, but industry has to find the way to save itself. It cannot count on government subsidies and trade barriers. America will restore its self-confidence not by "beating the Japanese" but by solving its own pressing problems.

Japan Moves Up to Number One in Total Car and Truck Output

By the first half of 1980, Japan had become the world's largest producer of cars and trucks. Preliminary statistics put Tokyo's total production at 5,460,000 vehicles for the first six months of the year, surpassing U.S. output during the same period by more than one million. The U.S. had invented the automobile and made it a household item, but it was no longer America's claim to fame.

America's automobile decline was not noticed at first. Imports were available but usually as luxury items: Rolls Royce, Mercedes and the like. Foreign cars began to be really popular when the cheap Volkswagen "bug" was introduced in the late 1950s. Still, imports were considered only a fad for the young or the rich. By the late 1970s, however, Americans were jolted to discover that imports had captured more than 25% of the domestic auto market. Politicians and the public, faced with the shutdowns of large factories at the very heart of the U.S. economy, demanded action against the import threat. It was a rude and late awakening.

Japanese automobiles were far and away the import leaders, outselling all other foreign makes combined. At first they were prized for their fuel efficiency. The oil embargoes of 1973 and 1978 forced drivers to keep more careful track of how they used fuel. As gasoline prices climbed to $1.00 a gallon and higher, the miles-per-gallon ratio became the first consideration in buying a car. The big, comfortable American car was stigmatized as a "gas-guzzler." Small cars achieved new respectability instead of being associated with young "swingers." The fuel shortage introduced Americans to Japanese cars, but skilled craftsmanship and reliability cemented the friendship. Detroit watched helplessly as American cars lost ground to foreign-made ones even when the price of gasoline stopped its dizzying climb.

Perhaps demographics has played a role. Big American cars were designed for big American families: parents, kids, neighbors' kids and the family pet. The 1980 census reported a significant increase in singles, in small households and in single-parent families. Who needed the station wagon?

THE DAILY OKLAHOMAN

Oklahoma City, Okla., December 26, 1980

WHAT would have been unthinkable a few years ago has now occurred: Japan has displaced the United States as the world's largest producer of automobiles.

Japanese production will top 10 million in 1980, compared to about 8 million by U.S. automakers — a lead that had been predicted because of domestic plant change-overs to smaller cars.

Losing the top ranking they've held since the dawn of the automotive age is symptomatic of the troubles confronting U.S. car producers. And it won't take long for the rippling effect of Detroit's crisis to show up in rubber, steel and plastics, of which the auto industry is the largest consumer.

Japanese takeover of the top spot, plus interest rates that have seriously depressed auto sales, cast further doubt on the survival of ailing Chrysler, which has already used $800 million of a $1.5 billion federal loan guarantee in a desperate bid to stave off bankruptcy.

Whether Chrysler can make it depends on many factors, the most immediate being whether the United Auto Workers can be persuaded to accept a wage freeze through 1982 aimed at saving $600 million. Without it, Chrysler chairman Lee Iacocca says the company can't draw on the remaining federal credit and will go belly up.

The only other possible option is merger with one of the Japanese or other foreign companies, such as the acquisition of controlling interest in American Motors by France's Renault.

Even with the freeze and another dollop of federal assistance, there is no assurance Chrysler can hack it. And it certainly can't unless there is a relatively rapid turn-around in the economy and a reduction of interest rates that will stimulate new-car sales.

Anybody who still doesn't believe U.S. basic industries are in a heap of trouble from lack of ability to compete in the world market has only to look at what's already happening to U.S. automakers.

The Seattle Times

Seattle, Wash., December 28, 1980

A DECADE ago, people would have scoffed at the thought. In the past few years, it was recognized as possible; more recently, inevitable. Last week it became official. Japan has replaced the U.S. as the world's No. 1 automobile producer.

Could what has happened in the auto industry be duplicated some day in the one transportation-manufacturing industry where Uncle Sam remains supreme?

In the manufacture of commercial aircraft, this country today enjoys the dominance once held by U.S. carmakers. Anyone inclined to be smug about that should give thought not only to what has happened to the auto industry, but to American industrial productivity in general.

Over a 20-year span ending in 1979, Japan had an average annual productivity growth of more than 7 per cent. West Germany, France and Italy averaged 4 to 4½ per cent. The United Kingdom had 2.3 per cent, Canada slightly less than 2 per cent. Among the major industrialized nations, the U.S. was dead last with 1.5 per cent.

In his annual year-end report, Karl G. Harr, Jr., president of the Aerospace Industries Association, had some pertinent thoughts on that depressing picture. Harr said:

"Clearly, we must reverse this trend or suffer continuing high inflation, further reduced international competitiveness, and abdication of our status as world industrial leader. There are many measures we could take to effect a productivity upturn, but paramount among them are these:

"We must substantially increase the level of federal research-and-development funding to spark innovation.

"We must provide incentives to stimulate private investment in R & D, to encourage capital investment for replacing obsolescent facilities with new, productivity-improving equipment."

In his campaign statements and cabinet choices, Ronald Reagan has shown understanding of the urgent need to restore U.S. industrial supremacy. Therein lies the strongest hope for the kind of changes Harr talks about.

The Star-Ledger

Newark, N. J., December 26, 1980

Adding insult to injury, Japan has outstripped the United States for the first time in yearly auto production in 1980. The Japan Automobile Manufacturers Association estimated that country's production at 11 million passenger vehicles, substantially ahead of the best estimates for the U.S. industry.

But the organization tactfully suggested that the United States might regain the lead once it fully adjusts to consumer demands for small, fuel-efficient cars. Hopefully, this possibility will become a reality.

BUFFALO EVENING NEWS

Buffalo, N. Y., December 27, 1980

It comes as a distinct shock to both the U.S. auto industry and the nation as a whole to hear that the United States, for the first time in history, is no longer the world's largest automaker.

With U.S. vehicle output down 33 percent so far this year and Japanese output up 15 percent, Japan is expected to far outrank the U.S. this year with a total output of 11 million to 8 million cars. While a measure of the degree to which the U.S. auto industry is faltering, this does not mean that the Japanese predominance will be permanent. A Japanese Datsun official correctly attributed the change to the sudden demand for small cars, which the U.S. automakers are only now beginning to make in a big way. The other major factor is that many people have put off buying cars in this year of recession, inflation and high interest rates.

These unusual economic conditions have been cited by Chrysler as one of the reasons for its failure, so far, to recover from the brink of bankruptcy. It has been a bad year to introduce its new K-car, the model on which Chrysler had banked to get back into the black.

* * *

Another impediment to sales of autos generally has been soaring prices. Prospective auto buyers are suffering from what dealers call "showroom shock." Many haven't been in the auto market recently, and they are shocked to discover that auto prices are several thousands of dollars higher than when they last visited auto showrooms. The average price of a typically equipped auto sold in the U.S. at the end of the 1980 model year was $7,795. For domestic autos only, it was even worse — nearly $9,000. And on the 1981 models prices have been raised by several hundred dollars more.

During any kind of economic uncertainty, buyers naturally tend to favor less-costly models. This has helped the foreign imports, especially since, with the high cost of borrowing, the extra $1,000 or more for domestic autos raises the overall financing cost to a prohibitive level for many.

The outlook is not entirely gloomy for the U.S. auto industry, however. Japanese auto output now shows signs of leveling off. The November output showed a slight decline for the first time in 20 months, and Japanese officials expect the 1981 output to show little change from this year. And with lower interest rates expected in the U.S. and with the Reagan administration expected to introduce new tax breaks for industry, the prospect is for an improved business climate that will give the refurbished U.S. auto industry the springboard it needs for recovery.

* * *

The alarming thing about Japanese competition is that they are beating us in a field in which we are supposed to excel — the use of technology. There is a misconception by many people that the somewhat lower Japanese wage scales are to blame for the flood of imports. That is a factor, but a more significant one is Japanese use of technology in modern plants. The average Japanese steelworker, for example, produces nearly twice as much steel as an American steelworker.

The U.S. auto plants now being rebuilt and retooled should be among the most modern and efficient in the world. In addition, there is a growing realization of the need for labor and management to work together in improving productivity and, in the case of a crisis such as Chrysler's, to seek to moderate wage costs, which are a major component of the auto industry's cost structure.

The auto industry is not out of the woods yet, but it has good reasons for hope as we enter the new year and, hopefully, a period of rising business activity.

ARGUS-LEADER

Sioux Falls, S. D., December 30, 1980

There's no joy in being the world's No. 2 automaker.

Japan is No. 1.

This is the first time the United States has not been the world's largest manufacturer of autos since it surpassed the French auto industry at the turn of the century.

Production of 7,824,728 cars and trucks in this country in 1980 is well under the 11 million built in Japan.

We hope that the New Year will see a turnaround for all American industry, and particularly in autos.

The lost production for U.S. manufacturers has meant fewer jobs and higher unemployment for factory workers. It brought record losses for the U.S. Big Three manufacturers and less business for dealers along auto row in American cities and towns.

The industry that gave the world Henry Ford, who put the common man on wheels and with his contemporaries in Detroit changed the course of history, should not be counted out.

The sophistication and problems of the times call for further renewal of the U.S. industrial plant. There should be new efforts to regain productivity and employ the mechanical savvy and engineering knowhow that kept the U.S. No. 1 in autos — until now.

Americans will be relying on personal transportation far into the future. That challenge should spur new determination in Detroit corporate board rooms to meet and beat the competition from overseas.

THE ATLANTA CONSTITUTION

Atlanta, Ga., January 7, 1981

The auto industry is one of the heart industries of the U.S. economy. Let's consider the good news and bad news about it.

The year 1980 was a disastrous one for the industry. Sales for the year — the final figures will be out today — will total about 6.6 million units, the worst performance since 1961 when 5.6 million cars were sold. To make the 6.6 million figure look even more gloomy is the fact that sales in recent weeks have been running at an annualized rate of about 6.1 million sales, which could bode ill for the first months of 1981.

Other statistics also show what a bad year 1980 was for the U.S. automakers. Financially, the year was the worst in the history of the industry with losses of $3.7 billion by the four domestic companies through September; final financial reports are still weeks away. Indefinite layoffs of almost 200,000 autoworkers are holding steady, perhaps even rising. This week, 13 of the 40 domestic car assembly plants are idle because of excess auto inventories. And Chrysler Corp., which is not too far this side of financial collapse, is asking the federal government, its workers, its bankers and its stockholders for additional help to stay afloat.

That's enough bad news. There is better news, little though it may be.

Reports released earlier this week showed that sales of Chrysler cars in the Dec. 11-20 period were up 19.8 percent from the same period a year ago. Chrysler was the only U.S. automaker with increased sales for the period, thanks, apparently, to its program of giving discounts tied to the rise and fall of the prime rate. For the year, Chrysler sales are off about 28 percent, but the recent rise is at least a hopeful sign for the struggling company.

For the industry as a whole, the U.S. Commerce Department is forecasting that sales of U.S.-made cars in the nation will rise about 9 percent to 7.2 million units this year and will rise to about 9.6 million in 1985, a level close to previous years' record sales. Also encouraging for the U.S. producers is the prediction that they will regain a considerable portion of sales lost in the past two-three years to foreign automakers.

For the short term, the most encouraging news is that interest rates are beginning to fall. The actuality of lower rates — along with the psychological effects of them — should result in increased auto purchases by consumers, increased "stocking up" by auto dealers and, if the trend holds long enough, the reopening of closed auto plants and the recalling of autoworkers. For sure, let's hope so.

THE BLADE

Toledo, Ohio, February 8, 1981

THE fact that the American auto industry faces hard times well into the 1980s has become increasingly clear in recent months. So, in view of that depressing outlook, a report issued by the federal Department of Transportation just as the Reagan administration took over the reins of government is especially timely.

The report, which makes more sense than most government studies, sets forth some ideas that have been conspicuously missing in government relations with car manufacturers. They deal, to put it bluntly, with the survival of the auto industry in this country.

Among the recommendations is that an import-restraint agreement be negotiated with Japan. This is being considered by the Reagan administration, but the study recognizes that restraints on imports are only part of the problem and that if any such agreement is negotiated, it would have to be partial and temporary. It would last only until U.S. manufacturers could gear up with an investment of perhaps as much as $70 billion over the next five years to produce more of the kinds of cars needed at a time of rising fuel costs.

The Transportation Department analysts left no doubt as to the magnitude of the job ahead. What will be required, they point out, is nothing less than "a new socio-economic compact between industry, labor, and government to retool industry, to increase productivity, to recommit to quality, to retrain and re-employ workers . . ." There are many facets involved for all three legs of this triad — government, industry, and labor — including tax and regulatory reform, restraint in wage demands, support for productivity growth and product quality, and intensified research for more fuel-efficient vehicles. If even one leg breaks, the recovery stool will inevitably collapse.

The report urges this country to emulate in some aspects the way the Japanese have built up their modern plants with high technology and a reputation for quality control. The two situations, of course, are not parallel in many respects, but certainly U.S. manufacturers could adapt some of these factors in their production lines.

The study is timely in one other respect — the fact that it was released just as the Reagan administration took office. Most of the recommendations seem to be in line with the approach already favored by the new Administration — tax and regulatory reform, reduction of arbitary federal requirements, and an effort by government to cooperate with rather than fight against the manufacturers.

The proposals therefore stand an excellent chance of being implemented. That could be good news not only for the industry but also for the national economy.

A House committee report recently suggested that Japanese industrial expertise is placing this country in the position of a developing nation supplying a more advanced country, that we might become haulers of wood and growers of crops in exchange for high-technology products from abroad. There probably is not much danger that this will happen, but the trend certainly will be in that direction if the Japanese challenge to U.S. industrial might is not met.

The Dallas Morning News

Dallas, Texas, January 2, 1981

IT MAY be only small comfort for the big American car manufacturers, but the runaway success of imported cars seems to be braking a bit.

The *Wall Street Journal* reports that the problem of picky, penny-pinching customers, plaguing U.S. car firms for months, is now beginning to show up in the import showrooms as well.

Import dealers who say they once were able to sell merely by taking orders and putting customers on waiting lists now tell the *Journal* that they have to get out and hustle hard to make sales. The imports, the *Journal* concludes, are not invincible after all.

With the credit crunch, both sides of the car industry are feeling the pinch. The American car makers and dealers certainly know the ropes in price competition and should be able to take on all comers there. It is in the quality race, however, that the foreign cars have really been making gains, at least in the eyes of customers.

Allen Pusey, in his recent, splendid series on the American auto industry, reported in *The Dallas Morning News* that the industry is trying hard to match the imports' image of high quality. Not only are inspectors and supervisors working extra hard to catch "dings," but they are cross-checking each other's work.

However, one quote may symbolize part of the problem. One supervisor, detailed to check a part close by the rear window, pointed out the part. But the window was smeared with a smudge of gunk. "That must be some kind of glue they used," the supervisor said. "That's not my problem, though."

Contrast this with the Japanese plants where, according to Commerce Secretary Philip Klutznick, every work station on the production line has a red button. Klutznick told a group of *News* editors a few months ago that every worker has the responsibility of pushing the red button and stopping the line if he sees any defect on the car before him. The Japanese philosophy, apparently, is that *any* defect on a car produced by a plant is *everybody's* problem. That philosophy pays off in the marketplace.

The manager of a large American car division told a newsman that "You have to be impressed" with Japanese cars when parts like rubber door seals are "in the right place, and the glue is under the rubber where it belongs."

American industry can match the technology and engineering that go into the imports, and we are encouraged by the progress in doing so. But getting that impressive ability to get "the glue . . . under the rubber where it belongs" may be somewhat tougher.

First and foremost, it will require that every American auto worker, from the chairman of the board to the newest production line trainee, must recognize the truth that the Japanese apparently already know: Quality of the product is everybody's responsibility, individually as well as collectively. And when anybody slips, everybody is likely to suffer.

The auto marketplace is now international. When a plant closes in Detroit because its products did not win the customers' favor, the jobs therein are likely to move not just across the city, but across the ocean.

THE PLAIN DEALER

Cleveland, Ohio, February 1, 1981

Japan has punctured the myth of the American auto industry's invulnerability. Last year it produced 11,042,884 motor vehicles, compared with 8,011,740 in the United States. Japan has replaced America as the world's largest producer for the first time in this century.

Japan has also largely discredited the economic theory that America's auto industry is so big and strong that it could forever profitably control the law of supply and demand — that even in bad times this country could still be the world leader in producing autos.

Partly, Japan did it by producing a better automobile, or at least one that has been considered a better automobile by many engineers and the people who drive and test autos for a living. Last year, for instance, in a survey by Ward's Auto World, a respected trade publication, a majority of 250 American automotive engineers said that Japan's cars were better engineered and of higher quality than American cars.

To the thousands of laid-off American auto workers, including more than 11,000 in the Cleveland area, the emergence of Japan is certainly disheartening. The changing shape of not only the American car but the entire auto industry is sure to mean that thousands of auto production jobs will be lost permanently.

It is difficult to fix specific blame for the sorry state of this basic American industry. Industry management, labor and the federal government share the blame, and there may still be time for them to cooperate to revitalize the auto industry and restore its declining competitive position in the world.

Neil Goldschmidt, former Department of Transportation secretary, has left behind an excellent and thoughtful report on the auto industry's plight. President Reagan and the 97th Congress should give it serious consideration in devising cures for the ailing industry.

Goldschmidt urges the federal government to negotiate a voluntary restraint with Japan on its auto imports until the U.S. auto industry retools and begins producing a new generation of smaller, fuel-efficient cars that are better engineered and better manufactured to compete with the Japanese autos.

Goldschmidt proposes that labor soften wage demands until the U.S. industry regains its lost competitive advantage. This is an essential sacrifice. While labor makes concessions, management needs to launch a major effort to boost productivity and improve its products. Auto industry management could balance wage concessions by guaranteeing profit sharing or other forms of compensation for labor.

The auto industry needs at least $70 billion in capital over the next five years to pay for the production machinery to satisfy market demand.

The federal government should fashion a program of accelerated depreciation allowances on new plants and equipment and other tax credits to aid capital investments. The industry also has complained loudly and bitterly over costly government regulations. Its position is overstated. But there is a need to ensure that regulations yield benefits worth the costs.

Most important is an atmosphere of cooperation among the federal government, industry management and labor. Japan over the years has cashed in on such relationships that are much less adversarial than in the United States. Japan's dividends of higher productivity and higher employment are at least partly because of government regulations that aid, rather than hinder, its auto industry. Its 1980 production record is the enviable result.

The Washington Star

and Daily News

Washington, D. C., January 16, 1981

Studies prepared by outgoing administrations tend to be superseded by the task force reports of incoming ones. But a report just released by Transportation Secretary Neil Goldschmidt on the survival of the U.S. automobile industry merits attention for its sensible advice.

Mr. Goldschmidt calls the preservation of the auto industry an issue of national security. If that seems exaggerated, consider that more than 14 million people — the holders of roughly one job in six — depend upon it. The industry uses 21 per cent of the nation's steel output, 60 per cent of its synthetic rubber, and spends about $40 billion for equipment and material each year. In short, when the economy slumps, the auto industry reels — and vice versa. So Mr. Goldschmidt states what has been obvious for some time and needs restating: that ways have to be found to counter burdensome federal regulations, escalating wage demands and intense competition from abroad.

None of the problems ought to be considered separately. The price advantage many Japanese cars hold reflects that country's higher productivity, lower wages, regulations which favor the industry — and corresponding U.S. weaknesses. The U.S. response, logically, would be to do more as the Japanese do — and Mr. Goldschmidt says just that. But that takes time, money and, in some cases, new laws.

With this in mind, Mr. Goldschmidt also proposes negotiating temporary restraints on Japanese imports, something the Carter administation has opposed. That is a delicate way of urging protectionism, but is it needed?

The Japanese certainly ought to practice some self-restraint, recognizing that free trade works both ways. They sell 1.8 million cars in the U.S. market, but virtually no U.S. cars are sold in Japan. Japan's aggressive and often one-sided marketing practices have already closed trade doors in Europe. But while Mr. Goldschmidt argues for "temporary" restraints, the imposition of import quotas could have harmful long-term effects. Restrictions would undoubtedly raise the costs of all automobiles, domestic and imported, and heighten inflation; they would increase energy consumption, by diminishing a source of fuel-efficient cars while the U.S. catches up; and U.S. car makers might feel less urgency about the future.

Such urgency ought, however, to hasten consideration of some of Secretary Goldschmidt's other suggestions, including changes in the tax code (which should apply to other U.S. industries) to alter depreciation schedules and credits on investments. More daring is the proposal that labor agree to a wage strategy designed to close the differential with Japan. The Transportation Department's analysis indicates that for every $1-per-hour increase deferred, the industry saves about $1 billion. Mr. Goldschmidt suggests that management offer some form of profit-sharing in return.

More immediately, he suggests replacing what he calls "the current adversarial regulatory process" with a negotiating system. At the very least, revision of a number of current regulations is called for. So is delay in implementing future ones. Roger B. Smith, the new chairman of General Motors, said that the equivalent of 26,000 GM employees spent last year complying with regulatory requirements and paperwork. Are things really so out of hand?

U.S. automakers are currently spending billions to improve designs and plant efficiency, in what has been called the greatest peacetime investment in U.S. history. Mr. Smith puts it this way: "We don't want the last word in [auto] plants. We want the next word after the last word." Such an approach, of course, is the most effective way to restore vitality to U.S. auto industry. Ultimately it is the only way.

The Boston Herald American

Boston, Mass., February 15, 1981

President Reagan, reflecting an altogether fitting sense of urgency, wasted no time in appointing a cabinet task force to analyze the ailing auto industry and recommend what might be done to resuscitate the Big Three.

Help for an industry upon which one American job in seven depends cannot come too soon. As noted, Chrysler is in deep trouble. Ford's financial condition is deteriorating rapidly and last year the number two automaker lost about $1.5 billion.

Even General Motors, long considered well nigh impregnable, registered a net loss of $763 million last year — its first red ink since 1921.

So, the task force will have no trouble getting Mr. Reagan's, or the nation's, attention when it releases its diagnosis and proposed cure sometime next month.

Valuable as the task force's work may be, it's apparent that President Reagan and at least some of his chief advisers already understand how and why the industry took sick and what it will take to restore it to health.

The President cited some of the contributing ills in his economic address to the nation: A critical shortage of investment capital needed for plant modernization and retooling, record high interest rates for both the industry and would-be car buyers, and stifling government regulations.

To the above, Mr. Reagan might well have added government controls that held gasoline prices at artificially low levels and thus subsidized the American public's demand for big, gas-guzzling autos during the mid and late 1970s when Detroit should have been switching to the production of more fuel-efficient models.

As it is, the industry must somehow generate roughly $70 billion by 1985 to finance the retooling essential to building whole new model lines of smaller automobiles. Moreover, it must raise this staggering sum at a time when the cost of borrowing is almost prohibitive and profits are either marginal or nonexistent.

Obviously, much depends on the Reagan administration's success in healing an economy that is nearly as sick as the auto industry. Curbing government spending would free capital for private investment, much of it in the auto industry. Reduced inflation and lower interest rates would spur new car sales, currently at a 20-year low.

But all of this will take time, and Detroit needs help now. Thus, the task force and the President will want to find ways to provide some immediate relief. Liberalized tax depreciation schedules for plant and equipment would help. So would an end to excessive government regulations that hinder productivity and add hundreds of dollars to the price of each new car.

Beyond these immediate measures, and the hoped-for general improvement in the nation's economy, the Reagan administration ought to consider a longer-range plan to disentangle the auto industry from the federal agencies that have nearly ruined it during the last decade.

There being no future in either protectionist barriers against imports or in government subsidies, a healthy American automobile industry must be one that can compete here and abroad. Regaining this competitive edge won't be possible until Washington gets out of the business of dictating design and marketing decisions that ought to be made in Detroit.

The Saginaw News

Saginaw, Mich., January 25, 1981

Gov. Milliken and six other state leaders say they will push an eight-part plan to revitalize the nation's auto industry.

At least two other parts are needed to complete a solid structure for recovery.

Our industry has to improve customers' perception of the quality of its products. And our government should offset the impact of high interest rates by giving buyers of American cars a specific tax break.

The elements of the governors' program are good, if somewhat obvious. For instance, who could quarrel with the call for a "healthy economy" and lower interest rates?

Some others are nebulous. The governors would prefer Japanese self-restraint in auto exports rather than U.S. controls. But that's up to the Japanese, and will probably work only if we show we're serious about controls as an alternative.

The rest of the program is more specific, and worthy of the joint efforts of the auto-producing states of Michigan, Ohio, Missouri, Wisconsin, Indiana, Pennsylvania and Illinois.

It seeks investment tax credits for innovation; quicker depreciation to encourage replacement of old equipment; cost-effectiveness analysis of regulations; and removal of anti-trust barriers to joint industry research.

Much of this involves industry's government burden. But the would-be car-buyer is saddled with the same burden, in the form of huge interest rates that discourage borrowing to pay for cars. What may be good for the economy as a whole is disastrous for Michigan in particular.

We don't expect the fight against inflation to yield to our state's parochial needs. But we do not expect to be forgotten entirely.

A wide range of tax policies serves the national interest by stimulating business activity. It is time for the auto industry *and* its sidelined customers to get the same shake.

If Congress needs a reason besides the plight of the nation's industrial heartland, it might consider that *nationally*, one of every six jobs is related to the auto industry. What happens to "Detroit" also happens to steel, rubber, aluminum, glass, machine tools, plastics and electronics.

We believe Congress should quickly approve a special income-tax credit for buyers of American-made cars. The credit should apply not just to smaller cars, as Gov. Milliken has proposed, but to all models. But its amount should increase depending on the fuel-efficiency rating of the purchase.

This policy would serve the national interest in both preserving a vital, basic domestic industry, and encouraging fuel conservation. Milliken and the other governors must make clear to President Reagan and Congress the urgency of the problem — and the need for this partial solution.

Neither states nor Congress nor White House, though, can do the auto industry's work for it. And a confidential General Motors report says that in the past year, especially, buyers rate imports higher in quality than domestically-made cars.

Most disturbing is the finding that this is more than just popular perception, or misconception; the GM study said American cars in many aspects of workmanship really don't measure up. There are just too many squeaks, rattles and misfitting doors.

We think our auto industry's management knows how to achieve superior quality control. We believe our auto workers are skilled enough to build better cars.

We don't expect American shift workers to engage in mass rah-rah exercises, Japanese-style. But management and labor need to review their goals and re-dedicate themselves to doing a better job that results in a better product.

Otherwise, the next noise you hear will be the death rattle of an international misfit industry. Lemons leave a bitter taste.

THE ARIZONA REPUBLIC

Phoenix, Ariz., February 22, 1981

FORD had a fourth-quarter loss last year of $316 million, bringing its total loss in 1980 to $1.54 billion.

This is the biggest loss of any U.S. corporation in any one year, but it's a record that will not last for long.

Chrysler will shatter it when it announces its fourth-quarter loss this week.

The company is believed to have lost $300 million during the fourth quarter, which would bring its loss for the year to $1.8 billion.

Despite a fourth-quarter profit of $62 million, General Motors had an over-all loss for 1980 of $763 million.

The Big Three automakers went $4.1 billion into the red last year.

Two questions arise:

✔ How long can the Big Three, once the pride of U.S. industry, continue to sustain such losses?

✔ And why are they doing so poorly?

To the first question, the answer is that Chrysler has just about reached the end of its rope. There are no further economies the company can make. Nor can it look for any more help from the government or its workers and suppliers.

Ford has not yet found it necessary to seek help from the government. It still is able to borrow money from the banks.

However, if Ford's fortunes do not turn, banks are bound to say, "Enough is enough."

To the second question there are innumerable answers.

Wages in the U.S. auto industry are so high, it's become extremely difficult for the Big Three to compete with foreign manufacturers, particularly with the Japanese.

Government safety, emissions and gasoline consumption standards put another inordinate burden on Detroit.

Stratospheric interest rates are giving potential car buyers pause.

Detroit failed to foresee the day when unleaded gasoline would be selling for $1.30 a gallon or more. It, therefore, moved into the small car market too late.

The Japanese also have developed a reputation for quality.

The Japanese are cutting into the European market just as they did into the U.S. market.

Once Ford was able to count on huge profits from Britain, West Germany and other European markets to finance its spending plans at home.

Ford is now losing money overseas.

Thus, everyone must share the blame for the sad state of the U.S. auto industry — management, government and labor.

WORCESTER TELEGRAM

Worcester, Mass., December 4, 1981

The report of the U.S. Commerce Department presented on the prospects of the beleagered American auto industry is grim.

It says the financial strength of the industry has been sapped. It suggests there is now a serious question whether Detroit, hit by weak sales and aggressive marketing from Japan, can restore its long-run competitiveness.

The report singles out as "most startling" the decline of the working capital of General Motors, Ford, Chrysler and American Motors from $13 billion at the end of 1978 to $300 million as of last Oct. 1. Working capital is money available for investment. Analysts say the industry must invest tens of billions more in modern factory equipment in order to catch up with the Japanese.

The report concludes that a revived economy is essential to help "remedy the industry's condition." But it suggests that even a boom in new-car sales would be only a first step toward solving Detroit's problems. It must boost productivity significantly in order to bring its fuel-efficient new models into direct price competition with comparable Japanese cars.

Robots and other efficient new equipment are not the only factors in solving the productivity problem. The industry also needs new methods. Some of that Japanese attitude of labor, management and government pulling together instead of against one another might help.

The companies also need to scale back the high wages and benefits that have made auto workers the envy of the rest of organized labor. That is a bitter prospect for the United Auto Workers, and it is disillusioning to the labor movement as a whole. The auto workers fought hard over the years to gain those high wages and benefits. But without some relief, the auto industry is threatened with drastic shrinkage and loss of jobs.

The task of restoring Detroit's competitiveness is difficult and chancy. So it is not surprising that the union and, at times, the companies have urged Washington to impose import duties to boost the price of foreign cars to that of comparable American models. The argument has been made that the American auto industry must be saved for national defense purposes — and that the alternative could be massive unemployment.

But the Carter and Reagan administrations have, in the main, resisted these urgings. High import duties could well start a trade war. They would allow Detroit to remain inefficient, raising questions of how well it would compete in war equipment if it can't compete in cars.

Moreover, higher prices for foreign cars would be inflationary and would be unfair to consumers, forcing them to subsidize Detroit's ills. And those swarms of little foreign cars have not been all bad. They have enabled the nation to cut its use of imported oil way down in recent years, helping the balance of payments on one hand even while hurting it on the other. Even the agreement forced on Japan earlier this year to cut auto exports to the United States by 12.1 percent — a Reagan administration sop to Detroit — was marginally anti-consumer and anti-free market.

The solution to Detroit's dilemma is the long, hard one in which the industry is already engaged — more automation, better work attitudes and increased savings in overhead and labor costs. Washington can best help by reducing regulation and encouraging savings and investment.

The Commerce Department's stern assessment of the American auto industry's plight may be too pessimistic. After all, the Detroit companies have been moving in the right direction. They now offer a wide array of good small cars, and they are working at ways to reduce prices in relation to those of imports.

Interest rates are now, finally, coming down. That should enable more people to afford to finance a new car. Many forecasters now look for an upturn in the economy next spring. That and pent-up demand could well lead to an upsurge in auto sales.

But there's no open road just yet for Detroit — just more ruts and potholes for miles and miles.

The Des Moines Register

Des Moines, Iowa, December 12, 1981

These are hard times for the U.S. auto industry. Car sales have been severely depressed because of high interest rates, "sticker shock" (prices), the recession and the competition of imported cars. Sales of U.S.-made cars are off one-third from their 1978 peak, while imported-car sales have risen 15 percent.

U.S. automakers are expected to lose a total of $5 billion in 1980 and 1981. The Commerce Department reported last week that the industry's working capital — money for investment — has fallen to $300 million from $13 billion three years ago. In mid-November, 179,000 auto workers were on indefinite layoff. Since 1978, total employment in the auto industry has dropped by more than 250,000.

Perhaps the most damaging consequence of these troubles is that auto companies have been forced to cut back or delay elements of their essential program to build more modern factories and better cars in order to meet the Japanese challenge. And what a challenge it is! Recent studies have concluded that the Japanese car industry can deliver a car to an American city for as much as $1,500 less than it costs the U.S. industry to get a comparable car to the same place. Other surveys show that a growing number of Americans feel that foreign-made cars are better than American cars.

One apparent solution to these problems would be to restrict severely the sales of imported cars. That would only force U.S. consumers to pay even more for new cars, would weaken the long-term competitiveness of the U.S. auto industry and perhaps would set off a trade war. A second possibility would be a major infusion of government aid, but that is unrealistic in the present Washington climate.

The government could help by changing the present depressing economic policies to provide the kind of healthy economy needed to stimulate car sales. Congress could alter Reagan's economic program to achieve lower interest rates, lower unemployment and a higher rate of growth. Government can help, too, in retraining the several hundred thousand laid-off auto workers who probably never can return to their old jobs.

The auto industry must take painful steps to cope with its biggest single problem: the fact that Japanese labor costs are much lower than Detroit's. Not only are Japanese auto workers paid less, they are more productive.

U.S. auto-labor contracts expire next year. The pay and productivity problems must be addressed in the contract negotiations. Therefore, it is encouraging that United Auto Workers President Douglas Fraser has authorized local bargaining councils to renegotiate contracts, saying "a rigid policy doesn't make sense anymore."

Japan Urged to Accept "Voluntary" Export Restraints

Reagan Administration officials held three days of talks March 23-26 with Japanese Foreign Minister Masayoshi Ito on the problems of the U.S. auto industry and the future of Japanese car imports. President Reagan was hoping to persuade the Japanese to voluntarily limit shipments of cars to the U.S. instead of facing U.S. import quotas. In talks with Secretary of State Alexander Haig, Defense Secretary Caspar Weinberger and special trade representative William Brock, Ito said his government strongly supported the "principle of free trade," and he warned that U.S. import quotas would have serious repercussions on the world economy. The U.S. had been warned also by West Germany that restrictions on Japanese cars could trigger similar restrictions on Japan by the European Community and lead to wholesale protectionist measures for other industries. Another obstacle to the Administration's goal was presented March 17 by Attorney General William French Smith, who warned that voluntary Japanese export restrictions could violate U.S. antitrust law.

A minor conflict arose over who would lead the auto talks with Japan. According to reports, Haig had assumed control of the talks, but Brock declared March 26 that President Reagan had given him "explicit assurances" that "any consultations, discussions, conversations or talks on autos or any other trade question will be under" Brock's authority.

Detroit Free Press

Detroit, Mich., March 27, 1981

THE JAPANESE are terrible teases and brilliant negotiators where trade is concerned, and the Reagan administration will be lucky indeed if the voluntary agreement it seeks as a way to hold down automobile imports is forthcoming.

Japan, however, has a moment of rare opportunity to defuse the trade issue in the United States and to avoid the dangers of legislated quotas. It should not take the Reagan administration's reluctance to yield to protectionist pressures as a sign of weakness. On this, the Reagan administration is trying to hold to its free-trading principles, as did the Carter administration before it, in spite of the grief that such a stand causes in parts of the American auto industry.

If the Japanese cannot be encouraged to reach a voluntary arrangement to limit the volume of imports, the effort to protect the American market may well get a second wind. Much of the country is unsympathetic to the idea of restricting automobile imports, both because there is still a considerable constituency for relatively liberal trade policies and because many consumers want to have the Japanese option open to them when they go car-shopping. But the distress within the auto industry is great enough that many defenders of the principle of a liberal trade policy, including a faction within the Reagan cabinet, have been hard-pressed to hold to their position.

The danger that the U.S. would resort to protectionism is real, despite the reluctance that President Reagan has shown. When unemployment is as high and persistent as it has been in auto manufacturing and related industries, the pressure for some kind of governmental response builds up.

Following his meeting with President Reagan on Tuesday, Japan's foreign minister, Masayoshi Ito, gave at least lip service to the idea that Japan should try to work some kind of voluntary agreement to hold down the volume of cars coming into the U.S. He expressed the hope that something could be worked out before the visit of the Japanese prime minister to Washington in early May.

The United States, because of its huge role in international trade, has a special responsibility to try to avoid resorting to protectionism itself and adding to the risk of an international trade war. But the Japanese have a responsibility, too, not to take unfair advantage of the American commitment to liberal trade policy — a responsibility they have not thus far faced.

Two very different American administrations have withstood a lot of heat to try to uphold the American side of that commitment. If the Japanese do not show a new conviction that trade should be fair as well as free, though, they could very well push the Reagan administration toward the legislated quotas it abhors. The next 30 to 45 days could very well determine whether those in the U.S. who have resisted protectionism will have solid ground left to stand on.

The Cincinnati Post

Cincinnati, Ohio, March 16, 1981

President Reagan is under heavy pressure from the auto companies, the UAW, governors of industrial states and even some members of his Cabinet to clamp restraints on imports of Japanese cars.

In the long-term interest of the U.S. economy and of the automakers themselves, Reagan should reject pleas for protection. Inevitably, a protected industry becomes an uncompetitive one.

The companies and their political allies want a "voluntary" agreement (meaning one forced on Tokyo) to cut Japanese car sales here "temporarily." And temporary may mean forever, since protectionism is a habit hard to kick.

If quotas are put on Japanese auto imports, a number of things will happen, all bad. First, the popular, fuel-efficient imports will see more demand than supply, so their prices will rise.

This will make it easy for Detroit to scrap pricing restraint and to follow the Japanese up. The result will be another burst of inflation, but of the most unfair kind because it will be at the expense of those least able to afford it, the beleaguered middle-income American taxpayer.

What fairness is there in forcing a $5-an-hour textile worker to pay more for a car to protect the job of a $19-an-hour auto worker and of a $900,000-a-year company president?

Protection also will permit the companies to put off steps they must take to survive. Their labor costs are out of line, almost twice as high as in Japan. They must resist cost increases from their suppliers and the UAW if they are to regain competitiveness.

All this is not to argue that the government should do nothing for automakers in their time of travail. They should be aided by relaxing regulation, reading the antitrust law less dogmatically and granting tax breaks for retooling.

Before doing anything, the president ought to telephone his friend, Prime Minister Margaret Thatcher, for advice. The British have protected and coddled their steel and auto industries for years. Both are high-cost, uncompetitive, failing enterprises that are pulling Britain down with them.

THE PLAIN DEALER

Cleveland, Ohio, March 18, 1981

President Reagan should follow through on his campaign pledge to try to get Japan to reduce its auto exports to America voluntarily. That would reduce the pressures by members of Congress who advocate stricter and more inflationary quotas. Among other problems, direct quotas would severely reduce competition and would lead to a spurt of higher car prices.

The timing is right for such voluntary restraints, which should have a specific time limit, after which the agreement would end. Japanese officials are wary of the growing protectionist mood in Congress. They are said to be willing to discuss curbing the flow of their small, fuel-efficient cars, which captured a record 23.8% of the American market last month.

The president, we believe, should listen to the advice of those cabinet members who argue in favor of voluntary restraints. They say the Reagan administration cannot risk a pure free trade policy on this issue.

A period of voluntary restraint — perhaps three years — by Japan would yield the auto industry time to retool and produce sufficient numbers of small, better-engineered, fuel-efficient cars to compete in the world market. Detroit needs to invest $80 billion over the coming years in an industrywide restructuring.

The president should insist on some conditions from the industry and the United Auto Workers union. The union should be required to exercise restraint in wage and fringe benefit demands. It has already done so in the case of Chrysler Corp. Every dollar saved in hourly labor costs can add up to $1 billion a year to the industry's cash flow. The auto makers, in turn, should provide incentives for restraint by labor. Profit sharing, for example, is an excellent incentive.

The administration also is considering asking Congress to approve tax credits for auto industry investments. These credits could improve the auto industry's cash flow much more quickly than the proposed faster depreciation allowances on new plants and equipment. The faster depreciation, endorsed by the president, needs some correcting if it is to allow the auto industry to make quick and efficient use of capital investments to improve productivity.

This auto industry package, with voluntary restraint from Japan and new tax incentives for investment, would give the American automobile industry three to five years to demonstrate whether it can compete in the world market. If it cannot, all agreements should end. This economic experiment is not too much to ask for a basic American industry that directly and indirectly employs more than four million Americans.

CHARLESTON EVENING POST

Charleston, S.C., March 19, 1981

Congress last year nearly passed a resolution empowering the president to negotiate an orderly marketing agreement with Japanese auto exporters. Thankfully the Senate held off — feeling that the incoming Reagan administration should decide how to handle the issue.

The U.S. auto industry has just endured the worst year in its history, losing billions of dollars, and the question of Japanese imports is still very much on congressmen's minds. If the decision is to restrict imports there will be cause to celebrate in both the boardrooms and the union halls of Detroit. The loser in such a decision will be the consumer.

With government controlling the competition, Detroit would be under fewer pressures to build the type of cars consumers want to buy. Controls would create an artificial shortage of Japanese cars and demand would drive up prices. Without foreign competition, Detroit's prices would rise even faster than they have been rising.

If quotas are imposed, other countries, seeing protectionism practiced here, will place controls on products imported from America. Then both U.S. workers and consumers lose. Government intervention on behalf of special interests is the root cause of many of the problems facing the U.S. economy today. Restrictive import quotas on Japanese cars would be a bold step in the wrong direction.

THE ARIZONA REPUBLIC

Phoenix, Ariz., March 27, 1981

JAPAN'S foreign minister has left Washington without any discernible agreement about restricting the export of Japanese motor vehicles to the United States.

It is possible, of course, that Masayoshi Ito hinted at some accommodation during his conferences with American officials up to, and including, President Reagan.

But all he said publicly was something about the advantages of free trade for both countries. American spokesmen were even more restrained in their remarks.

Ito was quite right about the advantages of free trade. But there is a limit to how far the United States will allow its auto and truck industry to be destroyed.

Prime Minister Zenko Suzuki is scheduled to visit the United States soon and he may well have decided to hold up any offer to limit Japanese exports to the United States until he can announce the proposal himself.

It should be clear to everyone that this country will not go on subsidizing a substantial portion of the American motor industry while Japan continues to skim off a very large part of the U.S. market.

For many years, when it could not compete with American manufacturers, Japan protected its own market from foreign competition.

The United States has every right to do the same thing during the years it will take for Detroit to tool up for efficient but inexpensive small cars.

If Suzuki is ready to deal with this problem when he visits Washington in May, the current economic strains between the United States and Japan should be promptly overcome.

The Times-Picayune
The States-Item

New Orleans, La., March 19, 1981

The decision of the Japanese government to voluntarily reduce the export of Japanese cars to the United States takes the Reagan administration off of a political hotseat and clears the way for smoother relations between Washington and Tokyo.

As was former President Jimmy Carter before him, President Reagan has been under intense pressure from beleaguered U.S. automakers to reduce the number of Japanese cars flowing into this country. Last year, the Japanese shipped 1.8 million cars to the United States and garnered almost 25 percent of the new car market.

The big growth in Japanese imports came while U.S. car makers were having their worst year and Washington was in the process of keeping the Chrysler Corp. financially afloat.

U.S. automakers argue that reductions in Japanese imports would give them a breathing spell in which to overhaul their plants for the production of smaller, higher-gas-mileage cars that will enable them to compete more successfully against the Japanese imports.

If the self-restraint of the Japanese gives U.S. automakers the break they say they need, U.S. car buyers could be the beneficiaries in the long run. In the short run, however, the reduction in Japanese imports reduces price competition to the advantage of U.S. automakers and dealers and to the disadvantage of the consuming public.

The Reagan administration has been reluctant to restrict Japanese imports because such action runs counter to its free market philosophy and probably would be inflationary. By voluntarily restricting his country's car exports, Prime Minister Zenko Suzuki has eased the pressure on the Reagan administration and on this aspect of U.S.-Japanese relations.

The Knickerbocker News

Albany, N.Y., March 20, 1981

Masayoshi Ito, Japan's foreign minister, is in Washington to exchange international pleasantries with our new president, Ronald Reagan.

The U.S. auto industry isn't in a pleasant mood these days, and Mr. Reagan shouldn't be, either.

It was Mr. Ito who pulled off Japan's quick duck-and-feint maneuver on auto imports this week. Japan's international trade minister, Rokusuke Tanaka, earlier agreed that the Japanese auto industry would be asked to restrain imports to the United States; Ito flatly contradicted him, mouthing trite words about "free trade" and such.

Mr. Ito's two faces were showing, however: Japan has had trade restrictions that have severely limited — virtually eliminated, to be frank — U.S. car imports into Japan for many years. Thus, U.S. automakers have never had the opportunity to build up a market in Japan, while the United States has been wide open to Japan.

Japan is apparently waiting to see whether the Reagan administration will insist on a cutback in Japanese imports, or would permit business as usual. Japanese automakers will quite happily continue shoveling cars into the United States without letup if permitted to.

We have consistently objected to the proposal by U.S. car manufacturers to impose an import quota on automobiles. We still object: Limiting competition merely encourages U.S. automakers to shrug off competitive pressures. In many cases, it has been automakers' own poor business decisions which has gotten them into their present pickle. Detroit continued to produce and promote chrome gas-hogs long after it was apparent that low-mileage, well-crafted imports were selling well and would sell better. Import quotas should never be used to protect the inefficient.

But the U.S. economy is skidding like an otter down a mudslide, and President Reagan must warn Minister Ito that if Japan continues its heedless, damaging import policies, excluding U.S. products from Japan while flooding the U.S. with Japanese automobiles and other goods, something will be done.

Free trade goes both ways, as Mr. Ito must realize. Japan has had everything its way for too long; we cannot continue business as usual. And we will not.

Portland Press Herald

Portland, Maine, March 26, 1981

No one questions that imported cars—and Japanese cars in particular—are hurting domestic auto sales. But this is not a sufficient reason to abandon our current free-market policies.

It is wrong to argue that imported cars constitute a total loss to America. Tens of thousands of our people are employed in the sale and service of imported cars. Many domestic-brand dealers also sell imports, and in some cases only the profits from those imports have allowed the dealers to stay in business in the face of declining domestic-car sales.

Moreover, Japan is in a strong position to retaliate economically against a unilateral import limit. That is why U.S. manufacturers of farm equipment, who enjoy strong sales in Japan, oppose unilateral action—as do some other American industries.

Even from the standpoint of supply and demand, the small Japanese cars are needed because Detroit cannot yet produce enough fuel-efficient cars to meet America's demand. The imposition of an arbitrary import ceiling not only would be against the national interest from an energy-consumption point of view but also might have the effect of slowing the domestic industry's conversion to small-car production.

And finally there is the question of quality.

It isn't necessary to argue that imported cars are better made than American cars. Since studies show people *believe* them to be better made, Detroit is battling to improve the quality of its own cars to erase the "perception gap." That too is a worthwhile goal.

On balance, then, it is in the public interest for the Reagan administration to let the free market do its work. The Congress, which is threatening to legislate an import ceiling, has pointed a political gun at the president's head, and the administration appears to be leaning toward negotiating a limit in preference to accepting congressional action.

But a political solution to this economic problem would be no solution at all. Indeed, since diminished import availability almost certainly would drive up the cost of American cars, a political solution might actually worsen the economic problem.

Any kind of import ceiling, whether voluntary or statutory, would treat only the symptoms of the disease, not the disease itself. There are more effective ways, such as using accelerated depreciation schedules, to cure the disease.

The domestic industry will certainly get the help it needs. But that help must take the form of encouraging productivity, not discouraging it. That is exactly what import restrictions would do.

WORCESTER TELEGRAM.

Worcester, Mass., March 22, 1981

Douglas Fraser, president of the United Auto Workers, let the cat out of the bag this week when he said there was no way that General Motors and Ford could get the same wage deal that Chrysler has been granted.

Instead, Fraser called for restricting imports of foreign cars, an issue that has the Reagan administration all tied up in knots.

It's understandable that Fraser wants to protect his union's inflated wage scales, now running 60 percent above average manufacturing wage scales in this country and 100 percent above wage scales of Japanese auto workers.

But it's not at all understandable why the Reagan team, or any part of it, is even considering quotas, "voluntary" or otherwise.

Economist Charles L. Schultze, former chairman of the Council of Economic Advisers, says "the course of inflation and unemployment in the U.S. over the next five years could depend as much or more on the auto decision as it does on the fate of the Reagan budget cuts and tax reductions."

Instead of cursing the Japanese automakers we should take off our hats to them. They have done more to straighten out our own auto industry than Detroit and Washington put together.

Japanese cars and trucks, arriving here at the rate of 1.9 million a year, have put the brakes on vehicle prices in this country. They have forced General Motors, Ford and Chrysler to sell hundreds of thousands of cars at discount prices. They have set a standard of quality and reliability that Detroit belatedly is trying to equal. Only within the past year have U.S. auto manufacturers begun producing cars that compare favorably with some aspects of Japanese cars. But Detroit still has a long way to go.

For this should our government punish Japan — our second biggest trading partner? Should our government take a step that, as night follows day, will result in higher prices for automobiles, both domestic and foreign?

No. Now is the chance for the Reagan administration to strike a blow for free enterprise, for competition, against inflation. Free trade is not only good for the consumer, for our economy in general and for good international relations. It is also good for the U.S. auto industry.

ALBUQUERQUE JOURNAL

Albuquerque, N.M., March 26, 1981

As a candidate, Ronald Reagan was dogmatic in his devotion to free enterprise. As president, Reagan is wavering toward the siren song of international trade barriers to assuage the auto industry's woes.

But any help that barriers to the importation of foreign automobiles would give Detroit would come out of the pocketbooks of American consumers.

Japanese cars sell well here because they're generally cheaper and more efficient. Any restriction on their importation would cause an increase in their prices — and in the prices of their Detroit competition.

Reagan, who declared inflation America's greatest domestic problem, would fuel it again in the transportation area if he surrendered to the import restriction temptation. Automobile prices would follow the upward trend of their Reagan-deregulated fuel.

The problems of Detroit are real. And it's little consolation that they are largely self-inflicted. The erstwhile world leader in automotive production is playing catch-up in the small car field because of a decade of tunnel vision.

Washington should look with favor on internal means to assuage the auto industry's grief. Selective relaxation of government red tape, or even postponement of environmental or safety standard tightening might be appropriate.

But the nation that has always been a free trade advocate in times of domestic industrial strength shouldn't reverse its stand when the free-trade advantage is going the other way.

The international retaliation that could follow any protectionist move in the auto industry could give Reagan a whole new batch of erstwhile exporting industries to rescue at home.

THE COMMERCIAL APPEAL

Memphis, Tenn., March 23, 1981

WHEN PRESIDENT Reagan asks the Senate Republican-leadership about "voluntary" quotas on Japanese cars, Howard Baker of Tennessee can stand up for the interests of his state and his country by advising against the plan.

Quotas are a bad deal all around, as previous efforts to "protect" U.S. industries so painfully show. In the case of steel, for instance, the result was disastrous for the industry in particular and the public in general. Not only did Americans pay higher and higher prices for goods made with domestic steel, but the steelmakers put off modernizing their plants, they rolled with high labor costs, and they postponed the problem of productivity. The unfortunate outcome is clear today in the plant closings, layoffs and high-priced investment the mills must now undertake.

Is this country prepared to do the same to its auto industry, when even the President's Council of Economic Advisers predicts a quota on cars from Japan wouldn't significantly help U.S. automakers but would drive up the cost of all cars substantially and, therefore, inflation?

The majority leader and the senators who follow him need to make plain to the President that the "breathing space" being contemplated can suffocate the car makers, themselves, along with the rest of the economy.

DESERET NEWS

Salt Lake City, Utah, March 18, 1981

The pressure is mounting for restrictions on the number of Japanese automobiles that can be imported into the United States. Such formidable advocates as the United Auto Workers, General Motors Chairman Roger Smith, Senate Democratic leader Robert Byrd, and some members of President Ronald Reagan's cabinet all are plugging for import curbs.

The union is concerned about the 600,000 unemployed in the auto industry. Smith, a strong advocate of free enterprise, finds himself on the other side of the fence in the argument over car import restrictions. He's understandably concerned about the $4.2 billion the domestic auto industry lost last year.

Transportation Secretary Drew Lewis and Commerce Secretary Malcom Baldridge are known to favor negotiations with the Japanese on a voluntary reduction of their car shipments to the United States. The cabinet members believe this would take some of the competitive pressure off the domestic auto industry.

Some of us continue to wonder why American consumers must now pay the price for the past failure of the domestic auto industry to see the need to move quickly and decisively into the small car market. Or for the exceedingly generous wage rates paid to auto workers. The best interests of the American consumers will not be served by import restrictions.

Still, some valid points may be made by protectionists. For example, the Japanese automackers may have a lighter tax burden thanks to the willingness of the United States to carry the full cost of Japan's defense program. In addition, U.S. industry can and should be protected against the unfair trade war weapon known as dumping — selling a country's exported products at less than cost.

Reagan must take steps to mitigate as much as possible any suffering among the unemployed. Industry tax burdens must be lightened. Possibly other incentives can be devised to assist the industry through a difficult period. But import quotas or other restrictive measures would only preserve inefficiency, keep prices high, and open the floodgates to similar protection demands from other industries.

THE LINCOLN STAR

Lincoln, Neb., March 26, 1981

With justification, President Reagan has walked very cautiously in the field of international trade. In meetings with the Japanese foreign minister, President Reagan has not pushed hard in the matter of Japanese auto exports to the U.S., simply explaining the U.S. auto industry problems to his visting dignitary.

The president would no doubt be delighted if the Japanese took it upon themselves to limit their U.S. auto-import business. This would solve for him what at the moment is a most difficult political problem. But hopefully, he will stick with his free-trade instincts, regardless of what the Japanese do.

The real problem was summed up by the Japanese foreign minister. He said: "We are not shipping cars to the U.S. because of a push from outside. There is a demand in this country. Consumers want Japanese cars."

His statement may not be totally accurate and is somewhat self-serving, but the desire of U.S. buyers for Japanese and other foreign-made cars is certainly a fact. It is that desire, the popularity of these cars in the marketplace, that has brought about the U.S. auto industry's loss of some 25 percent of its market.

Why this has happened can be argued forever with no definitive conclusion. Certainly, one-fourth of U.S. auto-buyers simply think the imported cars are a better buy, whether the reason be price, fuel efficiency, reliability, quality of production or whatever.

And while the U.S. industry now seems prepared to meet that challenge, it has been a long time arriving at that point. We believe quality control is now a matter of growing priority in the U.S. industry but quality control had been allowed to slip for a great many years, while the industry believed it had a hold on the market that could not be broken.

What Detroit did was take its customers for granted, exactly what a lot of other U.S. business and industry have been doing. Much of that is still going on today and to it, we believe, can be traced many of the economic problems the nation now faces. Disregard of market sensitivity to price and quality, a fault of both management and labor, has devalued the worth of countless products and fueled an inflationary spiral.

Such faults will not be corrected through artificial manipulation of the market by such means as import restrictions. Only industry itself can solve those problems and the quicker it proceeds to do so, the quicker it will regain the respect and confidence of the buyers it has lost.

THE ATLANTA CONSTITUTION

Atlanta, Ga., March 24, 1981

The chairman of the second largest car maker in Japan has indicated his firm, Nissan Motor Co., would comply with restraints on importing autos to the United States if they are imposed. That represents a reversal of an earlier stand and may mean an agreement is being discussed privately. It would be handled with the upmost diplomacy.

The Japanese executive, Katsuji Kawamata, said, "If the government calls for export curbs through a trade control law, our industry will not oppose it but will comply. It was not known whether he spoke for the Japanese auto industry as a whole.

The United States has been the champion of a free market and of competition, which tends to improve things overall and make them less expensive. There are those who insist the American car manufacturers can be on top again if the imported cars were kept out of the country for a few years. They are short-sighted. Even United Auto Workers President Douglas Fraser — whose union members are out of work in many places — says he prefers voluntary restraints to something mandated by law.

There are good reasons for that. If this country starts passing laws about how many cars can be imported into the United States almost certainly there would be retaliatory laws on U.S.-made goods. Japan, for instance, might say that it will go to France and Germany for its aircraft instead of Boeing, Lockheed and McDonnell Douglas. How far could that go? Our nation competes on the world market with a great many goods.

The Japanese are our friends. That should be kept in mind when discussions are held. Nobody holds a gun to the head of a prospective car-buyer in the United States and makes him purchase something made in Japan. That's his choice and it's just about as American as you can get.

The U.S. industry seems to be on the right track now with the kind of cars that can compete with any foreign country both in quality, gas consumption and price. Right now sales are down and that has something to do with the high interest rates. It won't be that way forever.

Perhaps the Japanese companies would be willing to build plants in the U.S. or even limit the number of cars they export as a sign of wanting to help a friend. But the U.S. government would be making a big mistake to try to force it down their throats.

Los Angeles Times

Los Angeles, Calif., March 25, 1981

As a believer in free trade, President Reagan doesn't want to see import quotas imposed on Japanese cars. But congressional sentiment for quotas may be building, as Senate Majority Leader Howard H. Baker Jr. (R-Tenn.) warned Reagan on Tuesday. If some in Congress have their way, Japanese car sales in the United States would be cut in coming years by at least 300,000 from their 1980 level.

Reagan's preferred course is to induce Japan to make voluntary cuts in its U.S. car sales. But that alternative, if adopted, would be a distinction without a difference. Voluntary or imposed, any quota system would harm consumers and the economy.

The harm would come first of all in higher prices charged for nearly all cars, foreign and U.S.-made. Force some popular Japanese cars out of the market while demand for them is high, and the prices of remaining imports would be pushed up. Reduce competition from overseas, and the prices of American cars would rise.

The U.S. International Trade Commission estimates that this restriction on freedom of choice would cost new car buyers at least $1 billion more a year. The Council of Economic Advisers says the result would be to bump up the cost of living by 0.5%. That in itself would lead to bigger pay increases in many sectors of the economy, including the auto industry

At best, the International Trade Commission says, 20,000 unemployed auto workers might be rehired if Japanese car sales were restricted. But, meanwhile, thousands of Americans employed in the imported-car industry would have to be laid off. Charging consumers an extra $1 billion a year to put less than 10% of the currently unemployed auto labor force back to work would be subsidization run wild.

Quota advocates claim that Japan is in good part responsible for the woes of the U.S. auto industry, though they have yet to produce any convincing evidence. In fact, U.S. car sales in recent years have fallen at a much greater rate than imported car sales have increased. They have fallen for three main reasons:

—*Management blunders.* The rising demand for smaller and more fuel-efficient cars was for too long not taken seriously by the U.S. industry. For too long the industry continued to put too many of its investment dollars into valueless annual style changes, too little into fundamental technological improvements. For too long it emphasized quantity at the expense of quality. In some years the industry, to its embarrassment, was required to recall for repairs as many cars as it was producing. Japanese cars, meanwhile, were gaining a reputation for quality and reliability.

—*Runaway costs.* Unit labor costs account for two-thirds of all costs. Quota advocates raise the dubious argument that the Japanese auto industry has a competitive advantage because of sharply lower labor costs. It makes more sense to turn that argument around. In the U.S. auto industry, hourly labor costs are now 60% higher than the average for all manufacturing. When foreign competition was minimal or considered unimportant, the auto companies went along with steadily soaring union wage and benefit demands, secure in the knowledge that the added costs could be passed on to, and swallowed by, consumers. By the time the companies woke up to the reality of foreign competition, they were stuck with staggering labor costs.

—*General economic conditions.* The typically equipped small U.S. car can now easily cost $8,000 or $9,000, while annual finance charges are typically in the 15% or 16% range. Fewer people are buying cars, and when they do buy they not unnaturally want the best quality and the highest fuel efficiency for their money.

Protectionism is no answer to the industry's problems of costs, quality and competitiveness. A few years ago the American steel industry, whose workers earn even more than those in auto unions, won restrictions of foreign imports. Did the industry make itself more efficient or bring its costs under control? No. It simply raised its prices, at the expense of all consumers.

American economists and European leaders have been expressing increasing concern that auto industry protectionism could set off an international trade war, to the potentially severe detriment of U.S. exporting industries and the world economy. Should that happen, should political expediency win in the fight over import quotas, millions of Americans—and not just new car buyers—would suffer the economic consequences.

The Seattle Times

Seattle, Wash., March 25, 1981

PRESIDENT Reagan met with Japanese Foreign Minister Masayoshi Ito yesterday in a White House session that must have been a big disappointment to hard-line protectionists in Congress, many of them strong Reagan supporters. For that, American consumers can be thankful.

The protectionists, of course, wanted Mr. Reagan to take a tough line with Ito on Japanese auto exports to this country, and threaten him with the imposition of mandatory import controls.

One of them, Senator John Danforth, Missouri Republican, implied that what Mr. Reagan did was "dance around the issue."

But the President held true to his general campaign stance, favoring free trade. He took a low-key approach with Ito, and White House officials said he is seeking agreement in Japan on some form of voluntary auto-export controls.

The importance of Mr. Reagan's refusal to succumb to heavy protectionist pressures can scarcely be overestimated. Charles L. Schultze, former chairman of the Council of Economic Advisers, sees the decision on auto imports as "a symbol of whether we mean to do what's necessary to fight inflation."

"An administration decision to refuse to grant import protection," Schultze says, "would send a signal to all industries that the government will not validate inflationary excesses in wage and price setting."

Even as the Japanese foreign minister was beginning his series of high-level visits in Washington, The Wall Street Journal reported that for the first time since late 1978, U.S. manufacturers have begun outselling the Japanese in the small-car segment of the market.

In the first five months of this model year, the newest U.S.-built small cars took a 36.6 per cent share, up from 31 per cent in the year-before period. At the same time, Japanese sales were static at about 31 per cent of the small-car market.

In view of this, there is reason to hope the day may not be far off when Japanese voluntary restraints — to say nothing of U.S. mandatory import quotas — would become a moot issue.

If and when that happens, all sectors of the American public, including the consumers, would benefit..

The Miami Herald

Miami, Fla., March 25, 1981

LAST year America's automobile industry suffered more than $4 billion in cumulative losses. Last month, 244,000 American auto workers remained jobless owing to the hard times afflicting Detroit.

While Detroit bleeds, Japan prospers. Japan is now the world's leading auto producer. Today when Americans buy new cars, one in four purchases a vehicle made in Japan.

The trend has yet to peak. Japan exported about 1.55 million cars to the United States in 1979, and 1.9 million in 1980. The January 1981 export rate, if maintained through the year, would exceed two million. Thus it is hardly surprising that cries for Government quotas to "protect" Detroit from Japanese competition grow louder.

President Reagan's Secretaries of Commerce, Transportation, and Labor support the American car industry's plea for shelter for at least three years. That would give Detroit time to retool to meet the new market tastes of the '80s, the argument goes.

Ronald Reagan himself promised jobless auto workers last fall that, if elected, he would act to stem the "deluge" of Japanese imports. Last week Senate Democratic Leader Robert Byrd joined the chorus.

The protectionist furor ignores the fact that Japanese cars sell so well simply because Americans choose freely to buy them. The great overlooked issue in this debate is consumer freedom of choice. If Detroit wants to recapture its lost market share, it should produce better cars, ones Americans prefer.

That was the point stressed insightfully by consumer advocate Ralph Nader last week in a letter to United Auto Workers president Douglas Fraser. The union should sell consumers on the safety advantages American cars have over Japanese ones, Mr. Nader advised, citing American cars' superiority in Transportation Department crash tests.

Despite such sage advice, the protectionist wave rises. The dangers it represents are immense. Trade barriers against Japanese cars easily could trigger responses in kind against American exports. Shelter for one industry could easily spawn protectionist coalitions for others. After cars, why not steel? Shoes? Sugar? Electronics?

Markets closed by nationalist protectionism encourage inflation by insulating industries from competition. Consumers are denied choices. Expansion abroad is thwarted. National hostilities grow. The history of the 1930s tells the story, and it is not a happy tale.

That is why the discussions now under way between Japanese officials and the Reagan Administration are so important. Japan says it wants to help, but it wants to see what Mr. Reagan proposes first. The initiative must come from the White House.

This poses to the Reagan Administration both a severe test and a great opportunity. It should reject protectionism, especially legislated import quotas. Yet some specific voluntary export restraints by Japan probably are the only politically feasible solution. Devising such targets within bounds that Tokyo, Detroit, and Washington all can tolerate requires the Administration to live up to its professed free-market principles while demonstrating a much-needed capacity for diplomatic finesse.

The Honolulu Advertiser

Honolulu, Ha., March 20, 1981

The scheduled visit of Japanese Prime Minister Zenko Suzuki to Washington to meet President Reagan in May imposes a deadline of sorts on the current trade clash over automobile imports.

And while Foreign Minister Masayoshi Ito, whose Washington visit starts Monday, intends mainly to see and be seen by new Reagan administration officials, he will no doubt find automobile imports on many minds.

BOTH GOVERNMENTS are divided on the question. Reagan's advisers are split between those who would establish restrictions or at least insist that the Japanese set up "voluntary" limits on auto exports and those who are opposed to trade restrictions.

The Japanese are split over which ministry, Foreign Affairs or International Trade and Industry, will deal with the problem, what the response should be, and what concessions should be asked of the U.S. if Japan does agree to cut its auto exports.

By producing over 11 million vehicles (as compared to 8 million in the U.S.) last year, Japan became the world's top auto manufacturer. Over half of Japan's production is for export and autos are the top foreign exchange earner.

THE REAGAN administration is still faced with the crisis in the American auto industry which had its worst year last year. About a quarter of a million auto workers are still laid off.

Last year, the U.S. International Trade Commission studied the problem and determined that the capture of 25 percent of the American auto market by foreign automakers was the result, not the cause, of Detroit's problems.

The commission's rejection of protectionism reinforced the inclination of the Carter administration. But by the end of his term even Carter's transportation secretary was calling for the U.S. to negotiate an import-restraint agreement with Japan.

Reagan is in principle a "free-trader" opposed to protectionism. But the pressure to appear to be doing something to help the troubled American auto industry is growing.

HOWEVER, REDUCING imports would mainly penalize American consumers, plus those who sell and service foreign cars. It would not guarantee any great relief to the American auto industry or really cure the autoworkers' unemployment woes.

And reducing the competition from the small and fuel-efficient (though not necessarily cheaper) foreign autos would not help force the American industry to make changes that are needed over the next few years if it is again to be as competitive as it can be.

Given the current climate in the U.S., however, Japan is sure to exercise "prudence," as the international trade minister put it, with government offering "administrative guidance" to Japanese auto makers.

Even the president of Nissan Motors believes that restraining exports for two months could help Detroit solve some short-term problems.

BEYOND THAT, Japan should look to increased exports to other areas of the world than the industrialized nations like the U.S. and Europe which have automobile industries of their own.

But ultimately, the solution to the American auto industry's problems must be found in America. President Reagan should avoid the quick-fix by saying no to new automobile import restrictions. They would do little more than invite retaliatory restrictions against American products abroad.

The Philadelphia Inquirer

Philadelphia, Pa., March 25, 1981

The Reagan administration, after several weeks of sharp disputation, has reportedly come up with a consensus on how to help the American auto industry out of its doldrums. The consensus involves "voluntary" restraints by the Japanese on their exports of automobiles to the United States.

The difference between "voluntary" and "compulsory" restraints is that "compulsory" would probably require legislation, opposed by the administration and in conflict with any number of international trade agreements. Otherwise, there is no difference, the effect being the same, except that "voluntary" limits would allow President Reagan to go on professing his adherence to the principles of free international trade while violating the practice.

Undoubtedly, Mr. Reagan will discuss the matter with Japanese Foreign Minister Masayoshi Ito, who is visiting the White House this week. The Japanese, for obvious reasons, want to get along with their major ally and trading partner, but they are not about to volunteer to curb the exports upon which they are dependent unless their arms are twisted out of shape.

Nor should they be asked to volunteer. Restrictions on foreign trade, "voluntary" or otherwise, are a bad idea. They are bad for the American consumer. They are bad for the American auto companies. They are bad for the American economy. Take those in order.

They are bad for the American consumer because they will limit his choices of — and drive up the prices he has to pay for — the kind of automobiles he wants.

They are bad for the American companies because they will relieve the pressure of competition, in terms of prices, efficiency, miles-per-gallon and so on, and in terms of labor costs. Make no mistake about it: Among the reasons the American companies are in difficulty is that for years they have been making overgenerous wage settlements with their workers and passing on the costs to the consumers. In 1980, the industry and United Auto Workers arrived at a settlement putting auto wages and fringes 60 percent above the average of U.S. manufacturing. In 1982, another round of negotiations will open. That "safety net" for Detroit will simply make it safe for the industry and union to agree on another inflationary wage and fringe package.

And inflation, it goes without further saying, is bad for the economy. Going to the aid of the U.S. auto industry with import restrictions would encourage other distressed industries to demand protection so they too could avoid the pressures of competition on their efficiency and their prices.

Mr. Reagan is quoted by his press secretary as saying that the federal government has contributed to the auto industry's troubles and "must bear part of the responsibility for helping the industry back on its feet."

There is a certain inconsistency in the administration's attitude. One could argue, with greater accuracy, that the federal government has contributed to the troubles of America's older cities, by building highways and through tax policies that encouraged the flight to the suburbs. One could make an excellent case that the federal government has contributed to the troubles of mass-transit systems and intercity railroads, by concentrating on highways and starving the rails. Yet far from proposing to help the cities, the mass-transit systems and the railroads, the administration wants to cut back or chop out programs developed over the years to assist them.

The administration would be hard-pressed to explain why it demands sacrifices in some sectors to fight inflation while giving the auto industry import protection in direct conflict with the anti-inflationary policies it proclaims. Indeed, import quotas, "voluntary" or otherwise, would undermine the political credibility of those policies as well as the economic rationale.

Tulsa World

Tulsa, Okla., March 21, 1981

THE JAPANESE have offered to reduce car exports to the United States as a balm to the troubled U.S. automobile industry. The Reagan Cabinet is said to be in general agreement that this voluntary approach is more desirable than official U.S. limits on imports.

If in fact some kind of protection is desirable, the Reagan advisors are right. Mandatory U.S. restrictions on a major import item could lead to reciprocal trade restrictions by the Japanese. American action against foreign competitors could trigger a new wave of protectionist policies throughout the industrial world. So if there is to be protection, let it be voluntary.

But it is not at all clear that temporary protection from foreign competition — either voluntary or mandatory — will be a good thing in the long run for U.S. auto manfacturing.

The Wall Street Journal suggests that import restrictions might even make matters worse for Detroit.

European steel companies agreed to voluntary restraints on U.S. sales from 1969 to 1975, the Journal recalled. During this period, steel prices rose rapidly. While the voluntary restraints gave American companies some "breathing room," the U.S. industry became less and less competitive. Sheltered from competition, the U.S. steel industry allowed its hourly labor costs to rise to a level 59 percent higher than the average for all American manufacturing, compared with 28 percent above average when the protection policy began.

Economist Charles L. Schultze wonders whether the U.S. automobile industry — already notoriously non-competitive in terms of wages and other production costs — might not grow even more so if encouraged by artificial limits on competition.

The auto industry has a worse record than steel in the matter of holding wages and other production costs to a competitive level.

In 1980, the automakers acquiesced to union demands which left the industry's wages and benefits 60 percent higher than the average for all U.S. manufacturing, according to Schultze. That goes a long way toward explaining why Americans have lost the competitive edge to cost-conscious foreign competitors.

Temporary charity from efficient foreigners can help only if the American companies use the respite to solve their own problems.

The Des Moines Register

Des Moines, Iowa, March 25, 1981

Should the overseas market for American farm products be jeopardized to protect the ailing U.S. auto industry from the threat of Japanese competition?

The question is being raised by a growing number of farm groups concerned that the drive to impose import limits on Japanese cars could backfire — leading Japan and other nations to retaliate against goods that Americans sell abroad.

Japan is the world's best customer for American farm products. Last year, the Japanese purchased $5.8 billion worth — including $1.4 billion of corn, and about $1 billion of soybeans.

Farm groups are worried that a U.S. decision to limit the sales of Japanese autos could lead Japan to seek other suppliers for its food needs, such as Brazil. It might also undermine the effort to open up the Japanese market to more American beef and citrus.

Farm groups are concerned also that a U.S. decision to violate free-trade principles could lead Europeans to impose harsher limits on the sale of American farm products in Europe, especially soybeans. There is considerable political pressure in Europe to protect the European farmer against American competition.

These worries led nine groups — including five farm organizations — to send a Mailgram to President Reagan last week. It warned: "A decision at this time to restrict imports of automobiles would not be a substantial help to the domestic industry and could undermine the fragile structure of world trade.

"Once the United States raises trade barriers as a special favor to a troubled industry, other industries will demand similar treatment, and other nations will swiftly follow suit. A protectionist response to the automobile problem could jeopardize future U.S. exports, penalize U.S. consumers, and, in general, retard growth in the U.S. economy."

Among the signers were the U.S. Feed Grains Council, the National Corn Growers Association, the American Soybean Association, the National Grange and the National Council of Farmer Cooperatives.

Robert Delano, president of the American Farm Bureau Federation, told The Register that he shares these concerns, and that the Farm Bureau would oppose restraints on Japanese auto imports.

Agriculture Secretary John Block's press secretary told The Register that Block privately has expressed concern that limits on Japanese car sales in this country — especially if they were imposed by Congress in the form of a quota — "could hurt our farm exports."

Many farm experts believe that the damage to U.S. farm exports would be minimal if restrictions on Japanese auto imports are "voluntarily" imposed by the Japanese. The experts warn, however, that even a voluntary arrangement could lead other troubled U.S. industries to seek protection from foreign competition. If the fever of protectionism spreads, it wouldn't be long before farm exports were affected.

It may sound good to talk about "protecting" the auto industry from Japanese competition, but such protection would be at the expense of others, farmers prominently among them.

The Oregonian

Portland, Ore., March 20, 1981

The proposed remedies for treating the depressed U.S. auto industry are about as encouraging as a dead battery. None guarantees that the flattened industry quickly can re-energize itself or that high employment and profits needed to boost the U.S. economy can be restored without a lot more pain.

Ralph Nader, who drove an early nail in the Detroit coffin with his attack on the Corvair's safety in the 1960s, now wants the auto industry to launch a campaign emphasizing government reports that U.S. cars are safer than Japanese imports. But past efforts to sell safety have not been successful.

The United Auto Workers is demanding that Japanese imports be restrained for five years. Some U.S. car builders want at least a temporary moratorium so Detroit can get some breathing space to produce large numbers of its new fuel-efficient models. The Japanese government indicated it would seek voluntary export reductions, then changed its mind.

Others in the industry do not want to imperil the U.S. leadership in the world free trade movement, arguing logically that import curbs are double-edged swords, since Detroit both imports Japanese cars itself and has factories abroad producing cars, engines and parts with cheaper labor. The labor-intensive Detroit industry is now remodeling its plants and installing automation devices that will cut labor costs and reduce job opportunities.

Meanwhile, the public again is showing an interest in large cars, having grown somewhat immune to high gasoline prices and bored with driving small boxes on wheels. The real big cars are probably gone forever, but public tastes can shift quickly from economy to luxury lines, a trend that even Japanese builders are seeking to exploit.

In all this, the Japanese are vulnerable to the criticism that they impose unduly strict rules on foreign imports, charging high inspection fees and making clean-air demands that are costly to meet. Whatever the truth, the fact is that only a tiny trickle of U.S. vehicles reaches the Japanese market to counter the tsunami of imports sweeping onto American shores.

There is no question that superior quality and lower prices helped Japanese sales, but the U.S. price now has become competitive and quality has improved. Yet, high interest costs, a part of the administration's program to cool inflation, have frozen Detroit as much as foreign competition in recent months, if not more so.

President Reagan promised in his campaign to curb Japanese imports, but his Cabinet is now badly split on the issue. Some, including Donald I. Regan, the treasury secretary, and David Stockman, the budget cutter, have proposed tax breaks for Detroit in lieu of trade barriers. Transportation Secretary Drew Lewis has urged trade restraints.

Congress also is divided, but it may enact some form of restraint if the administration backs such a proposal. What might be considered, in addition to some kind of a short moratorium, are guarantees that both Japan and the United States will play with the same rule book. Fair is fair, even in the marketplace. A clear understanding among friendly traders is a better way to clear the air than is the passage of punitive laws that can only make matters worse in the long term.

The Kansas City Times

Kansas City, Mo., March 21, 1981

There is nothing wrong with voluntary restraints on imported motor cars and encouraging signs are coming out of Washington and Tokyo that the United States and Japan are moving in that direction. The talk so far has been of a general nature—except for a bill in the U.S. Congress that would impose mandatory quotas on automobile imports, which many economists see as an opening gun in a world trade war.

Otto Lambsdorf, West German economics minister, predicted in Washington this week that mandatory quotas could signal a trade war that "would spread like fire once it gets started. Nobody is going to prevent it" then.

Facing the possibility of such a catastrophe, it is little wonder leaders in the two capitals are adopting a more cautious approach.

So far there is no official position in either nation and one is needed quickly to clear the air. Mr. Reagan's auto industry task force made its report Thursday, although no one has been willing so far to reveal its contents. This much is on the record: If any protectionist measure is taken, the president has said, "we would like to have it be a voluntary restraint."

Secretary of Labor Donovan put it this way in Detroit: The administration "will not put import restrictions on automobiles." The secretary added that voluntary restraints are "in the long-term interests (of the Japanese) and I think they will" go along.

Other American officials, including Drew Lewis, the secretary of transportation, and Malcolm Baldrige, secretary of commerce, have stopped short of mandated quotas enacted by Congress. There is no question that something needs to be done — and needs to be done in a hurry.

Whether Detroit is in trouble because of its own ineptitude, foreign imports, high interest rates, excessive prices or a combination of the four is debatable. But this basic industry is in bad shape and political pressures are mounting to do something about it.

Masayoshi Ito, the Japanese foreign minister, said in Tokyo recently that his government is "determined" to settle the dispute on an economic basis rather than invite a full-scale political controversy. Such a consequence would be dangerous for both nations—and for the rest of the world. Foreign Minister Ito arrives in Washington this weekend to begin the first Cabinet-level talks with the Reagan administration.

The Japanese are also leery of a legal approach. Rokusuke Tanaka, the trade minister of Japan, told the Diet this week that discussions were already under way on voluntary restraints there.

"For the moment, I have no intention of resorting to law," Rokusuke added. And for solid reasons this should not be done on either side of the Pacific.

Maybe it is only wishful thinking, but positive signals are now coming from both Tokyo and Washington suggesting a rational solution could be worked out this spring. A failure to listen to reason in either country could lead to mandatory quotas in the U.S. and the hideous specter of a trade war.

Sales of U.S. Autos Post Modest Recovery

The year 1981 sounded a more hopeful note than before for the U.S. auto industry. For the first time in more than a year, all three major automakers reported after-tax profits during the second quarter of the year. Chrysler Corp. said its second-quarter after-tax profit was $11.6 million; Ford Motor Co. reported an after-tax profit of $60 million, and General Motors Corp. said its after-tax profit was $514.6 million. All three cited significant cuts in production costs as a factor in their profits. Also, all three had embarked on much-heralded switchovers to compact cars bearing alphabet-soup names like "J" and "K." Widely publicized rebate offers also helped to spur sales.

Another boost to American morale came in September when the U.S. Environmental Protection Agency reported that an American-made car had achieved a fuel efficiency rate of 40 miles per gallon. The car was a new-model Chevrolet Chevette diesel and was the first American car ever to place among the top ten in terms of fuel economy. Faith in American know-how was renewed—but the Chevette was equipped with a Japanese-made engine.

The Kansas City Times

Kansas City, Mo., May 8, 1981

It is a bold and costly venture that the nation's car makers have undertaken in completely redesigning their product line to meet the realities of a market that demands small, fuel-efficient and quality-built cars. Yet it is not so much a calculated gamble on Detroit's part as it is a forced response to huge balance sheet losses and the rising competition of foreign imports, mostly Japanese. With the grudging agreement of Tokyo to modestly limit car exports to the U.S. — and forestall congressional imposition of import quotas — the car makers are given some breathing space in which to launch the great comeback

General Motors, the industry leader, expects to spend $40 billion by 1985 on the conversion retooling, but then the Big Three recorded losses totaling $4 billion just last year, a trend that could not be allowed to continue. GM's market blitz with its new front-wheel-drive "J" cars, (not "J" as in Japanese, the company says) is seen as the big gun of the effort, following the earlier success of the "X" car two years ago. The various model "J" cars, to be followed by other alphabetical types in the years to come, will be considerably smaller and lighter, with markedly greater fuel economy than the traditional lines they replace.

To counter the nagging charge that American vehicles do not match their foreign-built competitors in "fit and finish," GM will greatly expand the use of robot functions on its assembly lines and is pushing a "Quality of Work Life" program (which does sound Japanese) among its employees to encourage greater participation, pride and productivity.

Ford has been and will be coming along with its versions of fending off the foreigners by giving the people what they want and Chrysler and American Motors are doing the same. But the last two also are forging links with foreign car makers — Mitsubishi and Renault — to help then stay in the chase.

Considering the size of the American automotive industry and its direct effect on the national economy, the hope that Detroit will succeed is fervent. A great many jobs and incomes are depending on it — in Michigan, in Missouri and in Kansas — every corner of this country where wheels touch pavement.

Sunday Journal and Star

Lincoln, Neb., May 17, 1981

Flag waving is not something done in this space with any skill, the belief being that real patriotism is marked far less by show than by quiet and steady performance.

But we're kind of pulling for the success of General Motors, which this week begins selling its J-car subcompacts.

We had — and maintain — the very same view when Chrysler brought out its K-cars, and Ford commenced marketing of its subcompact Escort and Lynx models.

An arrogant reading of world events on the part of GM, Ford and Chrysler managements, combined with a healthy helping of lust for immediate profits, put Detroit in a weak competitive position. The Big Three trailed in the small-car market contest against well-made Japanese and German compacts and subcompacts.

The Big Three now are fighting back. For reputation and market shares. Aided, of course, by the protectionist strong arm the Reagan administration put on the Japanese government, compelling Japan to "voluntarily" limit exports of new cars to the United States to 1.68 million this year, and not much more the year after that.

The net result of leaning on the Japanese is to raise the cost to American consumers. That is a price the administration apparently thinks necessary to help Detroit survive, even as a no-growth industry, repeating Transportation Secretary Drew Lewis' forecast.

Perhaps the price is required; time will tell. Much depends upon the dependable quality of Detroit's products, their fuel efficiencies and the sustained enthusiasm by which they are merchandized.

An early reading of the J-car by Chicago Sun-Times auto expert Dan Jedlicka is hopeful. After testing a model — and finding some faults, such as a bit too much weight for the four-cylinder engine — Jedlicka concludes that the J-car "is the best small car GM ever has built."

Most buyers of Chrysler's K-cars and Ford's subcompacts reportedly express like satisfaction.

Detroit is not down and out. Not yet, anyway.

the Charleston Gazette

Charleston, W. Va., May 12, 1981

IN what might be described by some as a patriotically overblown account, the current *Newsweek* shows readers how General Motors is responding to the foreign car threat.

The story is mostly about the new "J" cars, soon to be marketed as 1982 models. But what struck us as significant was the revelation that GM has multiplied its use of assembly robots and otherwise has adopted the techniques of foreign manufacturers whose products have captured the fancy of so many Americans.

It is strictly a coincidence, perhaps, that a recent *New Yorker* review of Japanese industry dwelt lovingly on the extensive use of robots on Japanese automobile assembly lines. The inference was that these robots — which can weld, for instance, with precision humans cannot match — are the principal reason Japanese cars are thought to be put together better than U.S. cars.

Some of the alleged superiority of foreign cars is folklore, of course. But competent authorities — Consumer Union among them — consistently have given Japanese cars good ratings in this regard. The frequency-of-repair charts in *Consumer Reports* always show Toyotas, for instance, at or near the top.

General Motors, according to the *Newsweek* story, has taken an "if you can't lick 'em, join 'em" stance, and we think it is about time. After all, foreign competitors didn't hesitate to adopt American manufacturing techniques and ideas when Detroit commanded the market. Robots now are doing most of the welding on the new "J" cars whose parts are cut so it is impossible to join them if alignment isn't perfect. And various programs are aimed at making assembly line work less onerous to the worker.

Newsweek's own test drivers have pronounced the new GM cars — small but reasonably roomy — competitive with the imports that have plagued American carmakers for so long. This augurs well for a resurgence of American carmaking superiority. We assume, of course, that GM is not the only American manufacturer breaking away from yesterday's conceptions.

We are scornful of the proposition that American workers cannot turn out a good product. American airliners are the mainstay of many a foreign fleet, and the durability of American household appliances makes them sought-after by householders everywhere. There is no reason that a properly motivated American work force, turning out an imaginatively designed product, cannot revitalize the American car industry.

The Saginaw News

Saginaw, Mich., November 22, 1981

Faster than anyone might have expected, Michigan's automakers have caught up to their imported competition in two categories most important to buyers — mileage and quality.

The improvement also may be faster than buyers have realized. The problem facing the auto industry is getting its message out.

The message, as reported by the Environmental Protection Agency, is that U.S. makes now get better fuel economy in 14 of the 16 model classes where they are in direct competition for the consumer decision. One group was rated a tie. In the other category, the Ford Thunderbird, for example, would lose out to the Mercedes 240D.

More startling is how good the mileage is no matter what car you buy. The Ford Escort, for instance, averages 36.4 miles per gallon, and the equivalent imports get 34 mpg.

Really mileage-conscious buyers would still have only imports to choose from. U.S. automakers don't compete in the three lightest car classes. But those kind of cars, aside from their driveability across long U.S. distances, also don't support the options many Americans still favor, such as air conditioning and automatic transmission.

A separate report indicates that the U.S. industry is also catching up to imports in "fit and finish," an aspect that makes a tremendous difference in perception of surface quality. This study concedes that the best imports are still slightly better than the best domestic makes. But the worst imports are far worse.

This means both the industry and its employees have been doing what's necessary to recover: Designing more economical cars, and building them better.

Eventually, the word should get around to buyers. The trouble is that the U.S. industry needs help right now.

It simply has not done a good enough job of selling itself.

For years, the imports have hammered away at their economical qualities. The approach of their advertising has been classic. The volume of it is enormous.

Yet the heavily promoted Fiat Strada, for instance, loses to both the Plymouth Omni and Chevette in its mileage class.

For their own good, American car buyers ought to start making some objective comparisons. They've been giving their own auto industry a bad rap for long enough. It used to be deserved. The industry should get the word out that they might be surprised at what they'd find now.

The Dispatch

Columbus, Ohio, July 27, 1981

THERE WAS good news from the business community last week and it came from — of all places — the U.S. auto industry.

Chrysler and Ford both reported second quarter profits, signaling a sharp turnaround from the same period last year when they both reported losses.

The other good news was that new car sales by U.S. automakers rose 6 percent in mid-July from the same period a year ago. This upturn ended three straight reporting periods of declining sales that reached 20-year lows.

Chrysler's profit was a mere $12 million and it hardly indicates that the huge company has turned the corner and avoided collapse. But there is hope that the $1.2 billion in federal loan guarantees have helped stabilize the company long enough for it to mount a recovery. This was the first time since 1978 that the firm was able to post quarterly profits.

Ford reported a $60 million profit for its second qarter. In the same period last year, the company suffered a $468 milion loss.

Officials at both companies point to new model sales as reasons for their improved outooks. Ford's Escort and Chrysler's K-cars represent direct responses to public demand for small, fuel-efficient automobiles — the kinds that, for years, were offered only by foreign firms.

Auto industry problems are far from being solved. But this news from Ford and Chrysler gives the first reasons for optimism in quite a long time.

If the U.S. industry has learned its lesson, if it continues to give the public what it wants, the summer of 1981 could be remembered as the start of a healthy new era.

Government Regulations Blamed for Auto Difficulties

President Ronald Reagan came to office in 1981 determined to reduce the extent of government regulation in all sectors of the economy. His proposals were welcomed by the auto industry, which had long complained that regulations prevented them from competing effectively against foreign cars. Reagan was sympathetic to their complaints, asserting in a February speech on the economy that federal requirements increased the average cost of producing an American car by $666. In April, he unveiled a plan to eliminate or reduce 34 environmental and safety standards for American cars and trucks. The administration calculated that such changes would save the industry $1.4 billion and save consumers $9.3 billion through 1986. The proposed changes included cancelling the stricter emission standards scheduled for 1984, reducing standards for bumpers to withstand crashes and delaying for a year the requirement that carmakers install air bags or other passive restraints.

St. Louis Globe-Democrat

St. Louis, Mo., February 2, 1981

Liberal members of Congress and environmental extremists scream as if the sky were falling whenever taking an objective look at federal auto safety regulations is suggested. They react as if the auto industry is determined to scrap most of the rules.

Such is not the case. Detroit is seeking the elimination of needless regulations that are costing the industry (and ultimately car purchasers, too) hundreds of millions of dollars annually. This fact emerged in testimony by executives of the top three automakers at hearings before the Senate Commerce subcommittee on surface transportation, chaired by Sen. John C. Danforth, R-Mo.

General Motors reported it spent $2 billion complying with regulations last year. Of that total, $500 million was expended on rules the firm believes are unnecessary. This $500 million represents but one-fourth of GM's expenditures on regulatory compliance.

Ford Motor Co. submitted a list of 17 federal regulations, proposed or in effect, that it says should be abolished. That small number is hardly a drop in the bucket compared to the vast regulatory quagmire that confronts U.S. automakers.

The prompt attention the new Reagan administration is focusing on the problem is a welcome change in Washington's pace. Transportation Secretary Drew Lewis has promised a review of the matter. Vice President George Bush called on auto manufacturers to send him examples of "needless regulations that stifle productivity, economic growth and the creative genius of American enterprise."

Finally, a role may have been found for a vice president. He would serve the country well as an ombudsman who could be the middleman in the efforts to revitalize the U.S. economy.

Some officials of the Carter administration finally acknowledged the problems confronting Detroit. But the admission came at the end of the four-year term — too late to help, even if they had wanted to take that approach. In those waning days, then DOT Secretary Neil Goldschmidt issued a report picturing U.S. carmakers on a collision course with a disaster that threatens to bankrupt all but GM within a decade unless drastic changes are made.

Japanese car manufacturers with their newer, more automated plants are ahead from the start. They have a manufacturing cost advantage of $1,000 to $1,500 over American cars due to lower wages, higher productivity and favorable government regulations.

U.S. automakers lost $4 billion in 1980. Despite their competitive disadvantage, they are faced with the necessity of investing more than $70 billion by 1985 to attain the increase in small-car production capacity needed to turn the tide on foreign small carmakers.

It is estimated as many as 500,000 manufacturing jobs could be lost in the next 10 years if imports continue to flood this country and governmental hostility toward Detroit remains unabated.

Obviously, this is not a problem that can be kept on the back burner. The new administration should act with dispatch to transform its words into effective action that will bring the title of world's No. 1 carmaker back to the U.S.

TULSA WORLD

Tulsa, Okla., November 3, 1981

GENERAL MOTORS officials estimate that they spend $2.2 billion annually to comply with federal regulations.

"'When you're making $10 billion a year that $2 billion is one thing," GM's chairman told the New York Times recently, "but when you're losing $700 million a year as we did in 1980, you look at it and say, 'that's the difference between losing money and making money.'"

That paints a stark picture of the effects of over-regulation. Governmental red tape has become so thick that it poses a severe threat to business survival.

When the Reagan administration assumed office 10 months ago, American auto-makers anticipated an end to bureaucratic over-regulation. But, so far, there has been only minor change.

Two major — and costly — regulations soon come up for consideration by the Administration.

One now requires automakers to install bumpers capable of protection at 5 miles per hour. Automakers want the figure dropped to 2.5 miles per hour.

Another seeks to ease federally mandated air pollution emission standards. GM officials estimate these regulations mean about $1,100 on the price of a new car.

Of course, there is a more important debate here than the cost of federal regulation.

Eliminating the air bag requirement for 1983 models will save the auto industry and automobile buyers millions of dollars. Airbag proponents contend this will be offset by $5.4 billion in medical and insurance costs and lost wages due to auto deaths and injuries which they believe could be prevented by forcing everyone to buy an airbag. Their figures have been challenged as unscientific and arbitrary. No one knows for sure just how well the airbag would work. Nearly every expert agrees that a seatbelt is a much more effective safety restraint — *if motorists would habitually make use of it.*

The ultimate question is who should bear what proportion of the risks of automobile driving and how much individual judgment should be allowed in accepting risk.

The Chattanooga Times

Chattanooga, Tenn., February 16, 1981

Given the sentiment in the Reagan administration and Congress against additional regulation of business, it is likely the fuel-economy rules imposed on the U.S. auto industry will not be extended beyond 1985. In fact, the extension will become unnecessary if Congress responds favorably to periodic calls by some to erect trade barriers for imported automobiles. Nevertheless, unless domestic manufacturers move aggressively to meet the imports' high-mileage attractiveness for car buyers, Detroit's staggering economic problems will increase.

Former Transportation Neil Goldschmidt proposed before leaving office that the United States and Japan negotiate a plan whereby the latter would voluntarily limit its exports to this country. If the increase in the sale of Japanese cars in this country could be slowed over the next five years, then Detroit, through the expenditure of more than $70 billion in research and retooling costs, could by that time become more competitive.

It's possible that such an agreement, under which the American car-buyer would still be able to choose either a foreign or domestic product, would benefit the national interest. But an incentive for the domestic automobile industry to meet, or perhaps beat, the competition would be removed if Congress erected a protectionist wall of import quotas or unreasonable tariffs. Further, if Congress also relaxed federal mileage standards, Americans would be forced back to the position of having to choose expensive — and fuel-inefficient — cars.

There are some, the National Highway Traffic Safety Administration among them, who believe that Detroit should be required to meet federal fuel-economy standards for the rest of the 1980s. The reasoning is that by then, the auto industry could be producing cars that get upwards of 50 miles to the gallon, thereby contributing to a reduction in oil imports.

But with the decontrol of oil prices in this country, and the inexorable increases mandated by OPEC for foreign oil, it's virtually certain that the steady rise in gasoline prices in the U.S. will force American car manufacturers to produce energy-efficient vehicles. Either that, or go out of business. Some Detroit analysts predict that U.S. companies will have to spend more than $100 billion over the next decade to obtain technology comparable to that already used by Japanese and European manufacturers.

Part of that cost will be underwritten by proposed depreciation schedules for machinery, as well as investment tax credits. As for mileage standards, the automakers — their minds concentrated on the $4 billion they lost last year — are perhaps best suited to make those decisions. One of them, Chrysler, is still hanging in the marketplace by its fingernails; it can do nothing else but try to compete with the low-mileage imports.

Since it's virtually certain the nation will never see cheap gasoline again, tougher federal regulations on mileage are probably unnecessary. If Detroit doesn't know now what it must do to compete with the economical imports, it never will.

THE BLADE

Toledo, Ohio, April 2, 1981

THE White House, thank heavens, is honing its machete for a slashing attack on the thick layer of federal regulations that has been choking the auto manufacturers in this country.

A task force under Vice President Bush is taking aim at the plethora of costly rules and codes that set standards for car safety and performance. This is a key part of the Administration's plan to nurse the car builders back to health.

The American automotive industry is one of the most regulated in the world. The price of this massive intrusion by the Government has not been cheap. A study in 1978 by Murray Weidenbaum — now President Reagan's chief economic adviser — found that federally mandated safety and environmental standards increased the cost of the average 1978 car by $666. Since then, of course, the problem has worsened; General Motors officials have estimated that the cost of meeting all the standards on the boards for the 1980s would be more than $800 a car.

Among the targets of this worthy regulatory reform are those mandating heavier bumpers than the industry believes are necessary, emission standards for certain pollutants, and — best of all — the notorious air-bag contraption which some consumer advocates have been attempting to foist off on car manufacturers and the public.

Success in this effort by the Administration will save future car buyers millions of dollars and remove an inordinate burden from the industry. One recent estimate placed the cost of meeting safety, emission, and mileage standards between 1978 and 1985 at about $70 billion. That huge additional cost would come at the very time American manufacturers are struggling to compete with imports and during a period when Detroit is laying out an additional $50 billion or more to tool up to produce new smaller and more efficient automobiles.

Anything that can be done to prune the regulatory thicket imposed by the Environmental Protection Agency, the National Highway Traffic Safety Administration, and the Occupational Safety and Health Administration will be a blow for a healthier auto industry and, indirectly, for a sounder economy. It cannot come too soon.

SAN JOSE NEWS

San Jose, Calif., April 10, 1981

THE Reagan administration's move this week to scrap a slew of auto pollution and safety regulations may have some value as a symbolic political gesture. But its beneficial effect on the floundering American auto industry will be well-nigh imperceptible, and the damage it will do to the environment by relaxing future emission standards for trucks could be significant.

The White House announced Monday that it is removing, relaxing or delaying 34 different regulations to save domestic automakers nearly $1.4 billion and consumers $9.3 billion over the next five years.

Some of the rules on the administration's hit list — for example, a proposed requirement that cars be equipped with warning indicators to signal low tire pressure — are almost frivolous; others, such as the requirements for passive safety restraints (air bags and automatic seatbelts) and slightly more crash-resistant bumpers, we can do without.

In relaxing certain air pollution standards, though, the administration is trading a sizable chunk of the public interest for a negligible and uncertain economic gain.

The Clean Air Act currently stipulates that, by 1984, all cars will have to meet the same stringent air pollution standards now in effect for high-altitude areas like Denver. It seems reasonable for Congress to relax this requirement, as the administration has requested.

Much more questionable, though, are the administration's proposals to ease 1984 emissions standards for diesel cars and light trucks and for gasoline-powered heavy trucks — actions that could mean an important setback in the struggle for clean air in urban areas. Also dubious is the White House's decision not to extend federal fuel efficiency standards beyond 1985. The idea is that the marketplace will force Detroit to keep making more efficient cars, but recent history demonstrates that the marketplace sometimes doesn't send clear signals — and Detroit sometimes misses them.

The deregulation package would be easier to accept if there was reason to believe it would pull the domestic auto industry out of the morass, but it won't. The administration's own estimate is that it will reduce the sticker price of the average car and truck just $150 — not much of a saving on a vehicle that typically costs more than $9,000.

Detroit seems delighted. But American automakers still face the challenge of proving to American buyers that cars made in this country are simply better than those made elsewhere. And government can't do that for them.

Sentinel Star

Orlando, Fla., April 11, 1981

THERE'S something to be said for the Reagan administration's prescription for the ailing domestic automobile industry — delaying or deleting 34 environmental and, safety regulations. But there are still some reservations.

Not many will lament the delaying for one year — or even forgetting — a requirement that all big cars be equipped with either an air bag or a seat belt that automatically grabs the driver when the key is turned. And who can't appreciate cutting the estimated $150 per vehicle the 34 regulations would have cost? Certainly an industry as sick as America's auto works can equate relaxing those rules to a miracle drug.

The Reagan move is expected, over the next five years, to save the industry a capital investment of nearly $1.4 billion and cut consumer costs by $9.3 billion. The theory is that this would improve auto sales and, coupled with tax programs, help America catch up with foreign competitors.

But that said, the fact that most of the industry's ills were self-inflicted — by both management and labor — continues to stick in the nation's craw. And when last week, in the face of lagging sales, General Motors replaced a rebate program with one of the largest sticker price increases ever, the sympathy curve went through the basement.

The domestic auto industry lost a record $4.3 billion last year, much of it because foreign makers grabbed more than a quarter of the market with more fuel-efficient models.

Auto men do a lot of braying about how they were delayed from entering the small-car market by buyer demands for big cars while foreign makers have faced high gas prices for decades. It's true that even after the 1973 oil embargo there have been periodic backlogs for V-8 engines, but such captains of industry have no excuse for not anticipating current conditions.

The truth is, car makers, like most of America's industry, are afflicted with suitcase managers, whose concern rarely goes beyond next year's bonus and who lack the entrepreneurial instinct. Of course, that's not the only problem — accountants replacing engineers in running the companies, rule by committee and too much concentration of ownership also get blamed.

At any rate, the auto industry shows a lack of sincerity in its fight to regain health by running hand in hand with its unions to plead for government intervention with imports, begging for regulation relaxation and seeking government-backed loans. And worse, just when new car sales had plunged to a near-recession low, the biggie of the bunch tacked an average of $350 per unit to price tags.

If any industry is overregulated, it is autos, and many of those relaxed regulations were overrated in importance or are now unnecessary in light of current economics. There's even a case to be made for going easy with the antitrust laws in such efforts as joint ventures in developing better emission control devices. But, let's face it, some of those regulations are important to living and would never have come about without government pressure.

Relaxing some standards in light of auto industry problems makes sense, but it would have been much more palatable had that industry been helping itself a bit more. And if the industry expects the American public to do anything other than send a get-well card, it had better start practicing more preventive medicine.

The State

Columbia, S. C., April 14, 1981

NOW THAT President Reagan has begun to get government off the backs of Detroit's carmakers and workers, he can rightfully expect that they won't try to ride on government's back.

Excessive regulation, controls and standards have been a major cause of the disease that has gripped the industry. But they are not the only cause. Poor planning by management and excessive demands by labor have contributed heavily.

However, the Administration's decision to relax or eliminate 34 pollution and safety rules will give carmakers a fighting chance. The Administration says the move will save manufacturers, who lost $4.3 billion last year, nearly $1.4 billion in capital investment over the next five years. Buyers of automobiles and trucks would benefit by about $9.3 billion.

The Administration will also speed up government purchases of cars and spend $100 million on them this fiscal year.

One major key to helping the industry will be controlling inflation and lowering interest rates. The Reagan people are working on that.

Environmentalists and safety champions are not too happy about Mr. Reagan's move, but Administration spokesmen say the safety benefits of regulations being changed are negligible and the environment will be not suffer significantly.

Now what else can Detroit expect from Washington? Not much.

The industry wants limitation of Japanese automobile imports, but the Administration opposes protectionism and wants to let the free market operate. However, it has agreed to "monitor" the effects of Japanese imports.

The Presidential moves put the burden on the automotive management and labor to be efficient. They must compete in a free market — if they can. Management must not succumb again to bad planning, such as mass production of gas guzzlers in the face of gasoline shortages. Labor leaders must continue to resist the suicidal temptation to demand more and more until they negotiate themselves out of jobs.

The road will not be easy. Some people suspect the automotive industry is a mature one economically, and that government aid to companies like Chrysler simply diverts resources from more fruitful enterprises. Certainly, the nation must be compassionate toward the severe problems of Detroit and understand Detroit's impact on the national economy.

But at this time, the Reagan determination to cut regulations and promote free trade deserves a chance. The next move is up to the industry and its workers.

Plant Closings, Layoffs Beset U.S. Steel

U.S. Steel Corp., the nation's largest steelmaker, announced Nov. 27 that it was closing 10 plants entirely and shutting parts of six others. The action meant layoffs for 13,000 workers out of a total workforce of 165,000. Corporate Chairman David M. Roderick said the plants had become "noncompetitive for a variety of reasons, including operating costs, unfairly priced imports or excessive environmental spending requirements." Roderick's explanation was challenged by government officials, who observed that the steel industry had approved of the trigger-price mechanism established in 1978 to combat imports. (Under the trigger-price mechanism, if imported steel fell below a certain price, an investigation would begin automatically to determine if dumping charges could be brought against the exporter. Dumping meant selling a product abroad for less than its price at home.)

Washington also rejected Roderick's complaint of excessive environmental regulations. Officials said U.S. Steel had insisted on fighting the clean-air rules, while other domestic steelmakers had complied with the rules without hurting their profits. Although U.S. Steel was the country's biggest steelmaker, it was not the most profitable one. In 1978, U.S. Steel accounted for 73% of total steel sales in the country, but its profits comprised only 14% of the industry total. The company's earnings in the third quarter of 1979 were $81.3 million on sales of $3.2 billion, compared with second-quarter earnings of $88.8 million on sales of $2.8 billion.

Steel industry analysts agreed with Roderick that government regulations were a burden for U.S. Steel, but they blamed the company for failing to modernize and respond to competition from foreign producers. Roderick pledged Dec. 5 that U.S. Steel would embark on a new strategy to improve productivity. In the past, U.S. Steel had been criticized for refusing to adopt new technology unless it had been developed by the company itself. Roderick said that, in the future, U.S. Steel would buy the best technology available. He pointed to recent Japanese assistance in building a blast furnace at one facility and an arrangement with Sumitomo Metal Industries to solve production problems at another plant as examples of U.S. Steel's new policy. Roderick added that the company would reduce its range of products, a major departure from previous policy of competing in all steel markets.

THE TENNESSEAN

Nashville, Tenn.,
December 5, 1979

IT IS depressing year-end news for some 13,000 workers and the communities involved that the United States Steel Corp. plans to close 15 plants.

The steel plant closings are the largest announced this year, but there are other plant closings going on as the corporate world eyes recession and begins to chop unprofitable plants.

Atlantic Richfield is shutting down a chemical plant in Texas. St. Joe Minerals is firing 1,500 workers at a zinc smelter. And there are others. Leaner more efficient operations are the trend.

In the case of U.S. Steel, much of the blame for the closings has been placed at the door of the federal government because of its trade, environmental and tax policies. Obviously there is some truth in this, but the larger cause is that the domestic steel industry has been ailing for some time.

While the Japanese and West Germans were shifting to modern facilities and more efficient processes, the American steel industry was lagging in making changes. Its plants were outmoded, its efficiency down and, in short, it was becoming less and less competitive.

The industry has begun to look at itself in terms of cutting fat out of the operations. Two years ago Bethlehem Steel went through cutbacks. The same year Lykes Corp. laid off 4,000 workers at its Youngstown, Ohio, plant.

Although the impact of the closings will be very painful to employees and communities alike, the changes may give the industry a more competitive edge ultimately.

The Virginian-Pilot

Norfolk, Va., June 24, 1979

Detroit, which produces rumors as readily as it does the deep-breathing highway behemoths so beloved of oil sheikhs, has a new imported model. The elves in Wolfsburg, West Germany, who make Volkswagens are said to be poised to snap up Chrysler, the least of the automotive Big Three.

True or not, the notion has a certain logic to it. Volkswagen bought a new but unused Chrysler plant in Pennsylvania and is cranking out copies of its Rabbit. Chrysler buys engine blocks from Volkswagen for its own subcompact cars. Chrysler's earnings have been pallid of late, and for the last quarter were off nearly $54 million. And investments in the United States by overseas holders of billions of shrinking Yankee dollars are soaring.

Notwithstanding local disappointment over a Swedish Volvo plant in Chesapeake that is yet to flourish, the Volkswagen-Chrysler rumor has a certain appeal. If West German cash and know-how would perk up Chrysler, the domestic benefit in jobs and cash turnover would be welcome.

Yet the thought of Volkswagen's coming to Chrysler's rescue has its somber side. It bolsters the feeling in some places that the world's industrial giant isn't getting around as well as it used to. The editor of *Production Engineering* picked at that sore in his June issue. He grumbled over news that a Japanese firm would provide the technical advice and personnel training for a steel mill to be built in Pennsylvania. The new process involved was patented by a French company.

While cheering the advent of "the hottest technology," the editor was dismayed that the United States had let its lead in steel-making slip to the point that Americans must import means and knowledge. The point is well taken. Foreign industrial investments here may yield demonstrable economic benefits. But they don't do much for native pride and prestige.

THE RICHMOND NEWS LEADER

Richmond, Va., November 30, 1979

When U.S. Steel — the nation's 15th largest manufacturer — announced a retrenchment the other day, its board chairman, David Roderick, was frank about the causes: "The operations being terminated at this time have been noncompetitive for a variety of reasons, including operating costs, unfairly priced imports, or excessive environmental spending requirements."

In short, the anti-business attitude that prevails in Washington has mounting casualties. U.S. Steel will lay off 13,000 workers — 3,500 of them in shell-shocked Youngstown, Ohio. Two years ago, Youngstown lost 4,000 jobs when a major sheet and tube mill closed. In January, the Lykes Corporation will close its Youngstown mill, with a loss of 1,500 jobs. And shortly after U.S. Steel's announcement, Jones and Laughlin Steel revealed that it will close down in Youngstown and lay off 1,400 workers. The jobless rate in Youngstown will move into double digits.

Jobless workers who are the ultimate victims of the government's crack-the-whip war on business can blame their benevolent Uncle Sam for most of their troubles. Many of the nation's manufacturing plants are obsolete and thus noncompetitive, owing (1) to federal tax policies that discourage plant investment, and (2) to perpetual budget deficits that gobble up much available risk capital. Environmental regulations require outlays of millions for anti-pollution devices that produce no advances in productivity, job creation, or plant modernization.

In U.S. Steel's case, other factors complicate efforts to compete. Demand for steel has been dropping, as aluminum has gained wider use — especially in Detroit's auto manufacturing plants. Ford Motor Company, for instance announced yesterday that it would close 10 more of its plants next week. That will leave only four of Ford's 15 plants in operation.

U.S. Steel also says that the government does not allow fully for inflation in setting import tariffs on imported steel. A Japanese manufacturer, for instance, can import coal from West Virginia and export steel back to the U.S. at prices lower than the cost to produce it here. American producers contend competition from steel imports that undercut domestic steel constitutes dumping.

American business has waged a courageous battle for survival in the face of government enmity and public hostility to reasonable profits. Manufacturers have provided millions of well-paying jobs that promote higher living standards. They give America goods and services that are the envy of the world, while providing their stockholders with dividends drawn from their profits. Their sponsorship underwrites many cultural programs and projects that enrich their communities.

Yet business has the undeserved image of Bad Guy, and profits are regarded as loot stolen from the public. The miracle is that so many businesses and manufacturers endure. Thousands don't: They smother under the weight of acronymic oppression from Washington in the form of OSHA, the EPA, the FTC, the ICC, the EEOC, and a host of other bureaucracies inimical to free enterprise.

No doubt the reaction in Washington to U.S. Steel's announcement was a shrug and a smirk: The bigger they come, the harder they fall. Big business is bad, right? No. For proof, ask Youngstown.

The Dispatch

Columbus, Ohio, December 24, 1979

ONE OF THE reasons why U.S. Steel is planning to shut down 15 of its plants is that it is being forced to ring out the old without ringing in much of the new.

Case in point: Its Youngstown Works, one of the 15 plants slated for closure, has a rolling mill powered by a steam engine installed in 1908.

One of the reasons the American steel industry is finding it difficult to meet competition is that German and Japanese mills are up-to-the-minute modern.

And where did the Germans and Japanese get all of this fine technology? For the most part, from America, much of it in the post-World War II Marshall Plan era. So efficient has been this modern technology that Germany long ago was able to repay all of its Marshall Plan loans and Japan is near the point of liquidating its obligations.

If modern American technology was able to give Germany and Japan such efficient steel production why have not American mills used the same American technology? Good question.

Part of the reason is because German and Japanese governments extended special tax incentives to their mills and part is because German and Japanese steel workers were less demanding than their American counterparts. In the last 10 years, wages and benefits in the U.S. have risen from $5.38 to $16.53 an hour. Worse, productivity growth has declined from 3 percent to 2 percent a year. America's steel noncompetitiveness is built in.

TULSA WORLD

Tulsa, Okla., December 7, 1979

THE U. S. steel industry, suffering from antiquated facilities and repressive Government policies on investment, is losing its competitive battle with foreign steel, which is working with relatively new plants and friendly Governments.

The extent of domestic steel's problem was illustrated by the giant U. S. Steel's decision to close all or parts of 15 unprofitable plants. That move sent shock waves through the economy.

But perhaps even more alarming is the giant firm's admission that its technology is no longer the best in the world.

Board Chairman David M. Roderick announced his company has retained the services of Japanese firms to supply technical aid and equipment to try to reverse the fortunes of several steel mills that have been kept open.

U. S. Steel, along with the rest of the domestic steel producers, have been accused of refusing to employ any innovations not developed within the country.

That criticism can at least no longer be made.

But it is not a development of which U. S. Steel or the public in general can be proud.

It is an admission that Japan, as well as West Germany and other countries to a lesser extent, have started from scratch after World War II and surpassed the once dominant U. S. steel industry.

It has been done initially with help from the U. S., but mostly through Government policies which encouraged, rather than discouraged, new plants, new equipment and research and development by foreign firms.

American steel firms have been denied those policies. The huge sums of capital needed to keep abreast of foreign competition instead have been diverted to wasteful environmental equipment far beyond what should have been reasonably required.

Construction of new plants and purchase of modern equipment has been held back by unrealistic depreciation laws.

The Japanese technicians no doubt will provide some help to U. S. Steel, but what American producers really need is an infusion of Japanese political talent.

Steel's Nagging Troubles Attract Government Attention

The steel industry entered a depression that persisted throughout 1980. Production figures reported June 10 indicated that factories were operating at less than two-thirds of capacity. Total raw steel production for the first week of June was 1.83 million tons, down from 1.86 million the previous week. In comparison, production for the same week in 1979 was 2.84 million tons, representing 93.2% of capacity. One industry analyst estimated that 60,000 steelworkers had been laid off because of production cutbacks. In an effort to spur sales, major producers cut prices at the beginning of July.

At the same time, a report released by the Office of Technology Assessment warned that the country was in danger of becoming dependent on imports of steel unless domestic producers once more became competitive. The congressional agency recommended initiatives from government to encourage industry modernization while averting the threat of massive unemployment among steelworkers.

Against the backdrop of a serious economic crisis in the steel industry and the pressure of an election year, President Jimmy Carter introduced a series of proposals to aid the industry Sept. 30. Carter proposed reinstating the trigger-price mechanism, which had been suspended in March, and extending the deadline for steel plants to meet clean-air standards. He stressed that his plan to liberalize depreciation allowances for businesses would, if implemented, be a major asset to the steel industry.

The Philadelphia Inquirer

Philadelphia, Pa., July 24, 1980

The American steel industry is in trouble and has been for quite some time. The question is how to get it out of trouble. Since the late 1960s, the steel men's answer has been to demand protection against allegedly unfair competition from abroad. Last March, U.S. Steel filed a formal complaint against seven European countries, charging they had been dumping steel, that is, selling it below their costs of production, on the U.S. market.

The Carter administration responded by suspending the "trigger price mechanism" (TPM), which it had launched in 1978. Under TPM, a government investigation is "triggered" automatically when steel products are imported below the cost of production calculated for the most efficient producer — Japan. In effect, TPM sets a minimum price for steel. The administration managed to persuade America's major trading partners and allies that TPM was the least worst of evils.

Now the Carter administration, itself in trouble, is looking for ways to help the American steel industry. One way being explored is a negotiated settlement of U.S. Steel's antidumping complaint with a strengthened TPM. An administration official said, however, that "the government will not attempt to deal with the problem of steel by ignoring trade." That is fine as far as it goes, but the pressures in this political year are manifest.

The direction ought not to be toward protection. That would trigger inflationary pressures as well as retaliation by Europe and Japan, which also happen to be American markets.

What the domestic steel industry needs is not a crutch but a carrot and a stick. The carrot involves government policies such as accelerated tax credits for purchase of new machinery or facilities (and not just for steel) to encourage capital formation. It also involves policies aimed at getting the whole U.S. economy out of trouble.

The stick involves government policies to encourage competition, from abroad as well as at home. The evidence that the domestic steel industry has been injured by imports is not very strong, as the Wall Street Journal reported recently. One could make a better case that foreign competition has been a prime mover in forcing the domestic steel industry to get off its duff and modernize. Instead of bleating about unfair competition, American steel men would do better to borrow a slogan from Avis and try harder.

THE PLAIN DEALER

Cleveland, Ohio, June 28, 1980

Congress' Office of Technology Assessment has sounded an ominous warning: A continued decline of the domestic steel industry poses a repeat of the American experience of heavy dependence on foreign oil.

Had such a warning originated with the steel industry, it might be discarded by some people as corporate tactics to gain support for tax, regulatory and other reforms. But coming from a specialized and technology-oriented office of Congress gives it even more credibility.

A two-year study by the office suggests that increasing dependence on steel imports could in periods of strong world demand put the nation in "a shortage and price situation similar to its dependence on foreign oil."

This trend would be hazardous to the nation's health. There are serious implications for both the economy and national security. Consequences of continuing dependency on foreign sources for such a basic material must be avoided.

Clearly, there is a need for a political consensus in Washington to preserve and begin modernization of the steel industry. Rep. Charles A. Vanik, D-22, who asked for the study, is correct in insisting the government needs a national re-industrialization policy for steel.

"There is no question of the validity of the study," Vanik said. He is chairman of a House trade subcommittee.

The study finds that big steel producers are not entirely blameless. They have helped fulfill their own prophecy of decline. They have allowed their industry to reach a point of obsolete plants and equipment. Profitability could be raised by shutting down old plants. But that would increase unemployment and further stimulate demand for foreign steel, increasing the threat of dependency.

A cure is not beyond reach. The study suggests it lies in a combination of remedies, including reform of tax incentives for investment, federal support for spending on research and development of new technology, write-offs for costly pollution control spending and strengthening government procedures for detecting unfairly priced foreign steel. All are worthy of consideration in Washington.

A positive government response would then put it to the industry management and unions to improve productivity and restore the ingenuity that made America's steel industry once worth emulating overseas.

The Salt Lake Tribune

Salt Lake City, Utah, September 16, 1980

When Congress was considering the Clean Water Act back in 1977 the steel industry tried, and failed, to get itself exempted from the act's provision by asserting that the resultant loss of jobs and resources would be against the national interest.

Congress didn't go along, primarily because it would have been unfair to the other 80 or 90 percent of industrial dischargers who, at that time, had already complied with anti-pollution standards.

Steel manufacturing is a notoriously dirty undertaking and to get that industry to clean up its act will be expensive, but not impossible. Recognizing this fact, Congress required industry, including the steel industry, be in compliance with the Clean Water Act by 1984.

Congress that same year passed amendments to the Clean Air Act of 1970 setting a last ditch deadline of 1982 for compliance.

The steel industry, generally, has fought these legislative provisions vigorously, for the most part refusing to agree to compliance programs outlined by the Environmental Protection Agency.

And the industry is still resisting EPA insistence that it comply with the congressional mandates, claiming compliance would be disastrous; wiping out whole plants and causing massive unemployment in steel towns. That resistance has borne fruit, of a sort.

The EPA has agreed to recommend to President Carter that he, in turn, ask Congress to grant steel firms three year extensions of the 1982 clean air and the 1984 clean water deadlines, until 1985 and 1987 respectively. Attached to these recommendations is the provision that the industry speed modernization of its aging plants.

The recommendations are keyed to the assumption that the steel industry doesn't have the capital to modernize its facilities and clean up its pollution at the same time.

Initially, the proposal sounds as if the EPA has "caved in'" to the steel industry. A look at the details says differently.

The recommendations would allow compliance delays on a plant-by-plant basis providing that:

— Delay is necessary to allow a steel plant to modernize its production equipment.

— Money that would have been spent on pollution controls is used for modernization.

— EPA is satisfied that the steel firm will have sufficient funds to meet pollution control requirements later on.

— The firm agrees to clean up any of its facilities not now in compliance with clean water and air standards.

Considering the fundamentally antiquated state of the American steel industry, as against that of its principal foreign competitors, West Germany and Japan, there is something very pragmatic about the EPA's willingness to concede to the industry's request for an extension.

While the national interest theme persists in the industry maneuver, it has been pleasingly muted by recognition that even the steel industry is obliged to help all Americans enjoy the cleanest air and water possible. This bit of late 20th century reality was arrogantly missing from the industry's earlier attempt to obtain total and complete exemption from the nation's anti-pollution laws.

It now appears that not only will the steel industry's plants be modernized, but so will its sense of corporate responsibility.

The Washington Star
and Daily News

Washington, D. C., July 18, 1980

The auto industry's problems have been more widely publicized, but domestic steel may be in worse shape. Both face competition from abroad, but the steel industry has complained that much of theirs is unfair — a view reflected in U.S. Steel's anti-dumping suits. Both need to modernize, but the steel industry, with less investment capital, faces an annual shortfall of at least $600 million.

And while the auto industry gears up to compete against gas-saving imports, the Office of Technology Asssessment reports that high-priced steel imports could acquire as much as 44 per cent of the domestic market by the late 1980s as against 15-to-20 per cent in recent years. The report, prepared at the request of the House subcommittee on Trade, also notes that while domestic steel production increased by 20 per cent, world demand doubled — and Japan's production increased seven-fold.

It is not good news; and, in light of OTA's findings, neither is it surprising.

The report found — and the industry was quick to agree — that domestic steel has been hurt by federal policies which often have been "uncoordinated, contradictory and inattentive to critical issues." For example, OTA cites the government's refusal to allow steel producers to take energy investment tax credits by adopting continuous casting — a process which replaces several steelmaking steps in one operation — even though it promotes energy conservation.

Indeed, the slowness of U.S. steelmakers to adopt continuous casting is a measure of its technological lag. In 1978, Japan reached a 50 per cent level and the European producers 29 per cent. The U.S. level was 15 per cent. OTA found that while domestic steel producers have a good record for *product* innovations, the record "does not extend to the internal creation of new *production* processes."

The report places the blame on both industry and government. It recognizes that the domestic steel industry is not monolithic. Many smaller steelmakers, in fact, are expanding and profiting from the use of iron scrap even as larger "integrated" steelmakers, using iron ore conversion, are closing plants.

OTA suggests that the industry can be revitalized if money is available for capital expansion and calls for federal policies to promote capital recovery, including accelerated depreciation schedules and incentives for industrial R & D.

A strong steel industry is clearly in the national interest; evidence pointing to further decline is cause for worry. Even the industry's profits, OTA points out, are often at the expense of long-range planning and technological innovation.

What is to be done? Steel is not now a candidate for federal "bail-outs." Nor will the Carter administration's two-year-old "rescue plan" — the main ingredient of which was a trigger system to prevent dumping of foreign imports — ever do more than forestall coming to grips with the central problems: too little investment capital, not enough R & D, the high cost of regulatory compliance.

The OTA report bespeaks the seriousness of the question. It observes that "steel remains the most important engineering material in American society"; and it adds that without modernization the U.S. may face "a degree of steel import dependence (and) economic and national security problems . . . not unlike those now encountered with petroleum."

Newsday

Long Island, N. Y., October 6, 1980

The worst we can say about the White House's plan to rescue the American steel industry is that it's largely a packaging job, wrapping pieces of President Carter's economic proposals of last summer together with revived import restraints.

The best we can say is that the pieces aren't such bad ones, and that the very modesty of the package is something that recommends it.

That isn't intended to sound like faint praise, either. The nation's steel industry isn't as robust as it once was, but we're not at all sure it needs to be. If foreign suppliers can sell steel in this country for less than domestic producers can, why not buy more of it from abroad and let U.S. industry turn to things it can do better?

American steelmakers aren't on the verge of collapse anyway, and their problems aren't unique: Like many other U.S. manufacturers, they need new technology and equipment to operate more efficiently. So it seems reasonable to offer them the same sort of tax incentives to make the necessary investments that Washington offers to other industries.

The major difficulty with this approach is that it threatens to increase next year's federal budget deficit. That's likely to push up interest rates, thus robbing steel and other industries of the incentive it's intended to create. Congress will have to deal with that problem when it considers tax cut legislation.

Two other pieces of the package do amount to special favors, although they strike us as tolerable ones.

One is the revival of a government safeguard against steel imports that are unfairly low in price. It will mean higher prices for domestic steel, but without some sort of compromise on import policy the American steel industry might have successfully pressed a lawsuit seeking far stronger restraints on imports from European countries that subsidize foreign steel sales. This could have inspired a trade war that would painfully damage U.S. exports.

The other favor is a postponement, assuming Congress approves, in enforcing environmental standards on some aging steel plants. The industry, which is a major polluter, has been slow to clean up its operations, complaining that the cost has blocked new projects that would increase efficiency.

It's too bad to drag out those delays, even briefly. But it's an acceptable trade-off if it will encourage investment in more productive, and cleaner, new plants. Now, at least, we'll see if the industry will put its money where its smokestacks are.

The Miami Herald

Miami, Fla., October 9, 1980

PRESIDENT CARTER has cut a deal with the nation's troubled steel industry. His promises of Government help are not without virtue, though some elements of his program do unfortunately stink of political tradeoffs made at the public's expense.

Still, political realities being what they are, perhaps it's lucky that the stench was not more noxious. There is no question that the steel industry is in bad shape. The recent recession has forced drastic cutbacks in production. The number of American steelworkers employed in July was the lowest since 1933.

Short-term, anti-recessionary aid will not cure the industry's problems, however. America's steel industry suffers from fundamental flaws. Its outdated plants and equipment are no match for modern, efficient plants abroad.

If America's steel industry is to remain viable, it must rebuild, modernize, and trim its own fat to meet international competition. Part of Mr. Carter's proposed relief program would help it do that.

The President would spur investment in new plants and equipment via the 40 per cent accelerated-depreciation schedule and a doubled investment-tax credit that he first proposed Aug. 28. That is a sound approach; Congress should go along.

The President also felt it necessary, unfortunately, to agree to seek delays in steel-industry compliance with clean-air standards. Mr. Carter would permit the Environmental Protection Agency to allow case-by-case delays in compliance.

Congress should take a cold look at that promise. The steel industry argues that money saved from delaying purchase of pollution-control devices could be invested in more-efficient plants and equipment. But the Government has an obligation to the nation to ensure that those new plants meet clean-air goals.

Mr. Carter also pledged to restore the suspended trigger-price safeguard against foreign "dumping" of unfairly priced steel on the American market. That concession will be somewhat inflationary and protectionist. Two factors partially offset those objections, however.

First, foreign producers at times have dumped their steel in the U.S. market below cost, earning a protective reaction. Second, in return for reviving the trigger-price protection, Mr. Carter got U.S. Steel Corp. to drop its suit against seven European Steel producers.

Those complaints had threatened to result in a European retaliation of sanctions against American exports of grain and textiles. Now, however, European Common Market officials have pledged not to contest the new steel trigger-price mechanism.

Thus Mr. Carter's nod toward a little protectionism for steel was conditioned upon rejection of worse protectionism in Europe. Mr. Carter's ability to engineer such a trade-off displays a previously unnoticed diplomatic finesse. And that is the best thing that can be said about his politically motivated steel-industry relief program.

Post-Tribune
Guarding Your Interests Daily
Gary, Ind., September 24, 1980

Developments of late last week hardly served to clear out all the fog of doom and gloom which has surrounded the steel industry in recent months. For those who weighed them carefully, though, they afford a clearer, more encouraging perspective.

The gloomsayers had targeted in varying degrees (1) the entire domestic American steel industry, (2) steel and other production in the Midwest, and (3) Northwest Indiana's part of the Midwestern steel output. It had two basic components. One was the recent drastic drop in steel demand keyed primarily though not exclusively to weakened automobile sales. The other was the claim of steel's harshest critics that its troubles where imports were concerned were due largely not to unfair foreign competition, but to failure to modernize.

Well, Inland Steel took a big step toward refutation of that latter argument in its dedication last Friday of the $1 billion largest blast furnace in the Western Hemisphere. It wasn't a surprise. It had been under construction for years, and its dedication was a media event.

It was, however, an important one. It underlined that right here in Northwest Indiana, despite what critics have said, there is an ongoing modernization of steel production facilities. This particular Inland innovation may have been the most important such step since completion of Bethlehem Steel's Burns Harbor Works a decade ago.

Meanwhile, the dedication brought out Secretary of Commerce Philip M. Klutznick, who, while withholding details, confirmed that the national administration was bringing forth a new steel program in a few days. It's important that the political sector realizes it is taking such steps against a background of steel helping itself.

But Inland's was not the only important area steel development.

Word came that Calumet Industries, Inc., was leasing the abandoned Gary American Bridge Co. plant, the closing of which was one of the sadder recent area steel stories. A steel fabricating firm, Calumet Industries is one of the type needed to keep the area's industrial growth in key with its basic product.

Meanwhile, Bethlehem announced it would probably put its plate mill and a temporarily idled blast furnace back into operation within the next two weeks.

The generally good news was not without what some saw as negative aspects. Some picketed the Inland dedication contending such modernization meant a cutting of jobs. In one sense that could be true. A prime reason for modernization is more economical production. However, only through such improved efficiency can the steel industry in this country and this area hope to stay competitive and provide jobs.

Some saw the American Bridge plant's leasing as the final nail in an area coffin. In a sense, of course, it may be. It's final proof that U. S. Steel meant what it said in telling workers it couldn't continue in the bridge business unless they went along with some cutbacks, as union locals did at some other Americn Bridge plants. That, too, is part of the productivity story.

Neither the Inland major improvement, nor U. S. Steel's plant leasing nor Bethlehem's revving up means steel has turned everything around. All, though, fit into a pattern of continuing area industrial health. Don't count steel out.

THE CHRISTIAN SCIENCE MONITOR
Boston, Mass., October 1, 1980

We tremble whenever the US government announces it is prepared to rescue an ailing American industry. We are especially uncomfortable when such plans are announced in the middle of a tight presidential election race. In this case there is good economic reason to give the US steel industry a lift out of its doldrums. But a more dispassionate consideration of the matter would have been possible after Nov. 4. As it is, there is no question that President Carter has unveiled his steel rescue plan now in hopes of picking up political support in the crucial states across the great industrial belt.

What Americans will wonder is whether the program, which will give the industry more protection from foreign imports and more time to meet environmental standards, is economically justifiable. Or whether this is just another and unwarranted slide toward greater government involvement in the private sector – something both Democrats and Republicans profess to resist.

There indeed is a case for short-term aid. The US steel industry has been buffeted by growing competition from abroad. Foreign imports, now at about 16 percent of total steel use, could climb to as much as 40 percent of the domestic market by 1990 if the industry does not get back up on its feet, a congressional study warns. Yet with an outdated plant infrastructure and with some of the highest labor costs in the industrialized world, the industry finds it hard to turn things around when faced both by the growing imports and the tough requirements of the Clean Air Act.

The Carter plan will give the industry more time to comply with environmental standards. It liberalizes depreciation rules. And it also reinstates and raises the so-called "trigger price" at which government investigates whether imports are being dumped at below-cost prices. The latter system is justified by the administration on grounds that steel is not being produced generally on a truly competitive basis abroad because of the tax concessions which governments give their industries.

These measures seem reasonable given the magnitude of the problem. It is argued, and persuasively, that the government cannot permit such a basic industry to go under. A strong steel industry is essential to the nation's military and economic security and therefore warrants special consideration. Imagine the situation which would arise in wartime, say, if the US were heavily dependent on steel imports and ocean shipping were disrupted.

On the other side of the ledger, however, is the US steel industry's own record of mismanagement. When Japanese and European steelmakers were rebuilding or updating their plants with the latest technology, US firms were more worried about short-term profits than modernizing their mills and maintaining long-term competitiveness. As a result, companies were not able to direct any excess capital to upgrading production facilities and, as a result, productivity fell.

If the government now begins stepping in with short-term relief, at the least it should require more competence in management of the industry. Such relief should be made conditional on the steel firms modernizing their production facilities and on greater labor-management cooperation to boost productivity. Fortunately, the Carter plan does require steel firms to invest savings from postponed antipollution measures in plant modernization.

What concerns us, however, is the widening pattern of government intervention in the private sector. We have seen a financial bailout of Lockheed and then Chrysler. Is the steel industry next? Where will the trend stop and what will this do to the internal free-enterprise system as well as America's ability to compete effectively abroad? Republicans and Democrats alike recognize the importance of free trade in a world growing more and more interdependent. Yet it is clear that if the US begins to protect its own industries and throw up higher barriers to foreign imports, it will simply begin to spark retaliatory measures against US exports – and then other American industries will be hurt. There is no way to win a trade war.

This is the crucial long-term consideration the government must not overlook.

FORT WORTH STAR-TELEGRAM

Fort Worth, Tex., February 2, 1981

As the automobile industry goes, so to a large extent goes the steel business. And neither is booming at the moment.

But even if the automobile industry makes a complete revival, the outlook for steel will remain slightly hazy.

With fuel efficiency the new order of the day, the marriage between autos and steel is on shaky footing. The reason is simple: steel is heavy, and if automobiles are to deliver more miles per gallon, they must lighten the load.

So the builders of American automobiles are turning to lighter materials, such as aluminum and plastic, much as foreign auto makers have been doing for years.

Unquestionably, there are drawbacks to this conversion. Occupants of automobiles made of lighter weight material aren't as safe from collision impact as would be occupants of cars made of steel.

But everything has its price, and part of the price motorists will pay for greater fuel efficiency in the automobile of the future is increased vulnerability on impact. But, technology being what it is, this problem no doubt will be overcome as the auto manufacturers continue to strive toward the ideal combination of fuel efficiency and safety.

And in that technology probably lies the key to the American steel industry's future. A recent report out of Pittsburgh, one of the hubs of America's steel industry, contained some encouraging information along these lines.

Steelmakers, the report said, are working hard toward developing lighter and stronger materials for the manufacture of automobiles.

If they succeed, then steel will become truly competitive with aluminum and plastic, and the automobile industry will benefit greatly.

And that means so will the driving public.

And that's just about everybody.

THE INDIANAPOLIS NEWS

Indianapolis, Ind., May 20, 1981

Despite recent predictions that the Frost Belt is doomed to decline, it would be premature to start packing for the Sun Belt. Those who take a reasoned approach to the future of the industrial Northeast-Midwest will leave their suitcases in the closet.

The steel industry is a case in point.

Signs of the decline of America's steel industry have been written like graffitti in the crumbling neighborhoods of Gary and other cities where steel was queen. Last year raw steel production was just 23.3 million tons — the lowest since 1946. The cities that staked their fortunes on steel and its related auto industry have paid in human terms for the losses on industry's balance sheets.

But the outlook now calls for renewed hope. And if the Reagan administration recognizes the industrial potential waiting to be encouraged, surely it will do so.

In the last 20 years, effective domestic steelmaking capacity has barely increased, while domestic steel consumption has risen 60 percent. As a result, domestic production has not met demand since 1960, and the United States has become dependent on foreign-made steel.

Now American steel appears ready to stand up and fight for its share of the market. With world demand for steel expected to continue at record-breaking levels, U.S. Steel and the other giants have been getting slowly to their feet.

In order for the American steel industry to maintain its share of the domestic market in 1985, it would need to increase annual capacity by about 25 million tons. A recent General Accounting Office report concluded about 25 percent of the industry's physical plants are too old to efficiently compete with foreign producers.

Unless American steel plans to relinquish its primacy in the domestic market, an era of investment and growth is essential. At U.S. Steel in Gary, where the works have just celebrated a 75th anniversary, signs of vigor as well as age are visible. U.S. Steel has spent more than $600 million over the past six years to modernize its plant. Last year it showed a slim operating profit.

To prove the predictions of mass exodus are wrong, now is the time for government to get behind steel in a coordinated effort to encourage reindustrialization.

In its recent report, the GAO criticized such efforts in the past as piecemeal, lacking a comprehensive policy. The industry contemplates a 25-year rebuilding program, for example; government has come forth with legislation looking five years ahead.

Essential, of course, would be more rapid depreciation of steel facilities for tax purposes. Government also must stop its meandering between free market reliance and protectionism through the trigger-price mechanism. In the long run it will find it must permit steel prices to respond freely to market forces.

America has no intention of abandoning its traditional industrial base or the cities that house it. The steel industry, where the prospect of increased demand invites investment and growth, is a good place to prove it.

The Houston Post

Houston, Tex., May 19, 1981

South Korea is making steel so efficiently and cheaply that Japan is buying it. This portent for the U.S. steel industry goes beyond what the Koreans are doing. Predictions are that most of the world's expansion in steelmaking between now and the year 2000 will come in the developing countries rather than in the United States, Europe or Japan. Meanwhile, the world's steel overcapacity is expected to continue for years.

Under such conditions the outlook for the U.S. steel industry would appear bleak. But domestic steel producers can compete with the foreign mills by continuing what they are doing — building state-of-the-art plants and closing or upgrading obsolete facilities. Short of subsidies, the industry needs and deserves the federal government's support by way of incentives for plant modernization and energy conservation and by protection from unfair foreign competition.

The emergence of South Korea as a steelmaking power shows what the U.S. steel industry will be up against in the years ahead. Under government sponsorship, the Pohang Iron and Steel Co. built a $3.3 billion integrated steel works on the east coast of South Korea — bigger than any in the United States — and plans are being made for another of comparable size. The second plant would raise South Korea's steelmaking capacity to more than 19 million tons a year, about equal to the projected domestic annual need in the 1990s.

However, Pohang is now, and plans to continue, exporting 30 percent of its steel production in order to raise foreign currency to import iron ore and coking coal and to pay off foreign debts incurred in building its plants. An increasing amount of that foreign exchange is coming from Japan where the Korean steel is competitive in price with the domestic product. South Korea's output is still small in comparison with that of any major steel-producing country. Japan, for example, has an annual capacity of 150 million tons. But for reasons similar to South Korea's, other developing nations are expanding their steel production to meet domestic needs. They include China, Taiwan, India, Mexico and Brazil. Those countries also need foreign exchange and are exporting some of their steel to get it.

Low labor costs or governemnt subsidies, or both, are often factors in the low steel prices that developing countries can offer. But modern, integrated and automated plants are the main reasons for their competitiveness in the world markets. Under a favorable domestic business climate, U.S. steelmakers can be just as competitive. U.S. Steel Corp. pointed the way by closing down obsolete and inefficient plants in 1979 and turning to high-technology, cost-effective facilities. Although its production was down in 1980, the company turned a profit of $58 million from its steel manufacturing operations compared with a loss of $102 million in 1979. Other steel companies are modernizing as fast as possible. This is the key to soundness not only for U.S. steelmakers but for any other domestic industry that is vulnerable to overseas competition.

Japan Moving to Forefront in Computers, Hi-Tech Areas

Bruised over the Japanese victory in automobile manufacture, America contented itself with thinking that at least in the field of computer technology, the U.S. was securely number one. However, even "hi-tech" was not immune from Japanese competition. In 1980, the U.S. ran a trade deficit with Japan in semiconductors, the memory chips used in advanced computers. Japan also was forging ahead in creating chips with ever-greater memory capacities at cheaper cost. The situation was serious enough for high-level discussion between Washington and Tokyo. During a trip to the U.S. in May 1981, Japanese Premier Zenko Suzuki "agreed in principle to a mutual reduction of semiconductor tariffs," according to special U.S. trade representative Bill Brock. At that time, Japan levied a 10.1% tariff on imported semiconductors, while the U.S. tariff rate on imported chips was only 5.6%. The tentative agreement would reduce both countries' tariffs to 4.2% by 1982, if implemented. Semiconductors were a particularly sensitive issue in the U.S. because they were the object of an open "target" policy by Japan to overtake America within 10 years. The strategy involved low-interest government loans to business to subsidize research and development while maintaining high tariffs to discourage the sale of imports.

THE ARIZONA REPUBLIC

Phoenix, Ariz., December 21, 1981

JUST as the United States lost preeminence in automobile manufacturing to Japan, it now seems to be losing the lead in high technology to the Japanese.

This should be of special concern for Arizona, where high technology industries have led the state's industrial growth.

Yet, the U.S. discourages the marketing of U.S. high technology abroad. Unless government and manufacturers devise more inventive and less cumbersome methods to meet overseas competitors, the U.S. will be a runner-up in the world market.

The United States and Japan are now in a race to develop new aircraft, lasers, industrial robots, computers, semiconductors and fiber optics.

The Pentagon underwrites 30 percent of basic U.S. research and development while Japan's Ministry of International Trade and Industry (Miti) pays only 16 percent.

However, Japan adds something else.

Miti helps Japanese firms become internationally competitive.

Miti allows firms to collaborate on specific basic research projects but forces them to compete intensively in the marketplace.

The Pentagon awards many large contracts with virtually no competition.

A Miti R&D project may span a decade or more, while Pentagon contracts undergo relatively sudden changes because of defense shifts and political dictates.

High-tech marketing requires a global strategy. Miti has the authority to assemble export finance and controls, industrial location and expansion, and R&D.

Miti encourages exports through low-interest financing, subsidies to enter new overseas markets, and subsidized insurance for foreign risks.

The U.S. discourages high-tech export expansion by imposing strict controls over deliveries abroad.

The Japanese have an open door to do so.

At the same time, the U.S. subsidizes Japan's economy by spending far more on pure defense costs.

While Miti centralizes Japan's high-tech activities, the United States disperses functions throughout the government — in NASA, the Export-Import Bank, the Department of Commerce, the U.S. Trade Representative and others.

The Pentagon cannot be blamed. Its responsibility is defense, and although high-tech advances are important, they are nevertheless only a spinoff of the military's activity.

The responsibility is that of President Reagan and the Congress.

Perhaps only Reagan — with his deep belief in the free market system — can see the need for exceptional action to meet the competitive challenge laid down by Japan.

WORCESTER TELEGRAM

Worcester, Mass., November 12, 1981

Ever since the Japanese humbled the once-proud U.S. auto industry, business analysts have been expecting the same in the computer industry. Lester Thurow, an economist of pessimistic bent, goes so far as to say that Digital, Data General, Burroughs, Honeywell, Apple, IBM, Wang and the others are doomed to the same fate that has hit Chrysler, Ford and General Motors. Thurow says a Japanese master plan, in place for several years and backed by the full capital resources of the Japanese government, makes it certain that the Japanese will dominate the computer industry before the century is out.

Well, The Wall Street Journal has been looking into the situation and finds it far less gloomy. While the Japanese have been making inroads with printers, display tubes, semiconductors and the like, they are running up against tough U.S. opposition in big computers and in software, which is the fastest-growing sector of the industry.

IBM still controls 70 percent of the large computer market and the big Japanese computer firms accept IBM products as standard. At the other end of the scale, software development, language is a major barrier to the Japanese, the Journal reports.

The Japanese have had a plan to invade the computer market for at least 10 years, and have made some progress. But their share of the total is still only 10 or 15 percent, and the market is growing faster than the Japanese segment. By 1990, some experts say, there will be a data-processing market in the $1.25 *trillion* range. Even if Japan seized 20 percent of that, it would still leave a $1 trillion market for U.S. manufacturers.

The encouraging thing about the Journal story is that the U.S. computer industry has very carefully studied what happened to Detroit and is determined not to let it happen to them. At this stage of the game, it doesn't look as if there will be any Pearl Harbor for American data processors.

The Seattle Times

Seattle, Wash., March 8, 1982

THERE are solid grounds for concern that Japanese competition may eventually do to electronics, which ranks high on the Puget Sound area's list of potential growth industries, what it has already done to the U.S. auto industry.

Sixteen major U.S. electronics companies are exploring ways of meeting the Japanese challenge by adopting some Japanese methods, The Wall Street Journal reported the other day.

Some 50 executives from top semiconductor and computer concerns met recently with representatives of trade groups and the Defense Department in Orlando, Fla., to discuss a joint research and development venture of unpredented magnitude.

The focus was on possible formation of a profit-making company that would sell basic and applied research. The high-technology collective would function similarly to those sponsored in Japan under the ministry of international trade and industry.

As William Norris, chairman of the Control Data Corp., said at the Orlando meeting, Japanese companies' collective research, tariff barriers, and partitioning of product lines have wrought "massive distortion of the world competitive scene, which can be dealt with only by extraordinary action."

In other words — assuming antitrust hurdles can be surmounted — "beat 'em at their own game."

THE CHRISTIAN SCIENCE MONITOR

Boston, Mass., December 18, 1981

There is no longer any doubt. The race is now on to see which nation – the United States or Japan – controls the space-age technology of the future, the computer and information processing fields. The stakes are enormous, since many experts believe that the total electronics-based society now coming into play will be the "second industrial revolution," adding up to sweeping changes in national defense, manufacturing, and living styles.

At present the US is far out front in computers, given its commanding leadership in software, the electronic instruction that tells computers what to do. But it is clear that Japan – through a maze of formal and unofficial working arrangements among its government leadership, industry, and banking – is determined to accomplish in the area of computers and information processing what it has already done in consumer electronics and automobiles. That is, to wrest outright world market dominance in the 1980s.

Japan will have enormous technical and social hurdles to overcome to win this competition – "battle" may be a more precise term in light of the almost singular concentration on strategy and targeting involved. With an aging work force and rising consumer and union demands, Japanese officials will be hard pressed to hold down costs while expanding output. But the ability of Japan to galvanize its literate and hard-working population to meet such a challenge should not be taken lightly by US firms or the American public. To do so would be to ignore what happened throughout the 1960s and 1970s, when Japanese consumer electronics industries one by one challenged and shot ahead of their US counterparts. That also finally occurred in autos in 1980 when Japan became the world's leading manufacturer, toppling the Detroit-based US industry that had been king of the carmaker hill for more than 70 years.

Although MITI, the Japanese Ministry of International Trade and Industry, has lost some of the vast power it had in the 1950s and 1960s (largely because it lost its right to license companies to deal in foreign exchange), it still has tremendous influence in the economic life of the nation. MITI is currently propelling the drive to shift Japan from dependence on costly resource-based manufacturing industries such as steel, shipbuilding, chemicals, and aluminum to dominance in the high-technology area. It is doing this through a combination of generous no-interest loan programs to industry, research on a fifth-generation computer that can "think" and follow oral commands, and computer software development. Ample amounts of research and investment funds are also available within the private sector thanks to Japan's high savings rate.

The reasons for this profound change in national strategy are clear. As noted by Business Week in a recent special report on Japan, Japanese political and economic leaders have concluded that the resource-poor island nation must shift away from industries where oil and production costs are high into the one area that can capitalize on the nation's skilled work force. That is high technology.

The key to Japan's potential success is precisely its sizable lead in semiconducters and consumer electronics, including personal computers. From such a base, assuming that Japan can achieve at least equality with the US in computer software, it may well be able to underbid US computer firms like IBM in the world market. Not possible? Remember, it did just that with General Motors.

There is, of course, nothing intrinsically undesirable about strong Japanese leadership in computers, nor in good-natured competition between Japan and the US. Such competition should keep both national industries on their toes, and spur new technology. Moreover, the two nations are close trading partners and firm allies. Americans have amply demonstrated their high regard for the quality and price of Japanese products through their eager purchases at the sales counter.

On the other hand, there is another aspect to the computer competition that must not be overlooked by the American political and industrial leadership. If Japan (or any other nation, for that matter) were eventually to win the computer contest, this would present enormous implications for the US defense field. Computers play a vital role in advanced weaponry, and the US cannot afford to fall behind in so strategically crucial an area.

Will the US lose the computer war, as it lost the consumer electronics and car wars? Not necessarily. The US is still far out front in basic technology and innovativeness. But the outcome of the contest will also depend on the degree to which the US political, business, and union leadership are willing to work more closely together to achieve common ends. More financial resources will have to be made available for research and development. US firms will have to be more willing to slash prices abroad to hold markets. Perhaps most important, management and stockholders will have to take a longer view than just that of short-term profits. Japanese computer-related firms are now planning in time frames of a decade or more – just as Japanese car firms did with their products.

Will American companies heed the lesson?

The Evening Gazette

Worcester, Mass., August 7, 1981

When it comes to computers, U.S. companies are world champions. They dominate international as well as domestic markets.

But the Japanese are mounting a serious challenge. Supported by the Ministry of International Trade and Industry, Japanese computer companies are already competitive with those in the United States in the production of computer hardware — the actual machines and equipment needed for rapid computations and information processing.

Now the Japanese are concentrating on software, the programming that makes computers work.

Nippon Electric Corp., Japan's equivalent of General Electric Co., plans to start selling personal computers in the United States. A new subsidiary, NEC Information Systems Inc., based in Lexington, will spearhead the attack on the U.S. small computer market.

As with other Japanese products, Nippon Electric personal computers will cost less than their American counterparts. A typical small business computer system will sell for $4,000 to $5,000, as much as $1,000 less than a comparable U.S.-built system.

Reactions from U. S. computer companies differ.

An Wang, president of Wang Laboratories Inc., Lowell, has banned Japanese visitors from his company's plants. Edward Lesnick, assistant to the Chinese-born Wang, said the Japanese want to learn the business "but it doesn't make sense for us to teach them."

Maybe not at Wang. But Data General executives feel differently. The Westboro company owns part of a Japanese computer company — Nippon Data General — and has sent some of its most talented engineers to Japan to teach employees there how to build Data General computers and related products.

Whatever the correct approach, computer manufacturers in Massachusetts and elsewhere should take the Japanese seriously. They are talented and determined. Evidence of that is particularly clear in Detroit and Pittsburgh, where manufacturers of automobiles and steel are scrambling to recover from earlier Japanese invasions.

Houston Chronicle

Houston, Tex., May 25, 1982

For years, American businesses great and small contented themselves with filling the needs of the vast and wealthy American market. The rules and regulations governing commerce in this country were shaped accordingly: They presumed a domestic marketplace in which American companies competed, largely, against one another.

Then came the Japanese "invasion" of television sets, automobiles and other goods. Nothing has been the same since.

Enlightened — and perhaps frightened — by the successes of the Japanese and others in penetrating the U.S. market, American businesses began to widen their own horizons to include other, appealing markets abroad.

It was only a matter of time before both the Japanese successes in this country and the new-found desire of American business to find markets abroad should clash with the old rules. And so it has come to pass. The need to compete against other systems which play by different rules has come face to face with one of the hoary tenets of the system of American capitalism, the nation's antitrust laws.

Oddly, the most direct request for relief from the rules of antitrust to date has come, not from the beleaguered industries of Detroit or Pittsburgh, but from the high-technology industries. These long were regarded as this country's best way to "fight back" against the successes of the Japanese and others. But that was before the Japanese system, which allows, even encourages, industries to cooperate to design new products, turned its attention to the silicon chip. The impressive technological lead which a number of American companies had built on their own soon was closed by the cooperative efforts of the Japanese.

The concern now in "Silicon Valley" and elsewhere is that the Japanese will corner the world market on the next generation of computer chips. Industry representatives have asked the Congress to exempt them from antitrust provisions which prohibit sharing research, in order to launch a "counterattack" against the Japanese. Legislation has been introduced in the Congress to make that possible.

This is a matter which deserves the most careful study by the Congress. Certainly, no American industry should be unnecessarily hamstrung as it seeks to protect its position in the U.S. market, and as it seeks new world markets. At the same time, the other factors which shaped antitrust in the first place, chiefly the protection of the consumer, must not be overlooked.

Some changes may be in order, but no overhaul is needed. The antitrust rules which govern this nation's commerce have served this country too well for too long to be scrapped.

The Seattle Times

Seattle, Wash., January 24, 1982

FAR too little attention has been paid to a recent House subcommittee report on Japan's challenge to what many Americans mistakenly feel is a comfortable U.S. world technological leadership.

Japan's economic prowess, production "miracles," and massive trade surplus with other industrialized democracies are a familiar tale, to be sure.

But the American public does not seem yet to understand what Japan's surge toward leadership in high technology portends for the U.S. economy. The report, by the Ways and Means Committee's subcommittee on trade, spells out the alarming news:

"America faces an economic crisis precipitated by the brilliance of Japan's economic drive . . . in the high-technology products that count — the products that will dominate the world trade and economy for the rest of this century . . .

"Whether the Japanese today surpass the U.S. in their mastery of these high-technology products is debatable; it is not debatable that the trend lines indicate that they ***will*** surpass the U.S. and that the gap will widen dramatically unless the U.S. responds."

Some of the subcommittee's findings:

- Japan now spends more of its gross national product on research and development than does the U.S.
- It is concentrating on such key products as semi-conductors, computers, pharmaceuticals, machine tools, robotics, and even commercial aviation, "fields in which American leadership has long been almost undisputed."
- Technological equality and production superiority in semi-conductors make Japan a serious challenger in computers and all other electronic products.
- Already in robots, "the ultimate machine tool," Japan has a strong lead in production and use, and is "creating a revolution in manufacturing."

Import restrictions, of course, are not the answer. "Each such restraint," the subcommittee notes, "will only further burden the U.S. economy, making it more sluggish and inefficient, and we will slip further and further into a 'British disease' economy."

The real answer, obviously, is multifaceted. But it perhaps can best be summed up by the subcommittee's observation that America needs "some far-reaching changes in management techniques, labor-management relations, and societal as well as business attitudes toward productivity and quality control . . . It needs to rejuvenate its economy and the cohesiveness of its society."

Newsday

Long Island, N. Y., March 22, 1982

Look at your watch: Is it Japanese? How about your car? Your camera? Your television set?

Japan's success in building technological products like these has posed a major challenge to American manufacturers, who have had trouble competing even in their home markets. And it worries Washington, which must try to deal with this country's huge trade deficit with Japan.

One engine of Japanese technology has been the Japanese government's practice of heavily promoting promising industries, in part through collaborative research projects sponsored by its trade ministry. Washington hasn't taken that kind of role, which conflicts with traditional patterns of business-government relations here.

But now several large American manufacturers of computers and semiconductors are talking about uniting to set up a vast research and development outfit, similar to the Japanese government-sponsored collectives, in response to an expected wave of competition for their domestic and world markets.

That's a constructive response to the challenge. Congress should try to foster such collaborations—without suppressing competition among American manufacturers.

The dozen or so companies exploring the joint effort could go ahead without congressional action. But they would have to thread their way through the antitrust laws, with few landmarks to guide them.

Those laws are essential to prevent companies from conspiring to shut others out of the marketplace. But Congress could specifically authorize businesses to plan joint research projects in fields where foreign competition is particularly strong.

Then American industry could put its energies into building more competitive products instead of worrying about whether its research efforts violate the antitrust laws.

Herald News

Fall River, Mass., May 24, 1982

Senator Tsongas says that Japan's next target is high technology, and that it will seek to compete with this country in terms of the high technology industries.

What is more, it is evident from the Senator's remarks that he expects that the competition will be stiff.

Certainly any American who has been at all observant knows that the Japanese have made very serious inroads in terms of this country's traditional dominance of the automobile industry.

Virtually every American also knows that while some of the success the Japanese have had industrially springs from their own native ability, some of it is the result of overhead expenses that from the standpoint of the U.S., are ridiculously low.

It makes sense that the Japanese would now try their hand at high technology industries, and the warning Senator Tsongas has transmitted is both accurate and timely.

He is especially interested in the threat Japan may pose to high technology industries because of the importance of those industries to this state's economy.

With that in mind, he urges closer cooperation between industrial management and labor. He views the confrontational attitude that is so usual in this country as damaging in terms of any effort to offset Japanese threats to the high technology industries.

What Tsongas seems to mean is that if management and labor in this country view themselves as partners instead of competitors, they are more likely to be effective competitors with the Japanese.

The position taken by Tsongas certainly seems to make sense.

American technology has made the development of many industries dependent on this kind of special knowledge possible.

It does not follow, however, that these industries will have a permanent monopoly of the international trade in the items they manufacture. If the Japanese turn out to be able to make the same items better and cheaper, then the American industries are bound to suffer.

This is what Senator Tsongas is driving at, and what he hopes his warning will help American high technology industries, including those in Massachusetts, to avoid.

Will his warning fall on deaf ears? It shouldn't, since it will come as no surprise to most Americans.

But it may, simply because the confrontational attitude within this country's industries, as the Senator indicates, has been taken for granted for so long.

Nevertheless, this will be unfortunate, since what is needed here is not only an awareness that Japanese competition in this field is virtually inevitable, but a change of heart in terms of the way American management and labor view each other.

Senator Tsongas has been indicating that this is his view for some time now. In this instance, however, he has been more explicit than usual.

What he has to say is of more than academic interest. His message should be taken to heart by the high tecnology industries along Route 128. It should also be taken to heart here, where there is so much interest in having a high technology industry locate in the near future.

This is the time for foresight, a quality Japanese industrialists certainly possess.

It is not the time to cling to obsolete views of management-labor relations, which are not only out of date, but are positively damaging to the prospect of American industries competing successfully with the Japanese.

Japanese Executives Charged With IBM Computer Thefts

An eight-month undercover investigation into the disappearance of documents from the International Business Machines Corp. culminated June 22 in the charging of twenty Japanese businessmen with conspiracy to transport stolen property in foreign commerce. The executives, sixteen of them from two of the leading computer manufacturers in Japan, were charged with paying a total of $648,000 to a Federal Bureau of Investigation agent for technical and product design information stolen from IBM. The world's computer industry leader had been cooperating since 1981 with the FBI's investigation into computer thefts in California's "Silicon Valley". The payments were made to Glenmar Associates, an electronics consulting firm set up by the FBI as a front operation for its inquiry into industrial espionage. Hitachi Ltd. admitted June 23 that it had authorized payments for confidential IBM computer information, and Mitsubishi Electric Corp. days later said it too had paid for "sample information" from Glenmar. Both companies maintained, however, that they and their employees had believed Glenmar to be a legitimate research firm selling information obtained through legal channels, a common practice within the computer industry. The charges caused considerable consternation in Japan, where the motivation of the FBI was questioned in light of rising trade tensions between the two countries.

The Salt Lake Tribune

Salt Lake City, Utah, June 28, 1982

Whatever the outcome of federal charges against those Japanese businessmen for allegedly pirating International Business Machines (IBM) trade secrets, the vaunted Japanese reputation for industrial innovation has suffered a telling setback.

If the charges stick, the Japanese miracle workers will be shown as no better than the Russians who have relied heavily on stolen western technology to keep abreast of military equipment advances ever since World War II.

Actually, what the Japanese are accused of doing is fairly common practice within the American electronics industry. "Borrowing" the advances of competitors or "adapting" another company's discoveries legally or otherwise has been documented in numerous court fights between domestic producers. That the Japanese, intent on making it big in the computer field, should adapt the same tactics is therefore not surprising.

What is enlightening, however, is belated confirmation that the Japanese reputation for producing goods more efficiently and cheaper is based, at least in part, on the expensive and time consuming development work of others.

In light of the admission by two Japanese computer firms involved in the charges, that they did indeed buy, rather than develop certain computer-making knowhow, one is justified in wondering how many Japanese auto, camera and other electronic product manufacturers used similar methods to establish their impressive records in these fields once dominated by other countries.

Rocky Mountain News

A Scripps-Howard Newspaper — Reg. U.S. Pat. Off. — Colorado's First Newspaper—Founded in 1859

Denver, Colo., June 25, 1982

The fact that the FBI has charged 16 executives of two giant Japanese companies, Hitachi and Mitsubishi, with conspiracy to purchase stolen IBM computer secrets should come as no surprise.

The battle for supremacy in advanced-electronics is nothing less than a battle for technological and industrial leadership in the world. And as in every other area of economic endeavor, the Japanese take this struggle very seriously. Too seriously, it turns out.

Hitachi officials paid $546,000 to an FBI undercover agent for stolen IBM documents having an economic value of "millions and millions of dollars," according to prosecutors. Separately, Mitsubishi employees paid $26,000 to the agent for the same information.

It's hard to believe, as Hitachi and Mitsubishi officials insist, that the Japanese didn't know the information was stolen. These were no technological illiterates on a shopping spree. At the higher reaches of the computer business, everyone knows, or can quickly find out, what is and isn't legitimately for sale. Moreover, as several industry analysts have noted, the size of the payments alone indicate extremely unusual goings on.

So far, the Japanese reaction is outrage: not with the perfidy of their own companies, but with us. The FBI's use of undercover agents, it seems, violates their sense of fair play. Never mind, apparently, fair play in business.

For years the FBI was criticized for ignoring white collar crime while it fed an obsession with petty bank robbers and feckless radicals. No longer. And if the agency's strengthened efforts against white-collar crime also uncover foreign nationals attempting to undermine our very prosperity, all the better.

The Seattle Times

Seattle, Wash., July 12, 1982

THE case of the 18 Japanese computer-company employees charged with buying stolen IBM industrial secrets is likely to be a major news story on both sides of the Pacific for months to come. We hope Americans will take care to put the matter in perspective.

The charges could pump new life into an old and comforting stereotype, still too prevalent in this country, of Japanese success in the world marketplace being based largely on an ability to copy the products of other countries and then undersell the original producers.

To whatever extent that image may once have been accurate, it presents a woefully inadequate picture of today's Japanese industrial-export machine.

Some Japanese firms may engage in a little hanky-panky, to be sure. And imitating competitors is a standard — and usually legitimate — feature of business life in any country.

But Japan's sensational rise as a world-trade giant has far more to do with organizational genius, societal cohesiveness, sound planning, and just plain hard work than any copycat practices, legal or illegal.

It would be costly self-deception for Americans to think otherwise.

The Detroit News

Detroit, Mich., June 28, 1982

Lately, it seems, the Japanese have acquired an image of industrial invincibility.

Many of their consumer products, such as radios, stereos, TV sets, videotaping equipment, and of course, autos, are now fierce competitors for large slices of the American market. Newspapers and magazines have been filled with stories noting that American engineers and businessmen, who once hosted visitors from Japan, are now touring Japanese plants and firms.

Still other articles describe how various Japanese labor and inventory management techniques are being applied by U.S. companies. Publishing houses have added such titles to their lists as *The Art of Japanese Management, Japan As Number One,* and *The Strategist: The Art of Japanese Business.*

And there is indeed much to learn from the Japanese.

But America still has the edge in some fields — particularly in developing new technologies. The best evidence of this is the FBI's arrest of a number of Japanese businessmen accused of attempting to steal confidential computer technology information from IBM.

The incident provides a little reassurance on two counts: The FBI is on the job, and there's apparently still something we can do better than our Japanese competitors.

The Japanese are wonderful copycats and efficiency experts. But when it comes to technological invention, they have a well-deserved reputation for relying on stolen ideas.

THE ATLANTA CONSTITUTION

Atlanta, Ga., June 28, 1982

In a way, it's "old business" — this controversy about the FBI catching, in an undercover operation, employees of two large Japanese companies making illegal payments for what they thought were trade secrets of IBM, the world's dominant manufacturer of computers. But, in another way, it's "new business."

There's little new about industrial espionage, whether in America or other nations. Most companies, to be fair, surely don't attempt to steal or to buy illegally information or technology from other companies — information or technology that is needed to remain competitive and profitable. Most companies work to obtain the information legally or develop the technology on their own. But not always.

Over the years, there have been numerous occurrences of "stealing" information or technology. Some of these were uncovered and reported. Companies, especially those active in high-technology fields, maintain extensive security systems against "spies" employed by other companies.

The industrial espionage spotlight — the old business — is now focused on the two Japanese companies — Hitachi Ltd. and Mitsubishi Electric Co. Their employees were the ones caught in the FBI undercover operation. The Japanese actions, however, illustrate the new business — the rapidly growing international business and industrial competition in a worldwide economy.

The old business is, of course, illegal, and to be deplored. But the new business is perhaps the most important.

There has always been trade between nations. In the past two decades, however, trade between nations has become so extensive and interrelated that national economies are weak indeed if they are not closely involved in worldwide trade. No better example can be found than Japan; it would be a third-rate island nation had it not striven, following its military defeat in World War II, to become a giant in foreign trade. Japanese businessmen, along with Americans and West Europeans, are to be found competing in most nations of the world.

There are six major Japanese manufacturers of computers, related equipment and programs. They are striving to grow in the international markets, but their combined worldwide sales last year were less than one-third of the $30 billion recorded by IBM. Thus, the attempt to buy IBM trade secrets; with such secrets, Hitachi and Mitsubishi Electric probably would have been in a better competitive position.

There will be some attempts to turn this into more ammunition for justifying U.S. trade sanctions against Japanese manufacturers, of whatever products. The Japanese, to be sure, make it rather difficult for Americans to trade in Japan, while the Japanese face few limitations in the United States. But it would be a mistake in this day of a world economy for the United States to become involved in a trade-restrictions war with Japan or any other of our major allies. One restriction would only lead to another; they would, in turn, lead to retributions by other countries. In the end, we would all be hurt far more than the pain of the original wound.

The Oregonian

Portland, Ore., June 27, 1982

The undercover FBI sting operation that has led to charges against 18 Japanese and involved two major Japanese firms cannot avoid harming friendly relations between the two nations. Because of the great damage that can be done, the government must prove clearly that the investigation was not politically motivated, as some Japanese newspapers are charging.

Officials for both the Mitsubishi Electric Co. and Hitachi, Ltd., two of Japan's most highly respected companies, have denied attempting to steal computer secrets from IBM operations in California's "Silicon Valley." The companies said they believed they were making legal payments to private contractors when amounts ranging from $26,000 to $622,000 were paid to undercover FBI agents.

The electronics industry is badly shaken by the charges, although there are schools reportedly operating to train persons how to steal industrial secrets. Widespread efforts of foreign nations, such as the Soviet Union, to steal secrets are well known. And U.S. firms even steal from each other. It is the extent of the charges against the Japanese, the reputations of the companies and the high level of officials involved that has surprised the world of silicon chips and secret computer hardware.

U.S.-Japanese relationships have been knocked about recently by the heavy imbalance of Japanese exports to imports and the long-term failure of U.S. firms to penetrate Japanese markets.

To stop further erosion of trade and political relationships, it is important that both nations move quickly to get to the bottom of the charges, putting to rest statements that the investigation was politically motivated by a U.S. government anxious to get even with Japan for its successes in the American marketplace.

WORCESTER TELEGRAM

Worcester, Mass., June 25, 1982

It will take time for authorities to sort out the legalities in the case of those two big Japanese companies accused by the U.S. Justice Department of trying to steal computer secrets from International Business Machines Corp.

Hitachi Ltd. concedes that it authorized payment of $548,000 for IBM information offered by an American consulting firm that turned out to be an FBI front. Mitsubishi Electric Corp. admits to a $20,000 payment. But both companies deny any wrongdoing.

The federal charge against the seven representatives of the companies arrested in California is that they conspired to transport stolen property — the IBM secrets — from the United States to Japan.

Buying trade information from consulting firms is common practice in the computer industry worldwide. The key questions under U.S. law seem to be whether the Japanese businessmen believed the information they paid for was stolen and whether, as Hitachi has already suggested, they were entrapped by the FBI.

The case is big, sensational news in Japan. Government officials there are worried that it will inflame trade tensions between the two countries, already high because of American unemployment due to imports of low-cost, high-quality automobiles and other products from Japan.

It may do that. The very fact that IBM got the FBI to cooperate in defending its trade secrets suggests that the giant of the U.S. and world computer industry takes seriously the ambitious program that government-backed Japanese companies have launched to challenge U.S. supremacy in the field — and that Washington takes it seriously, too.

Traditionally, the Japanese are considered experts at production but lagging in innovation. But playing catch-up, especially in a field like computer software, puts a premium on finding out how leaders in the field do what they do. So it is not surprising that the battle is joined in the area of industrial spying.

It is clearly not just IBM, U.S. kingpin in big computers, that stands to lose. Hitachi, Mitsubishi and other Japanese companies are now getting into small computers, too. That hits home in Massachusetts, home of Digital Equipment Corp., Data General Corp. and other leaders of the world minicomputer industry.

Those companies say they have what it takes to keep ahead of any rivals. But the Japanese have shown what they can do to other American industries. The upshot of those California arrests will be worth watching. Thousands of Massachusetts jobs could be at stake.

The Washington Post

Washington, D.C., June 28, 1982

THE CASE of the computer secrets, and the two Japanese companies that wanted them, is going to be deeply damaging. The diminished reputations of the two companies will be the least of it. The more important threat is to American trade policy, and to this country's tradition of an open field for international competition.

Employees of the major Japanese computer manufacturers, Hitachi and Mitsubishi, have now been charged with conspiring to steal technical data describing the workings of new IBM machines. Whether these charges are well founded, the courts will decide. But the two companies' initial replies will awaken memories of the Lockheed kickback scandal seven years ago. At the time, Lockheed's defenders argued that (a) the company couldn't be expected to know what its employees were doing so far away and (b) the business was all through middlemen and (c) foreign ways are strange and (d) anyway, everybody did it. That was never a very persuasive case in behalf of an American company operating in Japan, nor will it be persuasive in behalf of Japanese companies operating here. Even if the criminal charges should turn out to be unjustified, it appears that Hitachi, in particular, was paying a lot of money for proprietary information with very little concern for its source.

This prosecution belongs to that influential category of events that, fairly or not, seem to substantiate a stereotype. In this instance, it's the stereotype of the Japanese manufacturer that exploits costly research done elsewhere to mount a devastatingly effective export drive. Like all stereotypes, this one is inaccurate in important respects. But through a dramatic display of, at the least, poor judgment, people representing these two companies have now given additional momentum to protectionist impulses here that are already powerful.

The more familiar strains of protectionism arise among those industries—steel, for example—in declining markets. But there's another variant on the theme that involves the rising industries. While the Reagan administration has maintained a generally good record on free trade, it has repeatedly expressed anxiety about the rapid gains of Japanese companies in those American markets that it considers crucial to the development of high technology. The leading example is integrated circuits and the communications and data processing equipment based on them. The administration foresees a future in which world competition in these industries will be dominated by a few huge companies, most of them national flagships, as the phrase goes, strongly backed by their governments. There has been much discussion of the ways in which the United States ought to respond. The computer conspiracy—if it was a conspiracy—strengthens all of the least desirable, and more protectionist, possibilities.

THE DAILY OKLAHOMAN

Oklahoma City, Okla., July 3, 1982

WHEN the American people discovered that a number of our larger international companies were using bribes to purchasing officers abroad to make sales, it was irrelevant that thousands of Americans had good jobs making jet airliners, computers and other things because of those sales.

Bribery is immoral and illegal by U.S. standards, so Congress went into one of its periodic morality sessions and tightened the laws on bribery. Our international trade suffered. Arab sheikhs suddenly discovered they could buy oil field equipment and TV sets elsewhere. The faltering European Airbus airliner program suddenly received some orders and has been doing well ever since.

The lesson that we might consider the local customs when doing business abroad was not learned in Washington. What we call bribery is considered normal business dealing in many countries. There is something entirely different, however, about the latest brouhaha over doing business with other countries.

This time it involves something that is very much a problem within our own country: industrial espionage. In our major cities, executives and planning offices are equipped with what amounts to electronic countermeasures to foil electronic eavesdroppers. Shredders to dispose of confidential documents are standard equipment in many companies.

That is not the way two major Japanese companies were operating, however. When accused of paying for stolen trade secrets of the IBM Corp., the Japanese indignantly protested that nothing had been stolen. The next day, they clarified that: Every single secret they had acquired had been paid for. The total amount involved is at least two-thirds of a million dollars in cash.

One Japanese industrialist shrugged the whole furor off with the statement that the Americans were being foolish — "This is the way the real world is. Everyone is seeking the latest information."

So we are back to the old reminder: East is East and West is West, and never the twain shall meet." Well, hardly ever, when it comes to really understanding each other.

The Kansas City Times

Kansas City, Mo., June 24, 1982

The full story of the 18 Japanese and one American who have been charged with attempting to steal computer secrets from IBM has yet to unfold. But enough is known to shock and anger Americans, who expect better of their friends. And if the Japanese have a friend in this postwar world, they have only to look eastward across the Pacific to see who it is.

This is not your run-of-the-mill spy thriller, with crazies in dark alleys trading purloined secrets for cash. These charges involve senior engineers, researchers and managers of two giant Japanese electronic companies — Hitachi and Mitsubishi. Both have won respect throughout the world for the quality of their products.

The obvious implication is that neither firm could compete fairly and squarely on world markets, based on their own talents and research. So, it is alleged, their representatives paid $648,000 to an undercover FBI agent for confidential information produced by the brains at IBM.

Another implication is that these secrets would in turn allow the two companies to produce better products so they could further undercut the U.S. on the world market. Whatever the information, it must have been highly prized or, as the FBI charges, so much money would not have changed hands.

It is no secret that America's open society is a sieve through which military intelligence and industrial secrets flow to other parts of the world. A great deal of sensitive information is gleaned from scientific publications and trade manuals that are freely exchanged. That's our problem and not the concern of those who reap the rewards. What troubles us is when the gaps are filled by stolen secrets that change hands on a cash basis.

The Russians have proved masters of this art and the Eastern European bloc is not far behind. In Silicon Valley and the Los Angeles area in California, they have bought military secrets wholesale as if at a supermarket, and occasionally they find a member of the armed forces who is for sale. These often are people desperate for money and on the edge of a breakdown for various reasons. But it is dismaying when senior employees of two major Japanese firms are charged with indulging in the same practice. And the allegations come from none less than Attorney General William French Smith and FBI Director William Webster, who should understand the international ramifications of the acquisitions.

Coming at a time when the Japanese trade surplus with the United States stands at more than $16 billion, and could top $20 billion this year, these charges could further fuel the fire of protectionism on Capitol Hill. Reciprocity legislation is already pending to force the Japanese to remove more non-tariff and other trade barriers or suffer the consequences through retaliation in the U.S. marketplace. Any urge now to proceed hysterically on those bills should be avoided at all costs.

Punishing an entire nation, to say nothing of a close and important ally, for the alleged sins of employees of two corporations, who stand accused but so far convicted of nothing, would be unfair. It also could have profound consequences on world trade, as well as the vital political and military ties that bind the two Pacific giants so closely together.

Deplorable as all of this is, and we are convinced most Japanese are reacting in the same shocked manner, the Pacific alliance stands and nothing should be done in Congress or elsewhere to jeopardize it on the basis of what has appeared so far in this case. But Japanese industry and government need to be careful in such matters. Americans tend to be trusting souls, perhaps too trusting. They do not like to be played for suckers.

The Dispatch

Columbus, Ohio, June 29, 1982

THE FBI'S most recently revealed undercover operation worked like a computer in an expensive IBM commercial. FBI Director William H. Webster proclaimed it "a classic example of the value of an undercover operation designed to ferret out the theft of high technology."

With the cooperation of the IBM Corp., FBI agents set up a bogus computer consulting firm and sold "secret" IBM computer plans and documents to representatives of two large Japanese computer firms for huge sums of money.

Then the FBI arrested seven Japanese nationals and two Americans and filed warrants against 12 other Japanese businessmen. Of those arrested, eight were charged with conspiracy to transport stolen property to Japan, and one was charged with receiving stolen documents.

Hitachi Ltd. paid $622,000 for supposedly stolen information, and Mitsubishi Electric Corp. paid $26,000, the FBI said.

The evidence gathered was sufficient to prompt top officials of the two large firms to admit such payments were made. To save face, the officials claimed their representatives believed they were dealing with legitimate computer consultants and that the transactions were not illegal.

If the quest for IBM secrets was above board, why did one Hitachi engineer, accompanied by an undercover FBI agent, hide in a darkened office of an East Coast firm until the time was right to sneak past a guard and take photographs of an advanced disc-storage device on lease from IBM?

It is not for us to rule "guilty" or "not guilty." That will and should be left to a judge and a jury.

The whole operation, however, certainly points out that the spying business is not limited to government intelligence agencies.

The Orlando Sentinel

Orlando, Fla., June 26, 1982

The arrest of six Japanese and one American accused of stealing industrial secrets from IBM is already being castigated in the Japanese press as some kind of ominous plot to spoil Japanese-American relations.

Two Japanese megabusinesses, Hitachi Ltd., and Mitsubishi Electric Corp., have belatedly admitted that they did pay $548,000 and $26,000 respectively for industrial information that the FBI says was stolen from IBM.

The Japanese corporations deny any wrongdoing, claiming they were simply paying to secure information on technological trends. Warrants are outstanding for 12 other Japanese in the same investigation, but the U.S. State Department has not said if it will try to extradite them.

All legal defenses and Japanese journalistic carping aside, the Japanese should realize that if they want to participate in the American market, they have to play by American rules. Industrial espionage is big business in the world of big corporations where winning the race to the megamarket can mean millions in profits. But it is against the law in the United States.

As for a plot to spoil Japanese-American relations, that's xenophobic nonsense. It's American car workers, not computer experts, who are out of work. If Americans were out to spoil Japanese-American relations, we would be slapping import quotas on Japanese cars, not handcuffs on Japanese computer experts.

St. Louis Globe-Democrat

St. Louis, Mo., June 26-27, 1982

Some Japanese newspapers have reacted in a curious way to the revelation that employees of two of their top electronics companies have been charged with paying $568,000 to purchase stolen information about IBM technology. The Japanese made the mistake of trying to buy the secrets from a "sting" company set up by the FBI.

Instead of criticizing the management of Hitachi and Mitsubishi, the two companies involved, the Japanese editorialists attacked the use of FBI agents to uncover the technology-buying scheme. They contended the Japanese firms were legitimately trying to gather information about IBM and had been tricked.

The amount of money involved, however, makes it rather clear what the Japanese employees were trying to do when they were caught. Firms wouldn't spend this much money for information legally available. This appears to be a case where the two companies wanted to obtain secret information about IBM's latest computers illegally in order to cut years off the time required to turn out similar products. They were willing to pay more than a half million dollars for IBM's closely guarded secrets.

There is no way to justify stealing industrial secrets or paying bribes to acquire them.

Rather than try to excuse their countrymen for allegedly engaging in industrial espionage, the Japanese editorialists would do better to ask these firms why they don't spend more money on their own research and development rather than try to gain an illegal shortcut to IBM's technology.

ST. LOUIS POST-DISPATCH

St. Louis, Mo., June 25, 1982

Business bribery — in this instance the exchange of money for business secrets — is like the tango: It takes two. It takes a briber and one is bribed.

But in the sensational case leading up the the arrest of employees of Hitachi Ltd. and Mitsubishi Electric Corp., there is an interesting twist. The focus is on the bribers — Japanese companies willing to pay cash for information that IBM says involves trade secrets. But where are the ones being bribed? Who was corrupted by the offer of cash to provide the "stolen" materials?

It was an FBI undercover agent who accepted the money. And where did he get the information he passed to the Japanese? From IBM executives working with federal authorities. So what we have here is not a real case of theft or a real case of business bribery. What we have is a conspiracy — actually two: one private, the other official.

The private conspiracy, of course, is the effort by the Japanese firms to purchase information about a competitor's products, and not ask a lot of questions about how it was obtained. It is thoroughly unethical, but hardly unheard of in the business world.

The other conspiracy is the more interesting, because it involves the federal government working in cooperation with a major U.S. corporation to prove that some Japanese firms spend money to buy information. One need not feel sorry for nor excuse the Japanese firms to understand why their nation was embarassed by the incident and why some commentators there think it was staged for political and propaganda purposes.

It would be one thing if the national security was at stake. But it wasn't. So was it public or commercial interests that were served by the showcase federal operations?

The Sun Reporter

San Francisco, Calif., June 25, 1982

THE HIGH-LEVEL drama involving charges of theft of computer secrets that is now unfolding in federal court here has all the makings of an espionage tale conceived specifically for our high-technology times. And for today's justice that is so often dealt by the "sting." Nineteen men, many of them employees of two of Japan's biggest and most-respected firms, have been charged with conspiring to snaffle the wizardry behind IBM's 3081 electronic processor. The stakes were high, involving vital elements of a booming industry. The repercussions have been understandably wide.

In Tokyo, top executives of Hitachi Ltd., the major manufacturer of electronics and machinery there, have admitted the company paid $546,000 for confidential information on IBM's newest computers, but they denied any wrongdoing and suggested their employees were victims of FBI entrapment. And Mitsubishi Electric Corp. rejected any basis for the accusations against its personnel.

THE FACTS are still to be proved in court. But this was a field ripe for industrial piracy. Japanese companies have begun an ambitious effort to become major exporters of virtually all computer and telecommunications products. They have surpassed IBM in sales of large computers in their own country. Worldwide sales of the top six Japanese companies are still less, however, than a third of the $30 billion recorded by IBM last year.

And if the initial allegations stand up (there seems every indication that this was a precise, careful investigation), the least one can say is that, while industrial espionage goes on all the time, the Japanese should have known better than to try to buy their way into the secret world of IBM. Their trouble was that IBM quite properly turned the case over to the FBI.

This all comes at a time of strained trade relations between the U.S. and Japan. But it doesn't set well to grumble, as now do some Japanese newspapers and members of that country's establishment, that the charges constitute a politically-motivated effort to "get" Japan. This appears on its face to have been a classic "sting" operation. It is the wide and dynamic range of this lucrative market that caused the effects to be international in scope.

THE MILWAUKEE JOURNAL

Milwaukee, Wisc., June 29, 1982

Industrial espionage has doubtless been around since James Watt cranked up his steam engine. But the FBI "sting" involving two major Japanese electronics firms, more than $500,000 for hot computer data and 21 persons arrested or charged, is the stuff of unusually high — and potentially damaging — drama.

If a sizable segment of the US public should leap wildly to the conclusion that Japan's enviable economic prowess has as much to do with the nastiness alleged in the Hitachi and Mitsubishi case as with industrial efficiency, the pressures for swift and punitive legislation would be terribly difficult to resist. The occasional murmurs about an economic "yellow peril" could find a loud, degrading voice.

Across the Pacific, indignation is a principal reaction to the FBI disclosures — and that, too, could be damaging. The US has been pressing Japan to assume more of the burden of its own defense. US-Japanese negotiations on trade barriers are at a delicate stage. It would benefit neither nation if excessive emotion heavily intruded into those critical areas.

Unhappily, that seems to be happening in Japan. The undercover operation, some Japanese commentators suggest, might be collusion between the US public and private sectors — the FBI and IBM, in this case — to try to deflate Japan's competitive surge. It also is being wondered there whether the Japanese executives might be victims of US entrapment.

While both Japanese firms concede the money was exchanged for valuable computer information that IBM says was stolen, they contend that the portrayals of the information as being purloined was an effort to "hype" the price.

It will take time to sort out the facts, and Americans are aware from Abscam how convoluted such affairs can be. We think the US government should move rapidly to litigate this vexing case so the gray areas might be better illuminated. That could help avoid premature judgments on both sides of the ocean, judgments perhaps infused too deeply with national pride.

The computer industry, symbol of the global high-technology race, is a fierce competitive pit. The line between licit and illicit techniques can sometimes be indistinct.

It would be a mistake for Americans to be too eager to indict or the Japanese to exculpate — based on what is so far known.

THE INDIANAPOLIS STAR

Indianapolis, Ind., June 28, 1982

Charges of FBI entrapment are being heard again — this time with a Japanese accent.

Tokyo newspapers hint, not very subtly, that the FBI "set up" those gullible innocents from Hitachi and Mitsubishi who bought stolen IBM computer secrets. In the first place, the coverage alleges, they didn't know the information was stolen and they were the victims of high-pressure sales tactics.

That line isn't going to sell very well on this side of the Pacific, where trade relations are already threatened by Japanese policies which effectively exploit the U.S. market at the same time they barricade the Japanese market from U.S. products. Stiff competition is one thing, pirating another company's secrets is another.

The FBI sting underscores only one aspect of the international traffic in industrial espionage. The stakes are tremendous and American corporations aren't the only losers. U.S. preeminence in technological research and development is being jeopardized, with consequences that can affect the jobs and livelihoods of millions of Americans. Most importantly, U.S. security is being undermined.

Witnesses before congressional committees have testified that Soviet theft of sophisticated technical information and equipment from American manufacturers has been a crucial ingredient in the upgrading of the Red strategic weapons systems.

The outcome of any future war between the East and the West may well be determined by what's going on today in California's "Silicon Valley," headquarters of this nation's computer and electronics research.

With that in mind, we have little sympathy for the Japanese snagged by the FBI. Instead, we hope government officials have the good sense to set all the traps necessary. Piracy has always been a risky business.

THE COMMERCIAL APPEAL

Memphis, Tenn., June 25, 1982

WHATEVER else may be said about the arrest of Japanese businessmen on charges of trying to steal U.S. industrial secrets, the case demonstrates how important the United States is to Japan's economic prosperity.

It's an old story. The Japanese have used technological advances made in the United States to help build their country into one of the world's leading economic powers. That may pain Americans somewhat, especially when Japanese companies take business away from U.S. firms. But Japan's success has been based mostly on its ability to manufacture high-quality products at low prices.

No fair-minded person can begrudge the Japanese the profits of hard work. Besides, in the long run, the competition may make U.S. business more efficient, resourceful and aggressive. If it doesn't, Americans have only themselves to blame.

FOR YEARS, the Japanese have been able to obtain U.S. technology in perfectly legal ways. In fact, they may be said to have become masters at the business. They've set up "listening posts" in this country to study patents and talk to U.S. engineers. One of the most important parts of the process is what's called "reverse engineering," in which products are bought to be taken apart to determine how they work.

The difference in the present case, according to the Federal Bureau of Investigation, is that commercial theft is involved. The FBI reports read as though they might have been lifted themselves — from a television script.

- Hitachi and Mitsubishi, two of the world's largest suppliers of computers and computer components, are in a battle with U.S. firms for leadership in the electronics field.
- Independently, the Japanese companies approach an unidentified U.S. firm seeking information. Eventually, they're put in touch with an undercover FBI agent who's set up a bogus computer consulting firm as part of a major FBI project to stop the theft of industrial secrets.
- The agent delivers several items to the Japanese in return for $122,000. At one point, he or another undercover FBI man hides with a Japanese in a darkened office of an East Coast plant. Then, when all's quiet, they slip past a security door so the engineer can take photographs of an advanced disc-storage device leased from International Business Machines.
- After an offer of $500,000 is made for more information, the FBI steps in to arrest six Japanese nationals and one American and issue arrest warrants for 12 other businessmen in Japan.

In Tokyo, a Hitachi spokesman admitted his company approved the purchase of information about IBM computer technology. But, he said, the company was offered the information by persons it thought were legitimate American consultants. Such transactions, he said, are common practice. The company didn't know, he claimed, that anything illegal was going on, and he speculated that Hitachi employes may have been "entrapped" by the FBI. Japanese newspapers compared the FBI operation with Abscam, which involved the supposed purchase of favors by wealthy Arabs from U.S. elected officials.

IT'S UNDERSTANDABLE if the Japanese prefer, at least right now, to think of the FBI operation as a political action taken to undermine the competition and to get even over imagined trade inequities. But the facts, presumably, will come out in a court of law. While it may be unusual for Japan to be involved in illegal business deals, it wouldn't be the first time. The Lockheed scandal forced the resignation of a Japanese prime minister.

In any event, the Japanese can't deny they wanted — even were dependent upon — know-how that's made only in America. It's nice to know that the United States is really needed.

Patent Rules Revised to Spur Drug Industry

Technological advances are not confined to computers. Drugs, too, are an important indicator of scientific progress in America. Complaints over the diminishing U.S. lead in the drug industry were voiced earlier than for automobiles and semiconductors. The Food and Drug Administration was routinely criticized for its slowness. In 1979, a report by the General Accounting Office said the average approval time for a new drug was 20 months in 1975. The GAO added that some drugs approved by the FDA between 1975 and 1978 had been available in Europe for as long as 12 years. As a result, in the words of N. Bruce Hannay of Bell Laboratories, "the rate of introduction of new drugs in the United States . . . took a precipitous decline . . . [since] 1962. . . . [It] dropped off by a factor of four, and there was no change in Western Europe." The situation had improved somewhat by 1980 from 1978, when Hannay made his statement at a National Academy of Sciences forum. In 1980, the FDA approved 27 new drugs, the largest annual number since 1962. Efforts to speed up the process even more continued in Congress in 1981 with the passage of a bill by the Senate to increase the patent life of new drugs. The measure would add up to seven more years to the 17-year patent life of a drug, to take into account the long federal approval process. Drug companies had long complained that the review process ate up most of the life of a new drug patent and thus reduced the marketing time of the drug and the company's profits.

ST. LOUIS POST-DISPATCH

St. Louis, Mo., September 19, 1981

The House committee that will soon begin hearings on a bill to extend the patent life of pharmaceutical products should find useful a recently released study by the Office of Technology Assessment. Patent laws are designed to protect innovators from competition for 17 years. Because it takes time for tests to ensure safety and efficacy, the effective patent life for drugs is often much shorter than 17 years. Hence bills are being considered in Congress to exclude up to seven years of the Food and Drug Administration's approval period from the patent life.

The OTA analysis presents pluses and minuses for the proposal. Drug industry profits have remained high and research and development expenditures have not declined despite the short effective patent terms. But, the OTA notes that patent extension would make increased research more likely on some types of drugs — those aimed at a large market. Prices for new drugs may be higher under the extended patent term, but competition is still likely to have a moderating effect as the patent does not bar marketing variations of the new drug. And it is possible that more and better drugs may be developed if the patent term were protected from dilution by the necessary FDA review.

The number of new drugs introduced annually in the U.S. has been declining over the last 20 years, and the length of time required for FDA approval has increased The Patent Term Restoration Act could lead to more medical breakthroughs because of incentives for research and development. It also would grant drug companies nearly the same patent protection other types of companies receive. Hence, the bill deserves favorable consideration.

BUFFALO EVENING NEWS

Buffalo, N. Y., July 6, 1981

No one disputes the benefits of requiring the federal Food and Drug Administration to make exacting tests of new drugs for safety and effectiveness before they can be sold to the public. But this testing process has consumed increasing periods of time — and that, many observers believe, contributes to a slowdown in the development of improved drugs to heal the sick.

That is because the lengthening time consumed by FDA procedures cuts into the life of the patent rights enjoyed by the new drugs' inventors. Under a 1936 law, the exclusive period for inventors to enjoy the financial rewards of new drugs runs for 17 years after the patent is granted. But after the patent is granted, the FDA testing period begins. One study indicates that 20 years ago the period of effective patent life after the FDA approved the drug was 15 or 16 years. By 1978, this had shrunk to 10 years.

With less time to recoup development costs and make a profit, pharmaceutical companies have less incentive to pioneer in their field. Critics believe that this diminished incentive has slowed the number of new drugs reaching the market.

One simple, sensible response to this problem is a bill pending in Congress that would restore the longer effective patent life of new drugs. Sponsored by Charles Mathias, R-Md., in the Senate and Robert Kastenmeier, D-Wis., in the House, the bill would delay the start of the 17-year exclusive marketing period until after the drug obtains FDA approval, with the delay limited to a maximum of seven years.

This wouldn't affect FDA evaluation procedures in any way, neither forcing dangerous shortcuts nor, for that matter, producing any needed reforms. But the change would restore the equity and incentives intended by the original 1936 patent law. Although some argue that the change could strengthen a pharmaceutical firm's monopoly position and brand-name identification on a major "breakthrough" innovation, that very possibility would be likely to spur competitors to develop their own equally useful or improved product as an alternative, thus reducing the risks of a monopoly position.

Most important, those afflicted by illness and disease are certainly not benefiting from any slowdown in drug research. If this change in the patent law would help accelerate beneficial innovation, as seems probable, then Congress ought to translate the proposal into law.

The Cleveland Press

Cleveland, Ohio, May 8, 1981

If a new policy of the Health and Human Services Department holds, the big pharmaceutical companies will get richer and prescription drug users will get poorer.

The policy, instituted by HHS Secretary Richard Schweiker, makes it harder for small companies to market cheap, generic versions of expensive brand-name drugs.

When 17-year patents held by original manufacturers expire, the Food and Drug Administration has allowed small companies to make and market identical versions of drugs without duplicating long and expensive studies of their safety and effectiveness.

This has substantially lowered the cost of drugs to consumers. For example, the tranquilizer Librium was selling for $14.65 per 100 tablets, but when the patent expired other firms began selling the same size bottle of similar pills for $1.95.

The big pharmaceutical firms don't like that because it eats into their business. They also claim that it stifles incentive to do research on new drugs.

What the big firms want is for the government to continue protecting their business and their profits even after their patent protection runs out.

Not only does that go against the grain of the Reagan administration's policy to lesson government regulation but it also would decrease competition and would cost consumers hundreds of millions of dollars a year.

Schweiker seems to have been taken in, at least temporarily, by the large pharmaceutical houses and has halted the FDA's licensing procedure for generic drugs.

THE TENNESSEAN

Nashville, Tenn., January 31, 1981

EVERY time consumers win a modest gain, it seems, the industry goes straight to the drawing board to devise a counter.

That is happening now in the pharmaceutical industry. Lobbyists for the industry are seeking an extension of the patent rights on drugs. This comes at a time when Louisiana has become the 48th state to allow the substitution of generic drugs for brand-name ones.

One of the arguments advanced for the use of generic drugs — which are substantially cheaper in most instances — was that the patent protection of 17 years was reward enough for the manufacturers of brand-name drugs. That was enough time, it was said, for them to recoup their costs and make their profits.

With 48 states now permitting generic drugs, that was obviously a persuasive argument — along with the fact that generic drugs, according to the Food and Drug Administration, are therapeutically equivalent in most instances. The FDA has published a list of prescription drugs with information on the therapeutic equivalence of generic and brand-name drugs.

Another sign of the persuasiveness of the consumers' argument is use. According to industry sources, generic drugs were only 2% of the market three years ago. Last year that share was up to 10%, according to the industry. A study by *Woman's Day* magazine puts the market share of generic drugs at 15%. The trend is obviously up.

Thus, the industry is preparing to lobby Congress to extend the patent production. Its argument goes: the time to develop a drug can run from eight to ten years and thus the patent protection is really only for nine years or less. The industry wants more patent protection while the drug is on the shelf, making money.

A part of this argument is that the research costs are enormous, $70 to $80 million for one drug alone in some cases. If these costs can be spread over a longer time, i.e., if the patent protects more time while the drug is marketed, it will benefit the consumer, say industry spokesmen.

The industry's sentiment is touching, but hardly persuasive. Last year the industry spent $1.6 billion on research, according to its spokesmen. It also spent $1 billion in advertising, including from $200 to $300 million on samples to physicians. Advertising, as well as research, is obviously a goodly portion of the cost that consumers are paying.

There may be areas of regulation of the pharmaceutical industry needing reform, but protecting and promoting the industry's profits isn't one of them.

Chicago Tribune

Chicago, Ill., May 2, 1981

Patents are intended to give investors and creators of a new product 17 years of exclusivity to reap a return on their investment and make a profit from their discovery before it can be copied freely by others. But for developers of new medical drugs, it hasn't been working out that way.

Today, the process of getting a new medication approved by the Food and Drug Administration (FDA) has become so complex that, on the average, almost half of the patent life of a drug now expires before the product can be put on the market. In some instances, a manufacturer has only three or four years left to sell a new medication before the patent runs out and it can be copied by competitors.

With less chance to earn back their initial investment — it cost an average of $70 million to develop a new drug in 1979 compared to $6 million in 1962 — pharmaceutical companies are less motivated to invest in research and drug development and increasingly inclined to shift to non-drug products. Drug companies introduced an average of 53 new medications per year between 1959 and 1962, but only an average of 18 per year between 1977 and 1979.

So Congress is considering new legislation that would stop the clock from running on the patent life of any product that must be reviewed and approved by a government agency before it can be put on the market. The bill would add to the remaining life of the patent the time elapsed between the initial application for classification as an "investigational new drug" and final FDA approval — up to a maximum of seven years. If passed, the new law would also help companies developing new chemical products, although government approval time is not quite as lengthy for these substances.

Some objections have been raised to the proposed legislation because it would lengthen the time until a drug could be copied by the developer's competitors and marketed as a generic product, presumably at a lower price. But in the long run, we all stand to benefit much more from the discovery and availability of new medications. It is far less expensive to treat patients with drugs than with surgery or long hospitalization, which may be the only alternatives. And one of the most effective ways to cut health care costs is to develop new medications. Enormous savings, for example, could be made if we had more effective drugs for heart disease, cancer, genetic disorders, respiratory diseases, and a long list of other ailments for which better treatment is urgently needed.

On the average, scientists now screen more than 10,000 possibilities for every one new medication that is eventually approved by the FDA and put on the market. The proposed legislation would provide some inducement to pharmaceutical companies to continue risking their time and money on such long shots.

The Boston Herald American

Boston, Mass., March 6, 1981

Encouraging research and innovation in any industry is largely a matter of providing sufficient incentives. That is why adequate patent laws are indispensable to the development of improved products. Patents permit innovative entrepreneurs to reap just rewards by protecting their exclusive right to market a new product for a set number of years.

The patent life for drugs is fixed in federal law at 17 years, long enough to guarantee the kind of financial return that rewards entrepreneurs and benefits the public by encouraging more and more research.

But in recent years, this cycle of reward and progress has been increasingly threatened. Partly as a result of a stricter Food and Drug Administration law passed in 1962, the regulatory lag between development of a new drug and its ultimate approval for market by the FDA has stretched to an average of nearly eight years.

Consequently, by the time enterprising pharmaceutical houses actually begin selling most of their new products, the effective life of their patents has shrunk by almost half.

The predictable result has been a steady decline during the last decade or so in the amount of money, time and effort pharmaceutical companies are willing to invest in research on new drugs. As rewards for innovation have diminished, so has innovation itself.

The obvious remedy would seem to be an amendment to the patent law compensating drug companies for at least some of the years they lose to the FDA review process.

Not surprisingly, the pharmaceutical industry is suggesting just that in the form of legislation that would extend patent life to ensure a full 17 years of protection once a new drug had been approved for market by the Food and Drug Administration.

That strikes us as eminently fair.

The Dallas Morning News

Dallas, Texas, March 9, 1981

A little-known, but certainly significant, bill is being studied in the United States Senate, a bill that would give a boost to medical drug research in the United States. Without it, the discoveries needed to save lives and reduce suffering will continue to be stifled by red tape.

The bill, S 255, would amend the patent law so that drug manufacturers would not lose years from the life of their patents while government agencies like the Food and Drug Administration hold up approval of the drug.

A patent gives exclusive selling rights to the inventor for only 17 years. Yet federal review can stretch out so long that by the day approval is granted, the company has only half that time or less to try to recoup money to pay for its research and reward its success.

As testing has become more sophisticated, the review time has increased. For example, in 1962, it took around two years and $6 million to bring a new medicine from the laboratory to the marketplace. It now takes from seven to 10 years and $70 million to complete the required testing.

The result has been the introduction of fewer new products. From 1955 through 1962, an average of 46 new drugs were introduced annually in the United States; today that average is only 17 new drugs.

The remedy is simple: Exempt the patent life that previously has been consumed during the government review. Let the 17 years begin after final approval is given. The change should encourage innovation, aid American companies in competition with foreign manufacturers — and, most importantly, save lives.

THE COMMERCIAL APPEAL

Memphis, Tenn., September 3, 1981

ALTHOUGH THE Reagan administration is working to inject price competition into health care, one piece of legislation appears to do just the opposite. Congress has been asked to restore the 17-year life of patents for medicines and medical devices.

By definition a patent is a monopoly granted by the government. The purpose isn't to foster competition among prices and products but just the opposite. It is protection of patent holders to let them recoup their expenses over time and at a fair profit.

IS EXTENDING that privilege any sort of prescription for containing health-care costs? In this case, at least, all is not as it seems.

The country has come a long way from the old patent-medicine days, when drumbeaters made every claim they could in order to make the sale. Today drugs must undergo extensive tests to show their safety and effectiveness. Those tests, however, have cut into both time and profit for U.S. pharmaceutical firms. Last year, for instance, the average patent life remaining when a new drug is marketed was 7 years and 5 months. The return on investment has declined, as well, from 15 per cent a decade ago to around 12 per cent.

The companies' trade arm, the Pharmaceutical Manufacturers Association, admits this percentage still is slightly higher than that of most industries. Yet, the association says, profits are needed for research and development.

It has a point. A recent study by the U.S. Office of Technological Assessment found that passage of the Patent Restoration Act could lead to higher drug costs; those costs, however, would tend to be offset by the development of new and more effective drugs by an industry with a history of plowing back its profits instead of investing in unrelated fields.

Even some with good reason to resist haven't opposed the patent law change. The American Association of Retired Persons, an organization of older citizens — the group which consumes the largest percentage of the nation's prescription drugs — acknowledges the potential rewards as well as the risks, including possible price gouging. But in a nation where prescription drug therapy takes only 8 cents of every health-care dollar (in 1960, it took almost 14 cents), and where new breakthroughs promise to reduce hospitalization time and doctors' fees, the benefits could be substantial.

THERE IS A danger, however, only indirectly related to patent life.

When physicians prescribe trendy, newer drugs instead of reliable, older medicines that would do just as well and cost substantially less, patients and their pocketbooks suffer. Sometimes the public purse does, too, when government-sponsored insurance coverage picks up the "fair price" of a prescription.

There are some drugs for which there is no substitute: Tagamet, for one, the ulcer prescription that's the nation's biggest seller, which retails for about 26 cents a tablet. Its price tag is mild, however, compared to those of the newer antibiotics, which can cost more than $1 per capsule and which can run $100 and above for a 100-pill prescription.

Anyone under the weather isn't going to quibble about a few cents if it cures what ails. But don't patients have to swallow enough without a druggist bill that makes them sick?

The Washington Post

Washington, D. C., May 20, 1981

THE DRUG industry is said to be at the brink of a new age of medical breakthroughs. It now hopes to strengthen its chances for solid returns on its research investments through a bill reported yesterday by the Senate Judiciary Committee. The bill would assure the drug companies and other industries subject to regulatory review that the protection afforded by patent laws is not seriously eroded by the often lengthy period of testing and review required before marketing is allowed. This is a reasonable assurance to require, and the Senate should approve the measure.

For reasons we assume have nothing to do with the locust cycle, patent law deems 17 years the appropriate period for protecting inventors from copycats. Since 1972, when requirements for more rigorous testing of drugs were added to the law, the time required for such preliminaries has stretched from seven to 10 years. As a result, by the time a drug is ready for market almost half the patent life has elapsed.

Since drugs are very expensive to develop, the industry argues that the effective curtailment of patent life discourages new research. Against the arguments of consumer advocates that longer patent lives will increase drug prices by delaying competition, the companies respond that encouraging more research will increase competition and thus lower prices; that drugs, however priced, are far and away the cheapest form of medical treatment and that longer patent protection may discourage high initial price markups now needed for quickly recouping costs.

There are merits on both sides of the price argument. The drug companies, moreover, with their enormous and durable profitability, do not make anyone's list of neediest cases. But there are stronger arguments in favor of patent life assurance. One is simple fairness. If 17 years is the right period for protecting the exclusive rights of inventors, there is no reason why those subject to federal regulation should be denied it solely by reason of that regulation.

There is also the strong desirability of reducing unwarranted pressure on the regulatory process. You don't have to be in favor of mindless bureaucratic delay to recognize the tremendous importance of thorough testing of drugs before they are widely peddled as the latest miracle cure. Some risk may be unavoidable, but no one can want to increase the chances of producing deformed infants.

Stronger regulation not only has reduced that possibility, but it may also have had other beneficial side effects. The higher cost of introducing new drugs, it is said, diverted companies from trial and error research and from the marketing of slightly better products into the basic biological research that is now promising to produce real cures for ailments ranging from asthma to heart disease and cancer.

There are probably ways that the FDA could further speed up clearance of major drug discoveries without jeopardizing the testing process. But assuring drug companies of a substantial period of patent protection is a reasonable and fair way to avoid having the desire for such protection translate into an unhealthy pressure on the review process.

TULSA WORLD

Tulsa, Okla., July 28, 1981

WHENEVER someone discovers a drug to ease pain or cure a disease, it takes from seven to ten years to get it approved for use by the Food and Drug Administration.

The time span is longer than required by any other industrial country. It is, in a word, unacceptable.

Unfortunately, the problem isn't as simple as it seems. The benefits of half a dozen new medications could be offset in a few weeks by the introduction of a drug that turned out to be dangerous. And that seems to happen less often in the United States than in some other countries.

A Congressionally-appointed Commission will begin investigating the FDA's drug approval procedures today. With luck, the agency will recommend ways to cut red tape and delay without sacrificing safety and reasonable caution.

"Pointless and mindlessly excessive regulation by the FDA" has denied millions of Americans the latest of pharmaceutical advances, says Rep. James H. Sheuer, D-N.Y. But . . .

"The American consumer now takes for granted the safety and effectiveness of prescription drugs that are marketed in this country," adds Rep. Albert Gore, D-Tenn.

The trick, of course, is to eliminate the stalling that worries Congressman Sheuer, but to do so without losing the safety and confidence cited by Congressman Gore.

Common sense says the Commission should be able to recommend a system that will meet both tests — safety and confidence without long, unnecessary delays.

The State

Columbia, S. C., August 11, 1981

THERE ARE scads of stories about the bovine pace of federal regulatory agencies in Washington, but none is more exasperating than the Food and Drug Administration's drug approval process.

Happily, there is an effort underway to speed things up at FDA. And there is another effort to try to offset the dilemma the FDA's lassitude causes pharmaceutical companies who have patented drugs but can't market them.

The problem is this: The time lag between the discovery of a new drug and final approval by the FDA is seven to 10 years. Since a drug may be patented for only 17 years, a substantial part of the life of the patent is wasted as far as the manufacturer is concerned.

According to a recent study by Dr. Stuart Butler for the Heritage Foundation, the effects on the U.S. drug industry are harmful. For one thing, the number of new drugs on the American market has fallen severely in the last two decades. We lag behind major European drug-producing nations. How often do we read of new medicines Europeans use that haven't yet reached the American market?

American pharmaceutical firms have also cut their funding of research for new drugs and are emphasizing instead research to extend the marketable life of existing drugs. The United States, then, is expected to lag further behind foreign research and development of new drugs for patients.

Two things are presently underway to remedy some of the delay. A commission has been created by Congress to conduct a six-month study of the FDA's drug approval procedures and to recommend improvements by next Feb. 1.

No one, of course, wants to sacrifice public safety in the regulatory process, and the commission is directed to develop a surveillance system that can provide quick withdrawal of drugs from the marketplace.

Another attempt is underway to extend the life of the patent by the amount of time consumed in the federal regulatory review, up to seven years. That is fair, since the industry is compelled to have its products tested by the federal government and loses much of the "effective life" of the patent in the process.

Speedier processing also would increase competition among the drug companies, which would mean lower prices for the consumer.

We are optimistic that efforts to improve the performance of the FDA will be successful without compromising the public's health and safety. Something really should be done.

THE INDIANAPOLIS STAR

Indianapolis, Ind., May 17, 1981

Positive results of the surgery on Pope John Paul, like those earlier of the surgery on President Ronald Reagan, owe much to infection-fighting antibiotics.

No one knows how many lives have been saved by the constellation of wonder drugs discovered in recent decades.

Dramatically, new drugs keep offering blessed alternatives to grim old treatments: tagamet instead of ulcer surgery, rifampin instead of a tuberculosis sanitarium and anti-psychotic medicines instead of the mental ward, to name a few.

But today introduction of new drugs in this country is being slowed by a thicket of governmental complications that discourage investment in research and drive up many drug prices and health care costs in general.

Many companies that accomplished past breakthroughs are cutting back on research or quitting the business. During 1954-58, 51 independent firms introduced at least one new drug. During 1972-76, that number fell about 19 percent to 40.5 firms.

Present patent policy hampers drug development.

When a firm discovers a new compound it must file at once for a patent. Research and development costs a company an average of $70 million per new product entering the market. But before the product can be sold, the firm must go through an involved approval process taking an average of seven to 10 years.

By the time the new product enters the market, less than half its original 17-year patent life is left.

The negative effect this has on research and new product development hurts the public as well as the drug industry. In 1960 a $3.5 billion pharmaceutical industry with effective 16-year patent lives introduced 50 new medicines. In 1980 a $22 billion industry with effective patent lives of less than 10 years introduced only 10 new medicines.

A thorough, conscientious approval process is essential. But so is a restoration of incentive to develop and market new medicines.

Legislation pending in Congress — H.R. 1937 — would restore to patent owners up to seven years of patent life lost due to government requirements which must be met before a product can be marketed.

We think that would help restore needed incentives to the pharmaceutical industry, encourage research and increase the number of new drugs introduced.

This in turn could restore to health and save the lives of untold numbers of persons, as has the array of wonder drugs introduced during the peak period of research and development.

The Seattle Times

Seattle, Wash., July 30, 1981

SINCE 1962 a simple equation that adds up to a thriving drug-research industry has become increasingly distorted. The incentives that have produced innovations are minimized because pharmaceutical manufacturers, in effect, have only a half a patent life on their products. If the trend continues, the public loses, too.

Drug manufacturers are given a 17-year patent on new products. But before the drug reaches the marketplace, it must undergo extended testing to prove not only that it is safe, but — since 1962 — also that it is effective.

Consequently, it takes eight or nine years before a new drug can be sold to the public — that's the half of the patent life that is lost. The companies then have eight or nine years to earn back the money they've spent developing and testing the drug and to support research into new drugs.

Many companies have not been able to do that profitably and have diversified to keep their profits up. The biggest single moneymaker for the Squibb Corp., long synonymous with vitamins, is Bubbleyum chewing gum. Other companies have turned to the production of household cleaners and cosmetics. Such diversity may please stockholders, but it does not encourage the manufacturers to invest an average of $70 million in the development of a new drug.

Legislation now before Congress (House Bill 1937 and Senate Bill 255) would restore some, if not all, of the lost patent life. Companies could retain exclusive rights to their innovations for an additional seven years (maximum) to offset the time it takes to do the testing.

The major argument against the measure, which has wide support, is that longer patent terms would raise the cost of the drugs to consumers. The bills do not affect existing patents, however; it would be many years before costs to consumers could be accurately calculated.

Even if costs rise, however, that has to be weighed against the need to encourage companies to continue researching and developing drugs, and to maintain a brisk pace in bringing the innovations into the marketplace.

St. Louis Globe-Democrat

St. Louis, Mo., June 20, 1981

It takes an average of seven years to win Food and Drug Administration approval of a new patented drug due to increased safety standards, more sophisticated testing techniques and other demands that have been added to the process through the years.

Under the present arrangement, there is no provision in the patent process to offset the long bureaucratic delay in approving marketing of new drugs. Companies which pay an average cost of $20 million to introduce a new drug simply lose seven years of the life of their patent. This means that the average effective life of a new patented drug actually is only 10 years, rather than the 17 allowed under law and enjoyed by other industries not subject to the long delays of the FDA.

As a matter of equity, it is difficult to defend the apparent unfair treatment of drug companies. They are entitled to the full benefit of a 17-year patent just like a company in any other field. Taking away an average of seven years from a patent life is no small matter when one considers how costly it is to develop a new drug and the many risks involved in bringing new products to the market. A great many of them don't make it commercially even after a drug company has obtained the patent and gone through the long process to gain FDA approval.

The shortened patent term is affecting research and development and introduction of new drugs in this country.

The record shows that in the early 1960s, before long delays were imposed, the average effective life of an approved patented drug was 16 years. As a result, a U.S. pharmaceutical industry with sales of $3.6 billion was able to produce 53 new drugs a year.

When amendments to the federal drug law altered the approval process, the delays mounted until the average effective life of a patented drug fell to only 10 years. The result was that a drug industry whose sales had grown to $16 billion a year was able to produce only 18 drugs a year in the 1977-79 period.

More importantly, undue delays in producing new remedies could shorten or endanger the lives of countless persons. While safety is paramount, needless red tape that hinders the production of life-saving drugs is unconscionable.

The Patent Restoration Act, which has received unanimous approval of the Senate Judiciary Committee, would extend a drug patent by the amount of time equal to the length of the regulatory review period required to gain approval for marketing.

Drug companies would still be required to go through what appears to be a too-lengthy review process, but at least it would give them the full patent life to which they are entitled. Since the government is the problem, the least Congress should do is to remove the heavy penalty that the long FDA delays have imposed. Congress should pass the Patent Restoration Act.

Wisconsin State Journal

Madison, Wisc., May 21, 1981

When you get a patent on a product, it gives you 17 years of protection from your competitors, right?

Well, not quite right. At least, not in the area of drugs and medicines. As a result, bills are under consideration in Congress to rectify a situation that deprives pharmaceutical firms of much of the "patent life" on their products, a situation unique to that industry.

The problem facing drug firms has crept up over the years. Many years ago, drug firms had more or less the same opportunities to protect their "inventions" as any other line of merchandise. Such firms would undertake research, develop a new product — a drug — and put it on the market after it had been tested and found to be valuable.

In more recent years, the federal testing process has become more extensive. The Food and Drug Administration sometimes requires more than a decade of testing and research after a drug is "invented."

But a pharmaceutical firm must patent a new drug immediately to protect its research investment. The life of the patent — 17 years — starts at that time. However, the firm may be seven to 12 years away from selling the new drug.

Bills in the Senate and House of Representatives would extend the life of a patent. In turn, the pharmaceutical firms believe that by being able to obtain a greater profit from their research will be a greater incentive to do more research.

Opposition to this bill comes from a few consumer groups, specifically from elderly people's advocacy groups. They assert that the shorter patent lives mean cheaper generic substitutes are available sooner.

While promotion of the use of generic drugs is laudable, what that argument ignores is that all consumers benefit from a high level of drug research. The more research there is, the more quickly improved drugs will come on the market.

That is why these bills deserve to be enacted.

The Birmingham News

Birmingham, Ala., May 26, 1981

If a great, new breakthrough in medicinal drugs is coming, as some predict, a bill approved by the Senate Judiciary Committee last week should help speed its coming and at the same time help keep prices of any new miracle drugs at reasonable levels.

The bill would assure the drug industry and others involved in bringing in new medicines that they will have a reasonable time in which to recover the costs of expensive research, development and testing.

At present, drug companies must invest millions in the discovery, development and testing of new drugs before they can be approved by the Food and Drug Administration as safe and effective for marketing. And therein lies the rub.

Under current patent laws, the drug companies have exclusive rights to the manufacture or licensing of the manufacture of their new drugs for a period of 17 years which has been deemed a reasonable time to recover development costs. But since the years spent in testing — during which the drug cannot be marketed — are charged against the patent life of a drug, manufacturers many times cannot recover the cost of development and show a reasonable profit.

For instance, since 1972, when requirements for more rigorous testing of new drugs were added to the law, the time required for such preliminaries has stretched from seven to 10 years. As a result of this stretch-out, by the time the drug is approved for the market, more than half the patent life has elapsed.

Since new drugs are tremendously expensive to develop, manufacturers argue that the excessive reduction in the patent life discourages new research and the development of new drugs meant to treat some of the worst killer diseases.

The drug makers also point out that a longer patent life will permit new drugs to be marketed at lower prices, since the manufacturer can spread recovery costs over a much longer period of time.

There are other strong arguments for giving drug manufacturers a normal patent assurance. One of the strongest is simple fairness. If 17 years is the proper time for protecting the exclusive rights of inventors, there is little logic that those subject to federal regulation should be denied their rights solely because of that regulation.

This is a measure that will benefit consumers and deserves the solid support of Alabama's congressional delegation.

THE BLADE

Toledo, Ohio, August 18, 1981

ARTHUR HAYES, new commissioner of the U.S. Food and Drug Administration, has picked a matter of extreme importance to pursue as a major goal of his stewardship of that controversial agency: doing battle with "drug lag" by finding ways to streamline the process of approving new drugs for use in everyday medicine.

"Drug lag" refers to the worrisome situation in which drugs that represent important new therapeutic advances over existing medications often are approved for use abroad years before FDA permits their use in the United States.

Ironically, some of these medications are originated by American research or by U. S. pharmaceutical houses. And they typically are more effective, more specific, and have fewer side effects than anything currently available. Americans thus sometimes wind up being among the last people in the industrialized world to have access to the best medicines. Examples of drugs approved first in Europe and only months or years later in the United States are legion. They range from beta-blockers such as Inderal for cardiovascular disease to bronchial dilators for asthma.

Dr. Hayes, formerly a respected pharmacologist at the Pennsylvania State University's Milton S. Hershey school of medicine, is proposing several measures that could reduce the approval time for important new compounds. He believes, for example, that pharmaceutical firms now are required to submit too much test data for FDA review. Preparing reams of minutely detailed experimental data adds to the costs of drug development, and evaluating it line by line adds to the approval time and needlessly mires the FDA staff.

Instead, Dr. Hayes wants drug companies to simply submit summaries of test data with the understanding that full data must be available if the agency needs it.

He also feels that the FDA may have incorporated unrealistically stringent standards for safety into its guidelines and procedures for testing new drugs. Absolute safety is an unobtainable goal in life, and public misconceptions about this fact complicate many vital areas of national interest from the expansion of nuclear power to the approval of new drugs.

FDA guidelines must recognize that there is no benefit without a corresponding risk. And they must also recognize the degree of risk that Americans routinely accept in their life-styles.

As long as they are compatible with FDA's basic mission of assuring that drugs are safe and effective, Dr. Hayes' proposals deserve the strong support of Congress and the public. Drug lag has been of concern since the 1960s. The time for action is long overdue.

The Pittsburgh Press

Pittsburgh, Pa., May 31, 1981

For almost two centuries the U.S. patent system has been promoting "the progress of science and useful arts" by giving holders of patents 17 years of protection for their discoveries.

Lately, however, the system has been working against innovation in certain fields, particularly drugs and chemicals, which by their nature must undergo thorough government testing to prove their safety and effectiveness before they can be marketed.

For medicines, the process now takes about seven to 10 years. The result is that the effective patent life of a new compound can be much shorter than the 17 years guaranteed by law.

The drug companies contend that since drug-law amendments in 1962 lengthened approval procedures a decline in new drugs has paralleled the decline in patent life. They have a case.

For example, of 12 new drugs or chemical entities approved by the Food and Drug Administration last year the average patent life remaining was less than 7½ years.

This cannot serve the interest of the public health or welfare, for it discourages the kind of pharmaceutical research that leads to better medicines.

* * *

Fortunately, Congress may do something about it this session. A "Patent Term Restoration Act" has been introduced in both houses.

This would direct that a "regulatory review" period be calculated for each product subject to such review — whether a drug or anything else — and that an equivalent period then be added to the product's patent life, up to a maximum of seven years.

This would help ensure that neither the drug industry nor the public would be needlessly penalized.

The Courier-Journal

Louisville, Ky., June 7, 1981

PATIENTS facing staggering bills for prescription drugs may look coolly on the current effort to extend the period in which pharmaceutical companies are protected against lower-cost competition. But the industry makes a good case, both in basic fairness and in the interest of consumers, for passage of this legislation.

The problem arose as an unintended offshoot of the 1962 congressional decision, in the wake of the Thalidomide birth-defects tragedy, to require more testing before new drugs could be marketed. Such testing shortens the life of drug patents; in a few cases, the entire 17-year period of patent protection has expired before the manufacturer has been allowed to market a drug.

What this means is reduced incentive to invest heavily in research. One reason fewer new drugs are being introduced is because companies with insufficient remaining patent protection may not be able to recoup their investments. Once the patent lapses, other companies that spent nothing on research can make and sell the same drug at much lower cost.

The cost of developing a new drug now averages about $70 million, up from $6 million when Congress passed the 1962 legislation. And the manufacturers note that they produced 50 new drugs in 1960 but only 12 in 1980. At a time when the industry believes profound breakthroughs to be imminent, it's clearly in the public interest to encourage more research, not less.

One useful step, of course, would be to streamline the regulatory process. The Food and Drug Administration has reduced some of the time required for animal and human testing of new drugs. But the industry recognizes that this approach has limits. So it wants Congress to extend the 17-year period of patent protection to restore at least part of the time lost to protracted reviews. (Such an extension also should reduce industry pressures for unwise shortcuts in testing.) The proposal has been approved by the Senate Judiciary Committee, but a companion House committee hasn't scheduled hearings.

An example of why the industry wants this bill is the Merck drug, Timoptic, patented in 1972. During subsequent animal testing, one of its spinoffs was found to be a treatment for glaucoma. Glaucoma is a frequent cause of blindness, especially in the elderly.

Merck then started four years of human tests in late 1974. When it sought to market the drug, the Food and Drug Administration put the product on a "fast track." As the first new anti-glaucoma drug in 70 years, it seemed a medical breakthrough.

But though Merck then won final approval in only eight months, that left only 10½ years of the 17-year patent period to recapture its research investment, pay for the "dry holes" in this and other research, and make a profit for stockholders. And Merck did better than most. The 12 drugs approved in 1980 by the FDA had an average remaining patent life of seven years and five months. That's about half what it was in 1960.

The proposed legislation would add as many years to a patent as were lost in regulatory review, but in no case more than seven years. It has strong Senate backing: Kentucky's Huddleston and Indiana's Lugar are among its 27 sponsors. But its 30 House sponsors (including Louisville's Representative Mazzoli) so far have failed to persuade the House Judiciary Committee to hold hearings. If not started soon, it will be too late for action in 1981.

This measure makes sense, not only in fairness to an industry unduly penalized by regulatory delays but in the interest of citizens whose very lives may hinge on research now unwisely discouraged. Drug costs might rise, though the industry says increased research actually would heighten competition and reduce prices in the marketplace. But they'll still be a bargain compared to other forms of medical treatment — and especially compared to sickness or death.

THE SACRAMENTO BEE

Sacramento, Calif., March 1, 1981

One of the anomalies of the nation's patent laws is that, while Congress has provided 17 years of patent protection for the makers and inventors of new products and devices, the government's own clearance procedures often reduce the benefits of such protection by several years — in some instances to little more than half the period provided by law. The most common categories of such products are drugs, pesticides and other chemicals which, because of extensive review procedures under the regulations of the Food and Drug Administration, the Environmental Protection Agency or other government agencies, sometimes cannot be marketed until six or seven years of the 17-year patent period have already expired. That obviously is a disincentive to research and development of new products.

There is now legislation in the Senate that would change all that. With the strong backing of the Pharmaceutical Manufacturers Association (PMA), Sen. Charles McC. Mathias of Maryland and a group of other senators are sponsoring a bill, S 255, that would, in effect, extend the life of a product's patent for the length of time it takes for it to clear the government's review process and reach the market. The PMA claims that "investment of funds in research and development of products such as drugs and chemicals requiring lengthy government approval is discouraged by shortened patent lives. A decline in new drug introductions has paralleled the decline in patent life and must be reversed to bring about a new encouragement of technological innovation in the United States."

That may be an overstatement — the decline in new drug introductions unquestionably stems from a variety of causes — yet the argument still has merit: obviously the shorter the effective life of a patent, the less incentive there is for innovation and development. There has been considerable pressure from drug manufacturers in recent years for the government to speed up its review procedures and thus get new products on the market more quickly. In some respects, such streamlining may make sense, but it also involves serious hazards in a field where the damage resulting from casual review of the safety of new products has been extensive.

The Mathias approach makes much more sense. The review process should be careful and deliberate, yet there is no reason why inventors and manufacturers ought to pay the price in reduced patent protection or why the process as a whole should unnecessarily discourage the invention of new drugs. If there is any economic virtue in a shorter patent period, it should be established through careful study and deliberation, not through the accidents of a review process designed to protect the health and safety of the nation.

OKLAHOMA CITY TIMES

Oklahoma City, Okla., May 28, 1981

GIVEN the present anti-business climate prevailing among much of the populace, it may be hard to work up much sympathy for drug and chemical companies in the dilemma they face under current patent laws.

But the public should take notice and put pressure on Congress to provide some relief from the regulatory burden that now stifles incentives for research and development. It would be in the public's interest to remove the barriers to a continuing flow of new and improved medical therapies.

The vehicle for doing so is the proposed Patent Term Restoration Act, embodied in S.R. 255 and H.R. 1937. The Senate Judiciary Committee has looked favorably upon it, but the bill faces a tougher run in the House Judiciary Committee. Oklahoma's Rep. Mike Synar is a member of that committee.

The basic patent law gives all holders of patents 17 years of protection for their discoveries. The object is to encourage research and development of new products.

By their nature, though, products like drugs and chemicals require a lengthy review process by the federal government to demonstrate their safety and effectiveness before they can be marketed. That means they are kept out of the commercial market and, thus, are denied part of their congressionally guaranteed 17 years of patented life protection.

The bill would restore the patent life consumed during this review and approval process. It would require that a "regulatory review" period be calculcated for each product and that an equal amount of time be restored to that product's patent, for a period not to exceed seven years.

Several good arguments can be marshaled for this change. One is simple fairness, treating all patent holders alike by giving them sufficient time to recoup their research and development costs in the commercial market. It takes seven to 10 years to get a new medicine through the federal Food and Drug Administration's approval procedures, so that its effective patent life is less than 10 years. The average is 7.5 years. For chemicals, the patent life has been reduced to 12 years.

That discourages investment in research and development and, consequently, inhibits technological innovation. From 1963 to 1975, the percentage of medicine and drug patents worldwide that originated in the United States declined from 66 percent to 54 percent.

Particularly hard hit are small businesses, the most innovative segment of the economy and the most dependable source of new jobs. They are the ones most in need of patent protection in order to attract investment capital. Also affected are university research programs, which invent new compounds but, because of the lengthy review process, find they have only a few years of royalty-bearing life in their patents.

Passage of the bill would help the pharmaceutical and chemical manufacturers, but society would be the real winner because of the medical advances and the possibility of lower prices resulting from increased competition that would come from greater research activity

CHARLESTON EVENING POST

Charleston, S. C., July 13, 1981

A legislative proposal, currently pending in a congressional subcommittee, could revive the lagging pharmaceutical industry and ultimately benefit all medical patients. The bill would extend the length of time a manufacturer is allowed to hold the patent on a new drug.

The present 45-year-old patent law treats all product inventions the same, allowing the developer to hold exclusive rights for 17 years. However, the drug industry must subject its new products to numerous, time-consuming, federally-mandated tests and evaluations before marketing. This testing period now eats up much of the 17-year patent life during which the manufacturer can expect to enjoy the new drug's exclusive market. (The testing period also denies needy patients the benefits of the research, but that's another subject.)

The proposal now before Congress would have the effect of extending the patent term of such products by the length of time consumed by regulatory review, up to a maximum of seven years.

Not only would such legislation spur research into new drugs but the increased competition among companies would benefit consumers. The end result could be more product competition with less emphasis on marketing competition.

The bill deserves a favorable endorsement.

Worker Productivity Lag Begins to Alarm Industry

By the end of the 1970s, American labor productivity was in a disturbing decline. The last year in which productivity rose during the decade was 1978, and the increase was a pitiful .4%. In the next year, 1979, productivity actually decreased by .9%, the first drop since 1974, during the recession produced by the sudden increase in oil prices. Suggested reasons for the productivity lag were many: low business investment, more inexperienced workers such as young people and women, increased government regulation and a fundamental shift in the economy from manufacturing to service industries. Whatever the reasons, or combination of reasons, overall productivity growth had been weakening for the past three decades. The President's Council of Economic Advisers said productivity had grown by 3.4% from 1948-55, by 3.1% from 1955-65, by 2.3% from 1965-73 and by 1% from 1973-77. Aside from hurting U.S. economic performance on the world stage, the productivity slowdown added to domestic inflation. Less output per worker meant increased unit costs of production for the manager and hence, higher prices for goods. Most Americans, however, cared less about the economic aspect of the problem than about its moral implications. America prided itself on its respect for the "work ethic;" the figures for labor productivity made a mockery of that cherished belief.

THE ATLANTA CONSTITUTION

Atlanta, Ga., February 8, 1978

For scores and scores of years, the productivity of American industry and American workers was the envy of the world. The United States made giant steps forward in economic development and provided a standard-of-living for most Americans that seemed like riches untold to millions of persons in other nations. The economic promise, along with the promise of freedom, was a major reason for the mass migration to these shores.

Somewhere along the line, something broke down.

America's productivity has put on the brakes. And it's causing a variety of unpleasantries — inflation, huge trade deficits, weakness of the dollar, slipping standards-of-living.

The Conference Board reported earlier this week that the growth rate of "total factor" productivity — a measurement of output combining total worker hours and capital investment — reached its peak in the U.S. in the mid-1960s, and since has grown at a considerably slower rate.

The peak annual growth rate a decade ago was 3 percent. Since then, it has grown at a rate of 1.3 percent, down more than half.

Why did this happen? The reasons vary. Insufficient investment in modern plants. Too much "featherbedding" by unions. Too much assuming "we had it made" and sitting back to take it easy. But a fact of life is that, no matter in what endeavor, when someone stops producing at his best, there's someone else waiting to take his place and do better.

Already there are other nations which can out-compete the United States in several industries on the international market. Productivity and standards-of-living are rising in other nations. Thundering hooves are behind us, gaining. To stay at the head of the stampede will take more than complaining to Washington about unfair foreign practices and calling on the federal government for protectionism.

BUFFALO EVENING NEWS

Buffalo, N. Y., September 15, 1978

People didn't talk about productivity and standard of living indexes in the Middle Ages — life went on more or less the same for hundreds of years.

Today, however, we take it for granted that our standard of living will gradually rise over a period of years. We have been able to make this assumption because the U.S. economy as a whole has been more and more productive — that is, the average output per man-hour keeps rising.

But we shouldn't take this too much for granted. Through the 1950s and 1960s, U.S. productivity rose about 3 percent a year. In the decade from 1967 to 1977, the increase was 27 percent, which doesn't sound too bad until compared with that of other industrial nations. Britain, which has often been cited as a horrible example of economic inefficiency, also had a 27 percent increase.

In contrast, Japan had 107 percent, France 72 percent, West Germany 70 percent and Italy 62 percent. The most alarming aspect is that in the last few years, U.S. productivity — and consequently our standard of living — has been rising hardly at all.

Why? A national economy is incredibly complex, and it is often hard to pinpoint causes, and when you can, to do anything about it. The end of the era of cheap oil is one factor — although one that affects other nations much more than it does the U.S. There is also the waste in the economy attributable to governmental bungling, poor management, strikes, overly stringent environmental rules, inflationary wage costs and the general decline of the "work ethic."

There are no easy answers, but perhaps the most promising area on which to concentrate is the steady encroachment of government on life and business — at a grievous cost in the form of rising tax burdens. Private enterprise is finding it harder and harder to raise funds for the new technology to make the economy more productive.

Technology, after all, is what makes our society different from that of the Middle Ages. Without constant capital investments in new technology, we cannot expect our economy to produce at a more efficient level. The lack of such investment could put the U.S. on a no-growth course that would make us fall further and further behind our industrial rivals.

The next few years should tell whether the U.S. is moving into a stagnant era, or whether it will be successful in reversing the present ominous trend and moving into a new dynamic economic cycle.

The Dispatch

Columbus, Ohio, May 6, 1979

TIME WAS when the industrialized world knew automatically that America produced high quality goods rapidly and at low unit cost. But America is no longer the leader.

Why? Because U.S. productivity, defined as output per hour of work, has gone stale. Its annual rate of increase has fallen to about the lowest of the industrialized nations. And economists are in general agreement that until U.S. productivity improves, there will be no effective cure for a major ailment — inflation.

Many reasons are given for inflation, the chief being excessive government spending and regulation. But a major factor is nonproductivity caused by an absence of a spirited and prideful work force. Also responsible is a lack of innovative technology, a reluctance by business to invest and a lack of support for the private sector from federal government.

The result is dishearteningly clear — in the decade of 1967-77 productivity rose at an annual rate of 7.9 percent in the Netherlands, 7.5 in Japan, 5.5 in Germany, 5.4 in France, but only 2.4 percent in the U.S. — and only 1.0 percent last year. At that rate there would be zero growth this year.

G. William Miller, chairman of the Federal Reserve Board, is concerned not only about the erosive effect of inflation on the nation's economy but its linkage to nonproductivity. He is advocating a commendable proposal — allow businesses to write off investments for government-mandated environmental and health expenses in a single year, for processing equipment in five years and for permanent facilities in 10 years.

But that is only part of the need. Still to be attained would be a more inspired work force, more innovative technology (more than a third of patents are being filed by foreigners), less timid investors and more daring managers. The economists are correct: Until productivity rises more rapidly, inflation will continue to sap the economy.

Lincoln Journal

Lincoln, Neb., January 2, 1978

Folk wisdom has latched on to the belief that America is losing the knack, or at least the liking, for work. One reason Japan's economy is outstripping our own is because the Japanese work harder, or so the theory goes.

There may be something to it. Last week the National Center for Productivity and Quality of Working Life reported that the growth of productivity in the United States slipped in the past 10 years to an average of 1.5 percent a year.

That's less than half the 3.2 percent average for the 1947-1966 boom years. It's also considerably behind not only Japan but Western Europe, too.

In fairness, the death of the work ethic may be exaggerated. Other factors were related to the drop of productivity, too.

The productivity center, created by Congress in 1975, pointed to two considerations that seem especially important. One is the increasing share of our economy represented by service industries. Productivity growth comes harder in services, where automation and labor-saving techniques have less application than in manufacturing. The other is the greater proportion, recently, of inexperienced employees, youths and women entering the labor force for the first time.

The inexperience factor should take care of itself, as the population bulge of young people passes and the phenomenon of women newly trooping to the workplace diminishes. But there probably is not much to be done about the evolution of a more service-oriented economy.

Some things can be done, however, to encourage productivity. They are implicit in the productivity center's report, though not stated as recommendations.

Government can encourage a greater ratio of capital to labor, which should result in better equipment and plants and therefore boost productivity. And government can stimulate research and development, whose discoveries and technology often lead to more worker output. Both R&D and the capital ratio have declined. Their promotion, of course, is tied to tax policies, and there are signs the Carter administration wants to head in these directions.

If U.S. productivity does not increase — or, at minimum, if its decline is not arrested — the nation's future is bleak. As the productivity center puts it, the recent slide "should be taken as a warning that our economy may not be able to deliver on our expectations."

More specifically, productivity is tied to gains in our living standard, to competing in international trade, to containing inflation, to providing jobs.

Even if the best-case scenario develops, the United States — and all industrial nations — may have to revise downward their economic expectations, given the energy realities. But if the productivity slide continues, what we will have is a worst-case situation.

And where does the work ethic fit in? Its erosion, obviously, is not the only matter at issue. Yet Americans' taste and talent for work just as obviously can make a difference in productivity. Certainly an insistence on ever-higher pay whether or not there is higher productivity makes an immense difference in our economy and its role in the world. Working hard has not lost its value.

Our society may have moved away from the precept that those who don't work won't eat. But the biblical injunction retains a great deal of truth: "Every man shall receive his own reward according to his own labor." Every nation, too. As Japan knows and America is learning.

Wisconsin State Journal

Madison, Wisc., January 5, 1978

There was a time when the United States was the most productive nation in the world.

We bragged about it. We reveled in our leadership. We exported far more than we imported, primarily because the partnership of skilled workers, technology and management let us produce goods more efficiently (and therefore at a better price) than other nations.

This salubrious state of affairs led to shorter work days and work weeks and to higher pay, enabling us to enjoy more leisure time and to improve our standard of living.

From 1947 through 1966, productivity grew at the rate of 3.2 percent each year. The workers were using machines with such efficiency that more could be done in fewer hours than ever before.

There has been a marked decline in that productive pace.

The National Center for Productivity and Quality of Working Life said in its yearend report that the growth of productivity in the U.S. slipped in the last 10 years to an average of 1.5 percent.

Further, from 1960 to 1976, Japan led the industrial world with an average annual increase in productivity of 8.9 percent; the U.S. averaged only 2.9 percent.

True, it is easier to have larger percentage increases when starting from a lower base, but Japan, West Germany and others have sustained their productive gains to a point where they can outsell the U.S. on the world market.

This has led the AFL-CIO to demand protective tariffs as America is undersold by Japan and Western European nations, ignoring the adverse effect this might have on American industries that still rely on foreign markets. One of eight American workers still work for export industries.

Is is more important to face up to the facts presented by the national center when it says the slowdown in the productivity rate puts the U.S. "far behind" Japan and Western Europe in terms of output per employe hour.

We no longer are the world's No. 1 producers. We're well down the list.

What will be the result? The national center says flatly that unless productivity improves "our economy may not be able to deliver on our expectations."

It said unless there is improvement, forget about an improved standard of living, forget about containing rising costs and forget about checking inflation.

It's time to face up to what is happening in the United States.

The national center identified four trends leading to the slowdown in productivity: Less experience as more women and young people enter the labor force, greater emphasis on providing services instead of producing goods, a proportiate decline in research and development and a slowdown in the ratio of capital to labor.

Succinctly, this tells us to get off our duffs and start putting out more or to look forward to becoming one of the have-not nations, over the long haul, with an accompanying decline in affluence and in living the "good life."

Casper Star-Tribune

Casper, Wyo., August 27, 1979

There is cause for serious concern in the announcement that non-farm productivity decreased by an annual rate of 5.7 percent in the second quarter of this year. This is the largest quarterly decline recorded since this series of statistics started to be recorded in 1947.

In the first half of the year the productivity decline was 3.3 percent - not an encouraging figure.

The Joint Economic Committee of Congress sees a period of economic slowdown and a decrease in the American standard of living and less employment unless this trend can be turned around.

The committee of which Sen. Lloyd Bentsen, D-Tex., is chairman is powerful because it is composed of senators who also are the heads of other influential committees.

We agree with the committee that attacking the productivity decline is a positive way of fighting inflation. Allowing productivity to continue to fall is a sure way of increasing the price of goods.

The statistics are measured in output by workers per hour. Because most costs are fixed, including labor, it follows like night after day that if a worker produces less and costs remain the same that the price of an individual article must rise.

One way of increasing worker productivity is by the addition of new machinery which does a job better than an old outmoded one which now may be in use.

However, tax laws have stretched out the period of depreciation on machinery making it impossible for a manufacturer to write off rapidly as a business expense the cost of the new machinery.

This can be remedied by legislation shortening the period of depreciation.

Much of the troubles and noncompetitive status of the United States Steel industry has been pointed to as stemming from obsolete plants. Japan has followed a policy of encouraging its steel industry to modernize through quick write-offs of new equipment.

Productivity lag means that there are more dollars looking for fewer items which ss a further incentive to higher prices.

The Carter administration has been looking at the other side of the picture trying to hold prices down by ineffective guidelines and tight money.

The Senate committee will render a service to the people if it can put through legislation which will encourage American industry to plant modernization and greater efficiency.

THE BLADE

Toledo, Ohio, June 28, 1979

THE most recent statistics on worker productivity in the major industrial countries should give Americans some cause for concern. The figures show unmistakably that the United States has lost its competitive edge in industrial production that had been almost taken for granted.

This worrisome development is reflected in the average wages paid to hourly workers in these nations. As reported in The Blade's Behind The News Section Sunday, American workers in 1972 were the highest paid in the world, earning an average of $3.81 an hour. In 1977, the latest year for which data are available, that figure had risen to $5.63 hourly, but wages in four other countries had grown even more. In Japan the average wage in 1977 had reached $6.70; in Sweden it was $6.13; in Belgium it was $6.10, and in West Germany, $5.76.

Stated another way, productivity in the United States increased only about 0.6 per cent in the years from 1972 to 1977 as compared with 3.5 per cent in Japan and West Germany. That is nearly six times as much as the U.S. average during that period. The nub of the matter was put succinctly by Sen. Jacob Javits, New York Republican, when he commented that "right now the United States is living too high on the hog . . . it is living on a standard which it is not producing to afford."

The causes, of course, go further than just the productivity of the average worker, involving such things as the makeup of each country's economic sectors, technological innovation, and the relationship of government, financial centers, and industry. And even though individual productivity still lies at the heart of the problem, the fact is that the United States is falling behind in most of these criteria.

A variety of suggestions has been advanced as to how to cope with the productivity problems facing this country, including the establishment of an "incomes policy" which would coordinate wages and profits in an effort to boost real production. Also a National Productivity Council has been created to coordinate federal activities in this area.

Government, however, cannot do the job that needs to be done to restore American productivity to the levels it has formerly enjoyed. That is a task for every business, labor union, and individual worker in the country.

The Cleveland Press

Cleveland, Ohio, August 15, 1979

President Carter has cited a drop in U.S. productivity as one of the symptoms of a "crisis of the American spirit."

Whether it is that, it's incontrovertibly economic bad news.

A steady growth in American productivity — output per man hour — propelled the United States to the rank of the world's foremost industrial nation.

But there has been a slowdown of the rate of productivity growth in the past years; and in the first two quarters of 1979 there has been a sharp drop. Five industrial nations now surpass the U.S. in per capita output of goods and services.

What does the falling rate mean? More inflation, for it increases the cost of goods. Perhaps a lowering of the American standard of living, for it will lead to a stagnating economy and an inability of consumers to afford what they need or want. A declining ability of the U.S. to compete in world trade, and a deterioration of American influence in the world.

To the extent that the productivity slippage can be laid to a decline in the American worker's zeal and pride of workmanship, it might be considered a symptom of a "crisis of the American spirit."

But the greater blame might properly be laid to government: Its corporate tax policies discourage investment in new and more efficient plants; its tax and interest-rate policies discourage individual savings that supply money for commercial borrowers; its regulations increase the cost of doing business; and its import and subsidization policies prop up inefficient industries and protect them against foreign competition.

The Joint Economic Committee of Congress warned the other day that lagging productivity is a major peril to the American way of life. Congress ought to heed the warning. We know of no other institution, organization or individual that is in a better position to do something about it.

WORCESTER TELEGRAM

Worcester, Mass., June 14, 1979

Nearly a century ago, the United States began inching ahead of other nations in productivity — output per worker-hour. By the beginning of World War II, no other major nation came close to the U.S. productivity rate or to the high U.S. standard of living that it made possible.

But after World War II, it was Western Europe and Japan that made strides in productivity, often with American aid. The U.S. productivity gain leveled off at 2.6 percent a year from 1960 to 1977, while the figure in West Germany was 5.5 percent; in Italy, 6.3; in Belgium and the Netherlands, 7.4, and in Japan, 8.8.

Set against this background, the recent news that the U.S. is actually losing ground in productivity is alarming. What are we doing wrong that productivity — and thus our standard of living — should drop by 4.6 percent, on an annual basis, during the first three months of this year?

Part of the explanation lies in the business cycle. Productivity normally slackens during a boom, when the work force has expanded rapidly and includes many inexperienced people. Obsolete and inefficient equipment may also have been pressed into service to meet demand at a time of high production.

But economists say most of the problem is chronic, not just cyclical. Rising American productivity in the past has resulted not so much from people working harder or with greater skill as from their being given machinery and techniques that add leverage to their efforts. Economists say the biggest cause of low productivity in recent years may be the low rate of business investment in new factory equipment and in research and development.

Many companies, bankers and investors complain they are discouraged from installing new equipment by high taxes and doubt about the future course of an increasingly government-controlled economy.

What's more, much of the new equipment installed in recent years has been by government order, to reduce factory air and water pollution. Cleaner air and water make life more pleasant, but they don't automatically improve productivity. They may even reduce it in some cases. And money spent on anti-pollution equipment is money not spent on machinery to improve production efficiency.

A number of proposals have been set forth to improve productivity. Industry argues that corporate taxes should be reduced and "double taxation" of dividends eliminated so as to make a higher percentage of pre-tax profit available for capital investment.

The American Productivity Center says managements should take steps to improve worker morale and should make it clear that "smarter" work is what is needed, not harder work. The center also urges that government regulations be tested by "productivity impact statements" and that antitrust laws be changed to permit cooperative industry efforts to improve productivity.

Once the adverse "productivity impact" of the new anti-pollution equipment works itself through the economy, U.S. productivity may start gaining again. But most of the experts say we won't be able to count on regular boosts in our standard of living unless the nation restores investment and ingenuity to the American workplace.

THE MILWAUKEE JOURNAL

Milwaukee, Wisc., July 1, 1979

Productivity might not be a hot topic at the neighborhood bar. Yet it goes to the heart of America's prosperity and high standard of living. Today US productivity is slipping and the nation needs to do something about it.

Look at what has happened to productivity — output per worker — in the private sector in the last 30 years. Between 1948 and 1955, productivity rose at an average annual rate of 3.4%. Between 1977 and 1978, it rose only 0.4%. And in the first quarter of 1979, the rate actually fell well below zero! Surely the implications for the future are alarming.

It is increased productivity that allows Americans to earn higher *real* wages. US wage rates can go up, but if output doesn't climb correspondingly, workers reap only inflation. And the nation has seen plenty of that lately. Although wages outstripping productivity is not the sole cause of inflation, the inflationary impact of declining productivity is immense.

Similarly disturbing is the way America's technological lead over international competitors, such as Germany and Japan, is dwindling at the same time US productivity is faltering. Technology and productivity are closely linked.

The growing competitiveness of foreign producers is seen in the wide array of high quality imported goods in US stores today. If the technology gap continues to narrow while our productivity stagnates, American manufacturers could soon be widely outsold in the world's marketplaces with considerable loss of American jobs.

What's wrong with America? Has Yankee ingenuity dissolved? Have we become an indolent people, demanding pleasure and shirking work. To some extent, yes. But there's still ample technological ingenuity and plenty of willing workers.

What is often overlooked is the great change in the economy's makeup and goals over the last three decades. These changes help explain the drop in productivity gains, but they don't necessarily provide solutions.

For instance, the economy has become increasingly a white collar, service oriented one, attracting growing numbers of women and youths, many of whom work part time at minimum-skill jobs. This service oriented work tends to be less productive than blue collar, skilled manufacturing. As a result, the nation's rate of productivity growth has slowed.

Meanwhile, Americans have developed a new consciousness about the *quality* of their lives. Consumer and environmental protection as well as improved safety on the job are examples of this concern, resulting in a rash of new government regulations. Industry has had to make large investments to comply with these standards. And while beneficial to the nation's living environment, these expenditures seldom enhance productivity very much. Government regulations, at their worst, can actually hamper productivity with red tape.

The trend toward less productive, service oriented jobs appears inevitable as the US post-industrial economy matures. Nevertheless, steps can and should be taken to minimize drags on productivity growth.

One, put a solid brake on inflation. Rampant inflation drives up consumption and depresses investment, which is a key to productivity growth. Curbing inflation will require a degree of wage, price and government spending restraint that has yet to emerge fully in the US.

Two, significantly increase capital investment. Most of America's trade rivals have national savings and investment rates much higher than the US. Investment produces technological change and higher productivity. Spurring investment will mean a smaller share of national output — at least initially — for consumption. Encouraging investment will also require Congress to look seriously at tax incentives such as reducing or eliminating capital gains taxes if gains are reinvested; increased investment tax credits; and accelerated depreciation allowances.

Three, boost spending on research and development (R&D). As the accompanying chart illustrates, the trend has been in the other direction. As a special kind of investment, R&D is the life blood of innovation and new technology. Again it may take specific tax incentives or even significant government financial support for private R&D to achieve this goal.

Four, review the regulatory process dealing with quality of life standards. Essential rules should not be changed, gains not sacrificed, but the bureaucratic red tape should be cleared away as much as possible so that these goals can be attained with the least cost in lost productivity.

If all these steps are taken the US should be able to get back on the track of higher productivity and be in a vastly better position to preserve its high standard of living — or at least prevent major erosion. It won't be painless. It means some sacrifice of consumption now for future benefit. But it can be done.

Productivity Lag Continues; Special Study Panel Named

Productivity registered its second year of decline in 1980, although a significant increase in manufacturing productivity was noted in the fourth quarter of the year. The only positive aspect of the .3% decline was that it was smaller than in the previous year. Economists and the Reagan administration blamed high interest rates, which discouraged investment, and the general economic slowdown. Although real economic recovery was not in sight, 1981 was a little better in terms of productivity. The final figure for the year showed a gain of 1%, with manufacturing leading the other sectors with a 2.7% increase. President Ronald Reagan had pledged in 1981 to appoint a National Productivity Advisory Committee, and the 34-member panel began its work in January 1982. It was the latest in a series of committees. The first was the National Commission on Productivity, created in 1970. It "had few staff members and no authority to carry out its mission," according to C. Jackson Grayson Jr. of the American Productivity Center, in testimony June 1, 1981 to the congressional Joint Economic Committee. The NCP was abolished in 1978, to be immediately succeeded by the National Productivity Council, which "had no staff, was not funded, took no actions, . . . and met only four times for a total of about four and a half hours" according to Grayson. Reagan apparently was determined to avoid the mistakes of the past, for he held a White House ceremony to inaugurate his new panel. He told the members that their work was "vitally important," and he instructed them to prepare "concrete suggestions and specific recommendations" for improving worker output and capital formation.

WORCESTER TELEGRAM.

Worcester, Mass., May 15, 1981

Rising productivity — output per worker hour — is what gave the United States a constantly improving standard of living for nearly a century, until just a few years ago.

By and large, it was accomplished not by getting people to work harder but by giving them better machinery and techniques. The results were better working conditions, more leisure time and more goods of all sorts for everybody.

In view of the importance of rising productivity in this country, it is perhaps surprising that Americans have commonly had the idea that a program to improve productivity was a fancy name for pushing workers harder on the job in order to line the pockets of top management.

But there are signs now of a change in that public attitude. According to a regional Lou Harris poll made public at a conference on productivity in Concord the other day, most workers now think better productivity benefits workers and consumers, not just management.

The poll says that fully 80 percent of people in the Northeast now see declining productivity as a serious problem requiring urgent attention. Nearly as many believe that failure to turn it around will result in fewer jobs, declining respect for the United States abroad, a lower standard of living and an inability to meet social obligations at home.

There are probably a number of reasons for this improved public understanding of productivity. Its decline — and, partly as a consequence, the rise of inflation rates over the past decade — has led to much more public discussion of productivity than in the past. A lot of attention has also been paid to the Reagan programs to boost productivity — deregulation of industry and tax incentives to boost investment in efficient new factories and machinery.

This new public focus on the importance of productivity comes none too soon.

Productivity has been rising smartly in Japan, West Germany and other nations that are this country's trade rivals. Here, meanwhile — where the term was invented — it has gone into a decline.

It's time for a change.

TULSA WORLD

Tulsa, Okla., February 2, 1981

FOR the third consecutive year the average American worker's hourly production dropped, a condition that is generally acknowledged to be a prime factor in inflation.

The U. S. Labor Department reported productivity dropped .3 percent during 1980, following drops of .4 percent in 1979 and .2 percent in 1978.

What does this mean? Does it mean Americans are lazy, that they are less energetic and dedicated than workers in other countries whose productivity rates are climbing?

It means nothing of the sort, and that is both good news and bad news. It perhaps would be easier to remedy if the American labor force had simply grown fat and lazy, because severe economic conditions would encourage an end to sloth.

But the situation is more complicated and more difficult to cure than simply urging everyone to work harder.

Simply put, Americans no longer have tools and equipment equal to their foreign competitors; they work in plants that are outdated. Additionally, more and more of them are devoting their time to filling out endless Government forms to comply with rules and regulations that in the end produce nothing.

It's universally agreed that the U. S. industrial plant must be rejuvenated. The U. S. worker must again be given the tools with which to do his job.

That is a major priority of the Reagan Administration. One way to help is to remove meaningless, burdensome, unproductive Government regulation. Another is to allow faster tax writeoffs of equipment and plants to encourage their construction.

Given the proper help to obtain the tools, the American worker is easily the most productive and efficient in the world. He ought to be given the chance to again prove that, and in the process, much of the economic stagnation and inflation plaguing the country will disappear.

The Dallas Morning News

Dallas, Texas, August 29, 1981

Wipe that gloomy prediction from the book. The one that foresaw another decline in the nation's productivity in the second quarter of 1981. New figures released by the U.S. Bureau of Labor Statistics show the American workers actually *increased* their productivity at a 0.7 percent annual rate.

In the business sector, productivity increased at a 2.8 percent rate and manufacturing productivity rose 4.3 percent at an annual rate.

Farm productivity was the major drag, with an annual decline of 1.9 percent registered in the second quarter.

Increased productivity is a key to the nation's economic revival. American workers are still the most productive in the world, but in recent years, foreign employees have been improving their productivity at a faster rate and closing the gap.

Many factors, including inflation and high taxes, affect productivity, and some of the obstacles to progress can be removed by Congress. But despite other discouraging economic news, the increase in productivity is encouraging and must be repeated. It means a bigger economic pie for more Americans to share.

THE ATLANTA CONSTITUTION

Atlanta, Ga., April 21, 1981

Results from a Los Angeles Times poll shed some interesting light on American workers. The general consensus has been that workers in the U.S. are not as interested in work as they once were and would like to quit if they could. But the poll said that's not true.

The poll showed 70 percent of those interviewed said they would continue working even if they were able to get enough money to live as comfortably as they would like for the rest of their lives without working. That's a switch in attitudes thought to exist.

Growing competition from foreign countries like Japan prompted the belief that the U.S. is losing economic battles mainly because American workers are not producing like they should. That's true in part, but listen to other things the workers had to say:

Two-thirds of those polled agreed they were not turning out as much each day as they should, and more than half said they could accomplish more on the job each day.

A true sign that the work ethic is still strong in this country was shown when those surveyed said they would work even if they did not have to, and that they believe they should be working harder than ever.

There is an appealing part of the survey of which employers should take note. The poll showed that today's workers expect more from their jobs than just a paycheck. An "interesting job" was the most important factor in achieving job satisfaction. Next in importance is a sense of accomplishment on the job, followed by having work that gives some prestige.

Believe it or not, the almighty dollar (money) was ranked fourth by those responding to the poll as a means of achieving job satisfaction.

This points out that many workers understand the plight of the country regarding its economic woes. The survey also suggests that employers may be causing some of the problems that result in less productivity of the worker. Could it be that the employer and the worker are not communicating?

If all it takes is a general understanding of what is expected of each other to turn the economy of this country around — then we better start talking to each other. There no doubt is a communication gap between the employer and the employee. The gap should be bridged, and quickly.

THE PLAIN DEALER

Cleveland, Ohio, May 5, 1981

The American economy is sick enough to require drastic remedies, as President Reagan says. But some bright spots are sparkling amid the general gloom.

One sparkler is the sudden 3.9% improvement in productivity in the first quarter of 1981 over the previous quarter. Productivity is output per worker hour. It has been dropping the last three years. So its shooting up is spectacular.

Declining productivity has been a cancerous sore. It means fewer goods. That drives prices upward. It means higher priced products, less competitive in the world market.

Other nations have been winning away big chunks of what used to be America's share of world trade: steel, autos, textiles, machinery, electronics. And when there are fewer goods turned out, prices inflate. Consumers pay more for less.

To cure the sag in productivity, more money must be invested in new equipment, in research and development, in productive machinery and plant. Once the world leader in industrial technology, the United States now invests less of its gross national product in new plant and equipment than any other industrial nation, including Britain.

Tax incentives must be used to lure money into investment, into solving the problems of slow productivity, into better industrial methods.

Labor and management must learn how better to set output goals and reach them. Morale among U.S. industrial workers and owners must be buoyed up. The time is ripe for it.

A decade or two ago, some critics said American workers were "bored, coddled, goal-less, antagonistic," and that corporate leaders were overconfident and out of touch with the public and their own employes.

Many workers already have indicated, in recent surveys, that they were willing to give up some economic advantages in order to solve the low-productivity puzzle. In this kind of atmosphere some chance of headway is more likely.

the Charleston Gazette

Charleston, W. Va., May 2, 1981

AT the same time President Ronald Reagan is trying so hard to tame the inflation monster by goading a lethargic economy, his country — to quote Emma Rothschild, the MIT economist and sociologist, in *The New York Review of Books* — "is moving toward a structure of employment ever more dominated by jobs that are badly paid, unchanging, and unproductive."

The United States has gone further toward a service and retail economy than any other of the world's largest industrial powers, she says.

In 1978, the United States had 51 percent of its employed population in service and retail trade. Sweden had 48 percent, Japan 44 percent, Germany 37 percent. Doubtless, these percentages have climbed during the three-year interval.

Meanwhile, the percentage of workers holding jobs in manufacturing plants has dipped alarmingly. The United States has less than 23 percent of its employed workers so occupied, Sweden 25 percent, France 27 percent and Germany 35 percent.

With more than half of America's work force engaged in labor of low productivity and with less than a quarter performing in jobs of high productivity, cutting inflation with the usual economic panaceas is probably an impossibility.

The Times-Picayune The States-Item

New Orleans, La., June 1, 1981

Low productivity probably is the least understood of all the factors that contribute to inflation. Americans tend to blame the government, large corporations, high union wages — almost anything and anybody but themselves — for inflation. But in recent years, when inflation has been at its highest, often the victims of inflation also have been among its contributors.

Productivity is a measure of economic efficiency. It measures the amount of goods and services produced in a paid working hour. The efficiency of American workers helped give our nation the highest standard of living in history. In recent years, however, the efficiency of the American worker has declined and inflation has risen.

When productivity declines, prices are increased to cover the labor costs. For example: Company A makes chairs. Company A pays two workers $8 an hour each to make the chairs. Company A's workers are very efficient. In one hour they produce two chairs at a combined wage of $16. After materials and other factors are figured in, the chairs sell for $50, and Company A makes a reasonable profit.

Company B also pays two workers $8 an hour to make chairs. But Company B's workers are not very efficient. They spend a lot of time goofing off. As a result, Company B's two workers only produce one chair in an hour. So to make up for the fact that it only got one chair for one hour that cost it $16, Company B increases the price of its one chair to $33. Two of its chairs, just like those produced by Company A, then sell for $66 rather than the $50 charged by Company A. The inefficiency of Company B's two workers has resulted in a $16 price increase to the public for the two chairs, and that's inflation.

This is, of course, an oversimplified example of the effects of high versus low productivity on inflation. Other factors, such as the efficiency of management, the quality of equipment and working conditions figure into productivity. But the example should demonstrate the role productivity plays in inflation.

When workers do less than their best they are contributing to inflation and diminishing the purchasing power of their own paychecks. The same, of course, goes for company executives whose lunch hours may be too long or who fail to eliminate inefficient practices.

Thus, when the U.S. Labor Department announced this week that productivity increased at an annual rate of 4.3 percent in the first quarter of this year, the largest gain in 3½ years, it was another encouraging sign that inflation might at last be declining. It also provides hope that Americans have begun to appreciate the connection between inflation and their own performance on the job.

THE INDIANAPOLIS NEWS

Indianapolis, Ind., July 13, 1981

Former Treasury Secretary William Simon will have plenty of massive volumes to study as head of the new Productivity Commission established by President Reagan.

All sorts of experts have propounded wide-ranging theories as to why productivity has declined in America in recent years. Some blame the loss of the work ethic. Others blame authoritarian management. Others blame the Federal government and its abundant regulations.

The public certainly sees the productivity decline as a major problem, according to a recent Harris poll that contrasts quite markedly with similar polls 10 years. More than 80 percent of the public believes that a failure to improve production will lead to fewer jobs.

Our fear is that Simon and his staff will get bogged down in the detailed studies and miss another story that is an essential part of any legitimate focus on the productivity issue.

No one has provided a formal academic study of a man named Wayne Alderson or his experience as a steel industry executive in western Pennsylvania. But Alderson's story is well documented, on film and in a book, and it provides the missing element in the other studies of this issue.

Alderson calls his theory the "value of the person concept — love, dignity, and respect for the individual worker." Though it may sound a bit vague or corny, the test is the marketplace, where it works quite well. He developed the concept as vice president for operations for a strike-ridden steel foundry, Pittron, in western Pennsylvania several years ago. The company had lost $6 million in 2½ years. Most of the employes had been laid off before the 84-day strike started.

Alderson helped work out a settlement, including the value of the person concept, producing some startling results in the next few months. Productivity rose 64 percent, absenteeism and grievances were nearly eliminated and profits soared to $6 million. He was later fired by Bucyrus Erie, which bought the foundry from Textron in 1974, but he has been taking his concept to other businesses since then, with the help of a dramatic film, "Miracle of Pittron," and now a book, *Stronger Than Steel*.

The book, written by R.C. Sproul, tells a story that Simon ought to read before he makes any authoritative recommendations on the subject of productivity. Other businesses have found Alderson's advice a bit different from traditional productivity recommendations, but quite helpful. Some of Alderson's success must be attributed to his own personality — an odd mixture of toughness, compassion and absolute personal sincerity. But he has discovered and practiced principles that are essential to any productivity commission.

Simon will not be doing his new job properly unless he familiarizes himself with Wayne Alderson, his ideas and experience.

The Boston Globe

Boston, Mass., September 7, 1981

pro-duc-tiv-i-ty: An economic buzzword. The capacity to produce. (Latin *producere*, to produce).

From straightforward roots and following an increasingly emotive entomological spiral, "productivity" has become an ante-raising word in the economic poker game.

"Produce," the root-ancestor, has an even-handed, nonthreatening air to it, perhaps deriving from the country farmstand vision of "fresh produce." Moving from farm to battlefield does not destroy the sense. We might say "the company produces parts for the MX missile" with as little emotion as if the product were plowshares.

One step along is "production," not so thoroughly mired on the factory assembly line that it cannot be theatrical. The production manager might exclaim, equally of missile parts or plowshares tumbling off the line: "That's quite a production!"

With "productive" the word begins to acquire a judgmental role. Not too judgmental, as its commendatory tone is more vague and noncommittal than approving. "It was a productive meeting," says the politican who wants to be quoted but not too much. Equally noncommittal is the eulogistic remark (a favorite of clergymen at funerals of rarely-seen parishioners): "He was a productive member of the community."

"Productivity" leaves nothing to chance. It is all business with no room left for temporizing or benefit-of-the-doubt. It is the word that separates ascendant Japan from effete America, or secures an audience for the technology salesman.

When economists want to tell us what's wrong with us, they talk a lot about our low productivity. It is a word that can be used gracefully to chide us by Lester Thurow or John Kenneth Galbraith, or grimly by the Congressional Budget Office: "Continued weak growth in productivity could have profound implications for American society . . ."

It crops up in a Xerox ad (accompanying an ungainly photograph of an oblivious string quartet in a chicken coop). "Mozart has improved productivity in hen houses," it declares. "Now, what can be done for offices?"

Futurist Alvin Toffler has signed on to the productivity ethic as central to his Third Wave electronic future. The old steno pool, he writes, must be automated "unless we are prepared to accept a reversion to the past, to give up our standard of living, and surrender many of the freedoms associated with a decent standard of living."

Economic curative or road to utopia, the notion of "productivity" carries us too far from the comfortable family farm, past the comfortably anonymous assembly line, into an abstract future. It may contain the destructiveness of its own illogic: "How," asks the believing Toffler, "do you measure the productivity of an office productivity consultant?"

Houston Chronicle

Houston, Texas, November 3, 1981

Hard work, nose to the grindstone, sacrifice — once a basic formula for success — may have fallen on hard times. There is strong evidence that many Americans are unwilling to embrace these principles any longer. This reticence, coupled with continued demands by most Americans for a high standards of living, is playing a part in the stubborn inflationary-recessionary cycle the country is facing.

Social analyst Daniel Yankelovich, in studies of American attitudes toward the work ethic, has found creeping apathy about the idea that hard work always pays off. In the mid-'60s, his surveys showed that 72 percent of college students believed hard work equals success, but in the early '70s only 40 percent accepted the correlation. Among the general population, the same shift was taking place. Between the late '60s and late '70s those trusting hard work to produce success dropped from 58 percent to 43 percent.

In contrast, while many Americans were rejecting the hard work-success formula, they were also demanding the continued trappings of success. At the top of that list was a comfortable home with all the modern conveniences. The '40s bungalow and Cape Cod gave way to the ranch style house of the '50s. In the '60s we had split-levels and in the '70s demand for even larger two-story homes soared. Donald I. Hovde, undersecretary of the U.S. Department of Housing and Urban Development in Houston recently said the public has "Cadillac expectations" for the '80s in their home needs.

In all this there is a clear contradiction between the rewards many want and the work they're willing to do to get them. And somewhere in that contradiction may lie a clue to the dilemma of co-existing inflation and recession, and a steady drop in the gross national product.

It's time for those Americans who scoff at hard work to make a choice if they want to maintain their expected lifestyles. For those unwilling to step up to the grindstone again, the American dream may quickly fade away.

Fort Worth Star-Telegram

Fort Worth, Texas, November 30, 1981

Although the temptation to blame America's current economic problems on overregulation by government and/or labor union intransigence is well nigh irresistable in many circles, a visiting economist may have zeroed in on a more likely target during a lecture in Dallas last week.

Lester Thurow, professor of economics and management at Massachusetts Institute of Technology, told a University of Texas at Dallas audience that the prime culprit in the nation's declining economic pre-eminence is "the death of American productivity."

In 1960, Thurow said, bicycle manufacturing was the only major industry in which the United States was less productive than its foreign competitors. Today, by contrast, America also trails in the production of automobiles, electronic equipment and steel, and as a result, the United States no longer can boast the world's highest standard of living.

One does not have to take Thurow's word for the productivity slump. In October, the Federal Reserve's index of industrial production indicated a nationwide decline of 1.5 percent. That was on the heels of a 1.2 percent slump, in September.

Some of the downturn can be attributed to the blossoming recession and its accompanying reduction in consumer purchasing power, but some of it also can be blamed on a decline in the American labor force's ability or willingness to produce goods in a manner capable of sustaining a competitive edge in the world marketplace.

According to Thurow, the productivity of the American worker has fallen .3 of a percent in each of the last three years. During that same period, productivity rose 4 percent per year in Germany and 8 percent per year in Japan.

Thurow blames the decline in American productivity on poor management techniques. To illustrate, he told of an American-owned-and-managed television factory that was sold to a Japanese company. When managed by the Americans, the company produced 1,000 sets a day and had an average of 140 errors per 100 sets. After a year under Japanese management, productivity had risen to 2,000 sets a day with seven errors per 100 sets.

"The problem clearly is that American management doesn't know how to get a work force interested in improving productivity," Thurow said.

That perhaps is true, but the work force, itself, must at least share in the blame. A recent survey conducted by the head of a national financial, accounting and data processing recruiting firm turned up evidence of deliberate waste of time and abuse of on-the-job time that is costing American businesses billions of dollars annually.

The most common examples of on-the-job-time abuse cited in the survey included arriving at work late, leaving early, overly long lunch hours, unwarranted sick time, socializing with co-workers, attending to personal business on company time and operating other businesses on the side.

Management may indeed need, to re-examine its methods and alter its procedures in order to spur American productivity, but the above-mentioned survey indicates that a large portion of the problem is attributable to poor attitudes on the part of workers, and that is something society as a whole will have to cure.

The Birmingham News

Birmingham, Ala., August 11, 1981

Americans are working less these days and enjoying it less.

These conclusions are reached on the basis of a study of employee benefits paid workers in 1979. As incredible as it may seem, almost half the benefits paid out in wages or salaries were for *time not worked* — pay received for vacations, holidays, coffee breaks and rest periods, jury duty, etc.

The study, done by the U.S. Chamber of Commerce, says approximately 13 percent of payrolls or about $162 billion in 1979 went for *time off.* In all employers paid out $390 billion — or 31.8 percent of payroll for employee benefits programs.

The object or purpose of employee benefits, which is to enhance workers' well-being, has almost universal support from employers in the country. But these expenses, like other costs of doing business, add to the the price of goods and services which eventually must be borne by consumers.

During normal times, the economy has managed to absorb these additional costs. But inflation and exorbitant taxes have had a very negative effect on productivity, compared to productivity in other nations. For instance, the productivity growth between 1970-78 in the United States was 2.3 percent, while in the Netherlands it was 6.2 percent. Percentages for West Germany were 5.3; 5.1 in France; 4.8 in Japan; 4.5 in Italy and 3.8 in Canada. The only major country in the West with a rate lower than the United States was Britain with 2.2.

The loss of productivity has cost the American worker in real terms. For instance, during October 1980, the average married worker with three dependents had real spendable earnings of $82.92 per week in 1967 dollars. Over the past 12 months, real spendable earnings dropped by 6 percent — or to the same level for 1960-61.

There is not a single cause for lagging U.S. productivity, but an important part of the problem is the rate of capital investment which, in turn depends on how much money both corporations and individuals have left after federal taxes. Capital investment is vital to creation of more productivity — through new plants and equipment — as well as to the creation of more jobs.

So increasing productivity during the 1980s at least to pre-1967 levels appears to be one of the best investments of all for the 1980s.

The Dispatch

Columbus, Ohio, September 21, 1981

THE AMERICAN Institute of Industrial Engineers again is sponsoring a unique Productivity Improvement Campaign during the month of October.

The campaign is self-explanatory. Its aim is to increase productivity growth by calling attention, for example, to this nation's lagging productivity for a decade.

The institute, which has a 40,000 membership, is engaged in productivity improvement in business, industry and public institutions.

It points out, logically enough, that since productivity is linked with inflation, every citizen is affected by the production rate.

When input improves sufficiently to cover added cost, prices can be stabilized and inflation checked, the engineers say. But if output of products and services fails to keep up with the cost of input needed to produce them, then prices have to go up in order for a company to stay in business.

The institute points out that job security is a major benefit of improved productivity. It explains this benefit results because the company is able to compete in the world marketplace.

Improved productivity also is important in government, it contends, because taxpayers have lost patience with waste and inefficiency in government agencies.

It is something to think about. It makes sense.

DESERET NEWS

Salt Lake City, Utah, October 19, 1981

The nation's productivity lag will be a major problem in the coming decade, according to a recent survey made by the American Institute of Industrial Engineers.

Though industrial engineers deal with productivity day after day, productivity should be everyone's concern since it affects everyone's economic well-being.

The engineers credit the most significant productivity gains in recent years to investment in new plant and equipment. Improved technology and more capital investment may be expected to bring additional gains, but the engineers say real improvement will have to come through other avenues.

Rated in the survey as the greatest need of industry is a change in the attitudes and abilities of management. Next in importance is improved worker attitudes and abilities.

The engineers see a downward trend in the American work ethic, noting that people today are not inclined to work as hard as they did 10 years ago. Most of the engineers also feel that pride in workmanship, loyalty, and motivation to work are on the decline.

As a result of these slippages, the United States is going to relinquish the number one spot in productivity. The engineers expect that the successor in the next decade probably will be Japan or Germany.

The trend could be reversed if American management and labor would realize that their best interests are basically the same. If both sides could dedicate to a team effort only part of the energy they now expend fighting each other, America could become more productive — and more prosperous.

Lincoln Journal

Lincoln, Neb., August 3, 1981

Everyone talks about the decline in U.S. productivity. But when C. Jackson Grayson talks, people listen — or should, at least.

A former educator and one-time price commissioner under President Nixon, Grayson now heads the American Productivity Center in Houston. Its mission is to educate the public about productivity, devise better ways of measuring it and teach management and labor how to improve it.

Grayson says society can't rely just on numbers to determine productivity. Effectiveness, too, must be considered: "Efficiency measures if you are doing things right. Effectiveness concerns itself with whether you are doing the right things."

Grayson points to the U.S. auto industry. It was doing things right, being efficient. The problem was, it wasn't producing the right cars. The Japanese were — they were effective as well as efficient.

Or, Grayson asks, which is the most productive medical facility — one that treats the most patients or one that deals with fewer people but teaches them how to remain healthy? Further, which computer is used most productively — one that turns out the most data, which may go onto a shelf unread, or another whose output is put to good use?

Anyone can apply the productivity center's standards to a variety of situations. Which restaurant is most productive — a place that serves hordes of diners, rushing them through their meal, or a competitor that caters to fewer people but makes dining pleasant, relaxing and enjoyable?

Or which is the most productive gas station — a fill-it-yourself operation that sells vast quantities of fuel, or one that pumps fewer gallons but checks your oil, battery and tires, prolonging the life of your car and enhancing your safety?

Grayson makes sense, especially in an age when industry's output increasingly is services instead of goods. Efficiency's fine. But effectiveness — which really involves quality of life — is better.

The Evening Gazette

Worcester, Mass., August 20, 1981

Ever since the figures began to show a marked decline in U.S. productivity a few years ago, all sorts of explanations have been put forward.

Some said the American worker was no longer willing to work as hard as the Japanese or Korean worker. Some blamed the influx of women and unskilled minorities to the labor force. Still others put the finger on unimaginative management.

It's none of the above, says economist Martin Baily of the Brookings Institution, a think tank in Washington. Or, at least, none of the above have had more than marginal influence.

Baily says the slowdown is due to the soaring oil prices of the 1970s. The ballooning costs of energy, as he sees it, caused many inefficient plants to close down or curb operations. The U.S. economic system, he says, was "hit with some hammer blows over the head" and the result has been that "Major structural changes have taken place that have rendered old capital obsolete."

Baily cites the auto industry as a prime example. The soaring costs of oil made necessary a thorough revamping of the industry. Old plants were closed, antiquated assembly lines were shut down for modernization, workers were laid off by the tens of thousands. Given that sort of convulsion in that major industry and others, it was inevitable that U.S. productivity would suffer.

Although Baily did not say it, the implication is that, with new, energy-efficient plants and better management procedures, U.S. labor will begin to compete effectively once again in the world market.

The "reindustrialization" of the United States is a trendy catchword on some campuses. But it is a necessity. If the Reagan economic program frees up enough venture capital to rebuild the industrial plant in this country, the next decade may see the United States regain its old pre-eminence.

Shortage of Skilled Workers Affects Technological Growth

New machines alone cannot solve America's technology problem. The machines need skilled workers to run them. Experts have cited the shortage of skilled workers as one of the major reasons for declining U.S. productivity. According to Brian Usilaner of the National Productivity Group, "the human factor contributes between 10 and 25 percent to productivity growth." Speaking Sept. 9, 1981 before the House Science and Technology Committee's subcommittee on science, research and technology, Usilaner added: "A need for new skills will accompany any new jobs created by automation. . . . A machine, a process, or a system may be ever so brilliantly contrived, but it is no more effective than the people operating and managing it. . . ."

The Houston Post

Houston, Texas, October 15, 1981

As economists and industrialists stress constantly, the United States has an urgent need for capital investment in modern plants and technology if we are to compete with foreign manufacturers. But less is heard about an accompanying need. Modernized plants need skilled workers to run them. Evidence grows that the skilled labor supply will lag far behind demand unless adequate training programs are developed.

In the computer industry alone, a million more trained technicians will be needed during this decade. Yet schools and colleges cannot find enough qualified teachers in the computer field to meet demands. The industry desperately needs skilled computer technicians and operators, keypunch operators, systems analysts and programmers. Yet only a fraction of the need for trained workers is being met.

This example raises questions about the nation's ability, under existing procedures, to satisfy demands for trained personnel in many other industries such as engineering, tool and dye making, bookkeeping and accounting, nursing, transportation and communications. We need more information to determine if current fears of future skilled labor shortages are valid and, if so, what can be done. Sen. Lloyd Bentsen intends to find out.

Bentsen said he will conduct an in-depth study of the nation's skilled labor force and the national ability to provide training to keep up with technological advances in industry.

The potential of Bentsen's project is sufficiently impressive in itself, but equally impressive is his suggestion that any skilled labor crisis could best be handled by private industry with federal involvement held to a minimum.

TULSA WORLD

Tulsa, Okla., January 25, 1982

TULSA'S employment picture is bright. According to the U.S. Department of Labor, the city's unemployment rate in November was the lowest in the nation at 3.6 percent.

The business boom fed by the oil and gas industries play a big role in Tulsa's economic health. But there is nothing unusual about the city's employment picture. In times of healthy economic growth, workers are in demand and unemployment is low.

But that low unemployment rate could be misconstrued. There are jobs available, but that does not mean anyone can find employment in Tulsa. The keys — as always — to a well-paying job are education and training. Tulsa has an abundance of jobs, but only for workers with the proper skills.

Employment specialists here say demand is greatest for welders, draftsmen, geophysicists, machinists, data processors, geologists. Those are not the sort of jobs one gets without a high school diploma and at least some additional schooling or job training.

In short, as one labor analyst put it, "The demand is not for people needing to be trained, or for unskilled laborers."

The worker in least demand continues to be the high school dropout with no job skills. A booming local economy isn't going to help him one bit.

Tulsa is blessed with economic health that most of the nation does not have. We all hope that one day cities like Detroit are again the productive centers they once were. We all long for days of robust economic growth on a national scale.

But for tomorrow's workers to participate in that growth, they must sharpen their job skills today. Tulsa proves that the benefits of economic health accrue only to those with the necessary job skills. That's an important lesson for high school students considering dropping out.

ST. LOUIS POST-DISPATCH

St. Louis, Mo., December 12, 1981

While unemployment continues to plague the economy as a whole — and particularly the automobile, steel and housing industries — the disease that afflicts the nation's high technology firms is just the opposite. In that area, the nation is "people poor," as an executive from Itek Corp. put it. There are more jobs than there are qualified men and women to fill them.

An estimated 17,000 to 25,000 unfilled positions exist in the computer, electronics and telecommunications fields. Moreover, the shortfall is likely to continue and even grow at least through 1985.

One reason for the shortage of trained personnel is the rapid expansion of the computer and related industries in recent years. The demand for electrical and mechanical engineers, computer programmers and other specialists has simply outpaced the number of technically trained graduates from the nation's educational system. Another reason matters are likely to get worse rather than better is the massive military spending program being pushed by the Reagan administration, which is expected to gobble up thousands of engineers, scientists and technicians.

Escalating demands for workers with scientific and technical skills mean an intensified competition between firms that cater to the private market — that is, produce goods and services people can use — and firms that help the Pentagon meet the needs of generals, admirals and friendly dictators abroad.

A federal scholarship program or expanded retraining programs for the unemployed might be two ways to help produce more technically trained graduates. But those are the very types of programs that are being eliminated or curtailed by the Reagan administration — in part because of the $1.6 trillion the administration wants to spend on the Pentagon over the next five years. As matters stand, Japan, with only half the population of the United States, graduates 5,000 more electronics engineers a year.

Because a shortage of technically trained workers may threaten some Pentagon projects, a program of scholarships or other subsidies may be created in the name of "national security." However it is justified, such an approach makes sense and would be in the broader national interest. But another, more fundamental problem would remain.

As long as the U.S. continues to put an inordinate share of its human and material resources into military programs — significantly greater than the percentages invested by our allies and economic competitors, such as Japan and Germany — it should come as no surprise that the lead in so many consumer and business applications of science and technology has been developed by others, giving them ready markets here and around the world.

Giving priority to the Pentagon has given this country a formidable military machine, but it has caused a kind of brain drain that leaves not just the new technology industries, but the nation's economy, shortchanged.

Robots in the Workplace: A Threat or a Solution?

Robots emerged from the pages of science fiction onto the factory floor in the late 1970s. Japan made news in the U.S. with its fully automated factories and other technological wonders. In an America already suffering from record post-war unemployment, talk of robots might seem cruel. (One could argue, however, that the invention of the automobile was cruel to carriage makers and blacksmiths.) Robots, in the words of Lewis F. Hanes of the Westinghouse Research and Development Center, might help "American industry . . . to remain solvent and competitive and to gain a larger share of the international market." In September 1981, he told the House Science and Technology Committee's subcommittee on science, research and technology: "The demands of the new technology will create opportunities for improvements in workers' status and pay. A market for new skills will develop." He added that with the introduction of robots, "the release from dangerous, dirty and tedious physical work may improve worker morale and encourage the development of professional attitudes. . . . At present, manual work on a conventional assembly line is paced by the speed of the machinery. . . . The new equipment will place the worker outside the flow of the assembly line. . . . [T]he worker will be able to create his or her own work pace. The stress created by timed, repetitive movements will be reduced."

The Evening Gazette

Worcester, Mass., December 15, 1981

You may have read last week about the maintenance worker in a Japanese auto factory who got in the way and was crushed to death by a robot. News reports said it was the first time, outside science fiction, a robot had killed a man.

American reporters promptly sought out a robot expert at New York University and got the following quote: "The Japanese companies may be counting too heavily on the intelligence of the human operator." The implication is that, in the brave new world of automated factories, if you don't want a robot to kill you, you'd better be smart enough to stay out of its way.

The Japanese, however, have already worked out a solution to the safety problem taking human failings into account. Some of their newest robots are outfitted with electronic sensors. When the robot senses that one of those dumb humans is close by, it automatically stops whatever it is doing.

An even better solution to the problem has been developed by Yamazaki Machinery Works Ltd. in Nagoya. It is the ultimate in automation, a factory where the robots service themselves so they don't even need maintenance humans.

The Yamazaki factory is considered the most thoroughgoing example of the new "manless" Japanese factories. In it, robots make robots — or at least robot parts. During the day, a few humans are involved, directing cranes, programming the robots by computer, and checking finished products. At night, the robots go right on working, each at its own rate. The only human in the plant is a night watchman.

If such a factory still sounds more like science fiction than reality, it may be because this one is in faraway Nagoya, Japan, instead of Detroit, Mich. Indeed, most industrial robots are in Japan, to Detroit's loss. The Robot Industry Association reports that, of a world population of 21,500 programmable robots, 14,000 — nearly-two thirds — are in Japan. Only 4,100 — less than one-fifth — are in the United States, which has prided itself on being the industrial leader of the world.

What's more, those robots that have helped Japanese automakers produce cars so inexpensively they sell for less than comparable U.S. cars in America, are multiplying much faster than American robots. A single Japanese company, Matsushita Electric, intends by 1990 to be using 100,000 robots — nearly five times the present world total of robots. General Motors Corp., which has the biggest robot program in the United States, aims for only 14,000 robots by 1990.

Robots aren't really out to get people, as in those science fiction tales. But American industry should be out to get robots, if it hopes to catch up with the Japanese any time soon.

The Arizona Republic

Phoenix, Ariz., April 28, 1981

JAPAN is now producing robots that can procreate — no, not like rabbits, but like robots.

They are robots that can manufacture robots to replace men and women on the assembly lines. They're already at work in the Fujitsu Fanuc factory near Mt. Fuji, manufacturing 100 robots a month almost without human assistance.

For 16 hours a day, in fact, the factory is run entirely by robots. About 100 workers come in for eight hours each day to make sure they're doing all right.

By 1986, the robots will be turning out 400 robots a month with the help of 200 humans.

Eventually, it may be possible to replace all the humans.

Hitachi, the giant Japanese electronics firm, recently mobilized 500 scientists and engineers to develop a new generation of robots that can replace foremen.

They will be able to walk up and down the factory floor supervising the robots on the assembly lines.

The Japanese are leading the world in the use of robots.

There are now 60,000 of them in operation in factories across the country, turning out high quality automobiles, cheap electronic equipment and other products.

They can be seen bobbing and weaving like demented chickens, welding doors to bodies, painting and performing other chores around the clock — faster, cheaper, and far more efficiently than humans.

Japanese industry already is the most efficient in the world. With newer and more sophisticated robots, the Japanese expect to increase efficiency by 70 percent in the next decade.

In the United States, which has only about 3,000 robots in operation, unions have resisted them because of fear that they will create mass unemployment.

Japanese unions have welcomed them because they eliminate unpleasant jobs.

They have not created mass unemployment in Japan — the country's unemployment rate is only 2 percent — because the companies have found other jobs for the workers.

Says a Fujitsu Fanuc spokesman:

"Just as the industrial revolution in 18th century England freed (or perhaps pushed) people from the land to work in industry, our inventions will push people further into the tertiary services of health, education and entertainment — to create a better life for all."

And U.S. unions had better stop fighting the use of robots. If U.S. industry falls any further behind Japan, their members won't have any jobs.

The Providence Journal

Providence, R. I., March 29, 1982

The metal arm begins to move. It hovers over a bin, drops down and clutches a feather. It rises, shifts direction, and drops again, neatly depositing the feather in a cap which bears the letters URI.

That's what happened last week — figuratively speaking. The University of Rhode Island's industrial robot program was named by the National Science Foundation a national center for robot research, the only one in the nation. With that honor came a four-year $700,400 grant and the kind of prestige that every institution of higher learning in the United States would be proud to receive.

"We are delighted to have the center," said James W. Dally, dean of the College of Engineering. "It is a recognition of excellence and it does give us a better opportunity to make our program even more successful than it is now."

URI got in on the ground floor of robot development 11 years ago. The program's technical directors, John R. Birk and Robert B. Kelley, are internationally known. Thirty-two companies now support the research conducted there with $25,000 annual contributions, and Dean Dally is aiming toward a total of 60 companies by expiration of the NSF grant in 1986.

Robotics is one of the fastest growing fields of industrial research. The Japanese pioneered it, but American industry is catching on fast and that is likely to put URI at the center of a developing boom. Several years ago Birk said that once manufacturers are "educated" to the advantages of robots, these computer-controlled mechanical arms will spread through industry at a pace faster than the one set by computers.

Rhode Islanders have many reasons to be proud of their state university. This is one more example. Everyone involved in the project deserves warmest congratulations.

The Washington Post

Times Herald

Washington, D. C., August 19, 1981

EVERY GENERATION or so since the Industrial Revolution, the technological alarm has been raised. A new wave of automation, it is warned, will soon descend upon us, leaving millions of displaced workers and distressed communities in its wake. In fact, these inundations do occur, and the consequences for some people's lives are substantial. Northern cities, for example, are still to some degree trying to absorb the descendants of the workers displaced by farm automation decades ago. The larger economy, however, moves onward with scarcely a surface ripple.

Now we are told that a new revolution is upon us. At its heart lies the tiny computer—the microprocessor, as it is known in the trade—that can process and store vast quantities of data in areas no bigger than a dime and at a fraction of the cost of 10 years ago. In the leading companies, computers are beginning to invade every phase of the production process—equipment design, ordering of raw materials, production control, quality checking and so on. On the assembly line itself, thousands of robots are elbowing aside skilled and semi-skilled workers. The robots vary in sophistication from the glorified machine tools on big assembly lines to the borderline humanoids that can "see" and "feel" their ways through a variety of tasks previously reserved to the uniquely adaptable human.

No one is predicting how fast the revolution will proceed. However, even before the new tax bill was passed with its presumably stimulative investment incentives, capital spending for automation was up sharply. Within a decade it is possible robots will replace several million factory workers. The inroads into the service sector may be still greater as computers take over the jobs of bank tellers, clerks, stenographers and even computer programmers—computers can, after all, program other computers.

Before you apply your sledgehammer to the nearest computer, remember that automation can bring many benefits. Robots can do many backbreaking, dangerous and thankless jobs with whatever passes for the electronic equivalent of a smile. Higher productivity and better quality control can raise standards of living and protect other American jobs from foreign competition. Nor is a wholesale loss of jobs likely. As the demand for semi-skilled office and production workers falls, the need for a new class of workers will increase—the people who tend to the needs of robots, for example. Some of these new jobs will require advanced degrees, but many others can be done by high school graduates with some technical training. Continued strong growth in the service sector and a drop-off in the number of youths entering the labor market in this decade could fill the gap between old jobs and new.

The transition, however, will place a strain on everyone involved. Large employers may ease matters by retraining displaced workers, involving workers in the conversion process and redesigning jobs to ease the new boredom of baby-sitting automatons. But many laid-off workers and new jobseekers are likely to end up in lower-paying jobs or on the unemployment lines. Long before the new wave of automation it was clear that the country's education and training systems failed to provide many youths and displaced workers with the skills and discipline they need to function in a modern economy. Whether the robots now entering our lives bring with them new prosperity or new hardship for many people will depend crucially on how quickly and well this failure can be remedied.

Pittsburgh Post-Gazette

Pittsburgh, Pa., March 29, 1981

Efforts by Pittsburgh and other communities in Pennsylvania to position themselves for the industries of the future have been given a boost with a $50,000 grant from the state Department of Commerce to study possible designs for a robotics factory.

The grant goes to the newly formed Economic Development Committee of the Allegheny Conference on Community Development. The project for "Robotics and Pennsylvania's Manufacturing Industry: Designing the Factory of the Future" will be conducted by the committee in cooperation with Carnegie Mellon University's Robotics Institute and Aerotech, Inc. Wisely, there will be a concurrent study of the impact of robotics upon the labor force in a manufacturing environment.

The state Commerce Department made five other grants, including one for $50,000 to the Pennsylvania Association of Colleges and Universities for a project titled, "Current Capabilities and Potential for Research: An Inventory of Pennsylvania Resources." The inventory would include existing research facilities, equipment and personnel within Pennsylvania.

The effort is evidence of increased realization by state government, as well as the private sector, that as Pennsylvania's bellwether industries, such as steel, decline in terms of job opportunities, the state has to look in new directions. That can be developing new high-tech industries such as robotics or in the research areas which may turn up the technology of the future.

Usually one thinks of such efforts as concentrated in university and industrial facilities. Therefore it is interesting to learn that in California a school district in the so-called "Silicon Valley" south of San Francisco is about to launch a "high-tech" high school. The Los Gatos-Saratoga Joint Union High School District in Los Gatos has decided to open this September such a high school to help fill manpower needs in the rapidly growing high-technology firms in the area. Silicon Valley industries have agreed to furnish support.

Education Week reports: "The problem the school will address — shortages of engineers and skilled technical workers that threaten to become more serious as the electronics industry grows — has been well-documented. Moreover, both educators and industry officials believe it is imperative that students become 'computer literate' and develop some experience with, and understanding of, the devices that recent surveys estimate will be part of 90 percent of all jobs by the year 2000."

The point is this: If Silicon Valley with its head start on most of the rest of the nation is extending its training for future industries even into the high school level, that demonstrates what faces latecomer regions such as this one. The new programs being funded by the state Commerce Department are but a beginning step toward what needs to be done in the future.

THE COMMERCIAL APPEAL

Memphis, Tenn.,
December 22, 1981

YEARS AGO the United States developed — and then almost abandoned — what has turned into an industrial revolution. Now the outcast appears to be headed toward Memphis, among other places, after growing of age on the other side of the Pacific.

"It was a great shock to our own engineers," a Japanese scientist said recently, when the U.S. invention was unveiled in his country in 1967. It caught on there, he said, because most Japanese think whatever comes from the United States "must be good."

The scientist was addressing a conference in Chicago on the "Robot Revolution." The industrial robot, after being welcomed and nurtured in Japan, is returning to the land of its creation.

Japanese industry has installed an estimated 10,000 robots in its factories, while the United States has only 4,000 to 5,000 at work. But the revolution also is catching on here. General Motors, which last year had 300 robots on assembly lines, plans to increase the number to 14,000 by 1990. Sharp Manufacturing Co., a Japanese firm, said this week it will add robots to its Memphis plant in 1986 or 1987.

THESE ROBOTS don't look like 3CPO or R2D2 of "Star Wars." Most are no more complicated than machine tools, with a computer brain, a power source and two jointed arms.

Like a human worker, they do such things as pull a metal sheet in and out of a punch press, attach a metal rotor to a succession of lathes and drills, spot-weld an automobile chassis and spray paint a chair.

The big differences are that the robots don't make the kinds of mistakes that result in human injuries, don't complain about monotonous tasks, work more quickly and meticulously and cheaply, and don't need health insurance, Social Security, vacations or coffee breaks.

But won't American workers lose jobs? Won't unions resist the revolution? To the Japanese scientist, the answers were simple. If jobs are guaranteed and the profits of higher productivity through automation are shared, he said, labor resistance will die down.

That, at least, has been the Japanese experience.

The biggest change that's needed, he said, is in the attitude of management. "The Japanese manager wants to take new technology and put it to work, while U.S. managers are reluctant to accept technology. Here they tend to put robots in a laboratory instead of in the plant."

BUT EVEN IF assembly lines can be manned by robots without hurting U.S. workers, what about the future? Sharp officials said their Memphis robots would be "intelligent" — capable of being programmed to do different jobs. White collar jobs, too? Secretarial work? Drafting? What are the limits?

As long as the brand says, "Made in the U.S.A.," Americans should have as much confidence in the product as the Japanese did in 1967. The limits of robots will be determined, presumably, by human inventiveness and ingenuity, which also can be counted on to expand opportunities for human workers.

U.S. industry shouldn't try to throw the revolution away a second time. It would be humiliating for another country to bring it back again.

Memphis, Tenn.,
March 5, 1982

THE ROBOTS are here.

Not those Tin Woodman-type characters that we used to read about and see in the cartoons, but the real robots — the supermachines that look more like the Canadarm on the space shuttle Columbia. These are the devices that build things — automobiles, clothes washers and dryers, refrigerators and ranges and a lot of other things.

Cobo Hall in Detroit has been full of these weird machines this week, twisting in all sorts of contortionist movements, reaching out to pick up just the right screw or bolt and fitting it to the right place on the item under construction, or welding one piece to another, or drilling holes exactly where they are needed. Thousands of production engineers and executives visiting the robotic industry trade show claim these machines can do just about anything people can do, even reproduce themselves.

Such robots, of course, aren't new. U.S. engineers have demonstrated them for years. But they are becoming a big thing now because our engineers showed them to the Japanese, and the Japanese — unlike American industrialists — thought they were great and put them to use. One reason the Japanese have built automobiles cheaper — and some think better — than American firms and workers is that robots have helped build them.

So now U.S. industrialists are accepting the General Electric dictum "Automate, emigrate or evaporate." They are turning to robots, and even union workers seem to consider them inevitable. At least nobody has been throwing steel-toed Red Wing safety shoes into the electronic consoles that have replaced the primitive gears of the industrial revolution.

Executives of such firms as GE, Westinghouse and IBM say the robot is more than a super machine tool. It is, they say, a component of "the factory of the future."

Robots are going to replace people in the factories, there is no question about that. Why else would anyone buy them at an average of $70,000 each? Their advantages over people also are obvious. They are tireless, even though they are worked 24 hours a day, seven days a week. They can work faster than human beings, and with greater assurance of accuracy even when doing work that for people would be dangerous, uncomfortable or deadly monotonous. They don't organize into unions to demand higher wages, more vacations and holidays or paternity leaves. The factories that house them need only minimum conditions of heat, cooling and lighting, thus reducing utility bills. And companies can forget about huge parking lots for employes to park their Toyotas and VW Rabbits.

BUT ROBOTS also will bring new problems, raise new questions that somebody will have to answer.

If robots permanently eliminate those jobs from which workers now are being "furloughed" and there are no jobs opening up for all those people just coming into the work force, where are those people going to find jobs to earn the money to buy the stuff the robots produce? Robots don't go to the shopping malls.

Will there then be a demand to cut the regular work week for the people who still are working to 30 or even 20 hours a week, so the available work can be shared? And if that happens, what will workers do with all that new leisure time? How will that affect our society and our economy?

If robots replace people on the assembly lines and the cost of the robots can be written off in three to five years under the new tax laws, will the stuff the robots produce cost less? If not, where will those "savings" go?

And what will robots do to foreign trade?

If manufacturers in the United States use the same sort of robots as the manufacturers in Japan, Korea, Hong Kong or Taiwan, what happens to that "cheap labor" advantage that businesses in the Far East have had? There won't be that much difference in the cost of robots.

At first that may seem to be a real gain for the United States. No more competition from those cheap-labor imports. But what happens to U.S. sales overseas when every other nation that has the necessary capital or can borrow it buys robots and builds what the United States wants to sell abroad for the same price?

IT'S TOO SOON to be worrying about such things? Not when you realize that the robot industry passed the $150 million mark in sales last year and that the number of companies producing robots increased 30 per cent in the same year.

We may be able soon to leave the making of things to R2D2 and C3P0, but we can't leave to them the solutions to the problems that their introduction will create. That still will require human intellects with the capacity to anticipate and resolve such problems. It is time for those minds to be about that task, too.

Post-Tribune

Guarding Your Interests Daily

Gary, Ind., February 25, 1982

Now the Japanese are beating us in building robots. What next?

Their supremacy in robot technology will be on display at the fifth biannual Robot Automation Exhibit and Trade Fair in Milan, Italy, Monday.

Industry sources say the Japanese have 10,000 robots in use, twice as many as Americans do. Some 150 companies build robots in Japan, in contrast to about 30 in the U.S.

U.S. industry has been slow in putting robots to work, so U.S. robot makers have turned to Western Europe and Japan for sales. These countries say the use of robots keeps them competitive with U.S. manufacturers.

Government incentives also have spurred Japan's lead in the robot race. Last one across the finish line may be left behind to rust.

The Detroit News

Detroit, Mich., March 4, 1982

Workers in Japan say they love their robots. They name the machines after favorite baseball players, and union leaders jokingly say the robots should pay dues.

But there is deep concern below the surface. Many workers are worried about losing their jobs. Unions are making gloomy employment projections.

Japan claims to have 70 percent of the world's operating robots. Certainly, the country has gone further than most, having one factory in which robots design and build other robots, droning along in rooms where men are only rarely seen. The operation must be particularly eerie at night because the machines continue their labors in the dark.

Robots deeply trouble Kohei Goshi, the patriarchal founder of the Japan Productivity Center (JPC), the organization that taught management and labor the value of cooperation over confrontation, and the mutual benefits of strict quality control.

Mr. Goshi notes that Japan is accelerating the deployment of robots. And he worries about the probable result of the automation binge in a resource-poor country that relies on value-added manufacturing industries to keep its people busy and prosperous.

Consider, he says, what an awesome decision would be faced if Japan's competitors went on an even greater robotics drive, cut their costs low enough to drive Japan out of the export markets. What if an industrial giant like the United States raced past Japan?

Japan would have to choose — robots and competitive prices, or men and higher costs. "If it comes to that, Japan will choose men, and opt for the human values," he contends.

Yet the thought troubles him because this grim scenario could deprive Japan of the foreign-exchange wealth needed to buy resources, without which its factories cannot operate. Japan's ability to survive would be as severely tested then as it was after World War II.

The JPC has already begun intensive studies to determine how far automation should go, whether it should be checked, and when. Of course, such thoughtful planning has enabled Japan to prosper in spite of all the odds.

Isn't there a message here for U.S. industrialists?

THE SAGINAW NEWS

Saginaw, Mich., March 7, 1982

Michigan is economically sick with, at the very least, a mild case of depression. Yet, even in such straits, there is a new chance for a healthier tomorrow.

Milliken's prescription: A strong dose of an antibiotic called "robotics."

The machines which could help cure the very human maladies of joblessness and hopelessness were examined in Detroit at a four-day conference and exposition known as Robots IV. The test results indicated that they may be the best answer to the years-long search for economic diversification in Michigan.

Look at the picture. The auto industry is down as it has never been before. Technical modernization is essential to a comeback. It simply costs too much to make cars the same way as 50 and 60 years ago, with heavy human labor.

The automakers' commitment of hundreds of millions for new equipment is an early clue that even a revived industry will never again look the same in terms of payrolls.

For good reason, the prospect must comfort the companies. Machines don't take paid holidays. They don't get pensions. They don't strike. They don't get sick. They don't run up millions in foot treatment medical payments. They never complain about conditions on the shop floor.

That seems inhuman. So does the specter of men and women cut off from jobs and paychecks, from their sense of economic contribution and worth. And, as one very wise woman we happen to know well recently reminded us, "they (automated machines) don't pay into Social Security, either."

Robots are inhuman, yes. But perhaps not inhumane.

For the people of Michigan, salvation may very well lie in the intense effort Milliken urged to make the state the leader in producing the same machines that threaten jobs.

The idea is to take a seeming threat and convert it into an opportunity. It is the sort of thinking the old industrial regions need if they are to survive, and even thrive, in the coming new high-tech economy.

Along those lines, Milliken thinks big. He wants to make Michigan the world No. 1 in robot research and manufacture.

Why not? We led the world for decades in making autos. Yielding that status to Japan is no reason to surrender on other fronts.

As Milliken pointed out, Ann Arbor's new Industrial Technology Institute provides a running start. Ford is leaping ahead this summer with a new center to develop robotic and automation equipment.

Robotics would exploit the brainpower of Michigan's fine schools and the "skillpower," if you will, of its trained work force. It would shift jobs from the auto industry into robot research and manufacturing.

The prospect is scary — and sensible.

In the newspaper industry, we're well aware that survival depended on automated production. In other industries, too, human talent will still supply the creativity — but machines will translate it into the goods, services and information people need.

There's no need to fear the robots — if we make them work for, not against, us.

THE ATLANTA CONSTITUTION

Atlanta, Ga., April 14, 1982

UP UNTIL now, industrial robots have been installed in this country primarily in dirty, dull or dangerous jobs humans don't want, jobs such as forging, welding, spray painting, die casting, materials handling and the like.

But, predictions are that industrial robots are the blue-collar workers of the future, predictions that have alarmed industrial workers and their unions.

Not to worry, the Joint Economic Committee of Congress reports. Rather than causing job losses, industrial robots can create new ones and stimulate economic growth, the committee concluded. "History shows that labor saving techniques have led to improved living standards, higher real wages and employment growth," the report continued.

Development and use of industrial robots is of such importance to the nation's economic future that there can be no timidity in their exploitation. America needs to move boldly to develop them and to apply them to industrial tasks.

As we have stated before, America competes in a one-world economy. The factories of Japan and West Germany and, indeed, all the rest of the world are competing in a single market.

Japan has already embarked on a crash program to employ industrial robots. It has about 70 percent of the world's robots and about five times more than does the U.S.

Despite the numbers, however, the U.S. appears to have technological superiority. Industrial robots here are more complicated and tend to be more versatile.

A first major test of industry's willingness to employ them, and the workers willingness to let them, will come in the auto industry. The Joint Economic Committee report notes that 100,000 jobs in the auto industry alone could be eliminated and a million factory jobs could be lost by 1990 to robots.

While the numbers of jobs eliminated may appear frightening, the panel suggests that retraining programs can provide for the orderly transition of displaced workers to new careers. Because robots will increase productivity, employers and employees alike will benefit, the committee said.

The U.S. has lost the technological edge in many industries to other nations. The race to develop and employ sophisticated robots is one the U.S. can't afford to lose.

The Union Leader

Manchester, N. H., March 19, 1982

A congressional study, commissioned by Congress' Office of Technology Assessment, tell us that robots eventually could replace humans as workers freeing mankind to do more creative tasks and ushering in a new era when all work is voluntary.

Really? Well, as we fall behind the industrious Japanese in just about everything from necessities to gadgets, there are indications that too many people in the United States ***already*** regard all work as voluntary.

Now, thanks to the congressional study, they can say that they're not lazy —just ahead of their time.

America Seeks Explanations for Japan's Economic Success

How does Japan do it? Books attempting to explain the Japanese "economic miracle" were prominently featured in bookstores across the U.S. in the early 1980s. Barely five years before, the Japanese success was dismissed as merely a product of hard-driving sales techniques or worse, a conspiracy to keep imports out and foil honest competition. Only slowly did the suspicion grow that the U.S.-Japanese trade gap was a result of forces more fundamental than salesmanship. Theories about the remarkable rise in Japanese wealth and industrial might ranged from the logical to the mystic. Japanese industry was organized more efficiently—or Japanese culture produced more dedicated workers. Whatever the hypotheses, most researchers agreed that there was something special about Japan. Japanese goods were acknowledged to be of superior quality; Japanese workers were thought to take more pride in their work; Japanese labor-management relations were considered more harmonious, and Japanese education was praised for being more rigorous. The Japanese national character was analyzed for clues. The Japanese were praised for placing more stress on group responsibility instead of individual gratification. The Japanese had a tradition that emphasized harmony and consensus in decision-making instead of "winning the rat-race" over one's fellow-workers. Japanese were taught obedience to superiors, while Americans were encouraged to go their own ways. Theories multiplied, but answers were few. In facing the challenge of Japan, America faces a dilemma familiar to much of the developing world, one which Japan itself had faced—and solved—more than a hundred years ago: how to adopt an apparently superior economic system while remaining faithful to one's own national or cultural values.

THE KANSAS CITY STAR

Kansas City, Mo., September 14, 1979

As we watch with increasing uneasiness the Japanese overtaking us in gross national product, we are reminded of someone we used to know — the old-fashioned American capitalist, willing to take chances, innovate and quick to seize an advantage — the men and women who built the railroads, power companies and other industries that came to symbolize the Industrial Revolution throughout the world.

To be sure, there still are some fire-eating companies. But more and more it seems that capitalism, American-style, is approaching a senility of numbers, percentages and odds that avoid even modest risk to produce a steady but unimpressive rate of return. The Japanese, on the other hand, are turning research and development into a fine art as they run about the world grabbing markets and introducing products that people want in an unprecedented example of coordination between private industry and democratic government.

It is ironic then that the solution to our Western problems was suggested recently by a man who symbolizes a most conservative Eastern thought, a man who was seeing the Empire State Building for the first time — the Dalai Lama, the 14th reincarnation of Avalokiteshvara, the patron saint of Tibet.

He was explaining that Buddhism has learned from the West how to throw off its detailed rituals and dogma. "We were becoming more Buddhist than Buddha himself," the Dalai Lama laughed.

But perhaps the American capitalist, with all his no-risk investments and computer-supplied percentages, not to mention government pillows and protection, is becoming far less capitalistic than the original capitalists who knew how to take a chance.

THE INDIANAPOLIS STAR

Indianapolis, Ind., February 11, 1978

The world has long marveled at the Japanese economic miracle.

James D. Hodgson, who was U.S. ambassador to Japan from 1974 to 1977, has some pertinent observations on the Japanese working world. Formerly a U.S. secretary of labor, Hodgson has culled some pointers from Japan that could do much to improve American management-labor relations. They are especially relevant now, at a time when strikes and threats of work stoppages are disrupting the economy.

The Japanese labor force takes for granted benefits that workers in this country have sought for years. With unemployment rate of 2 percent, and youth unemployment even lower, Japan is a worker's paradise. Eighty percent of the working Japanese report that they are satisfied with their jobs. Older workers enjoy lifetime job security and are paid on a seniority basis. Unions — usually made up of the workers of just one company — routinely win substantial wage and benefit increases without resorting to strikes.

Japan does well by business management, too. The Japanese worker normally starts his career with a thorough background in basic language and mathematical skills. Workers generally stay with one company their entire working lives, providing a stable work force. The Japanese government rarely interferes in the private sector and almost never in labor-management relations. Former ambassador Hodgson points out that while the Japanese have freely imported American technology, they have chosen not to import our expensive, restrictive government regulatory systems.

The Japanese status quo has its drawbacks, of course. Rigid hiring and firing practices are built in and workers receive little reward for outstanding achievement. Equal pay for equal work is not common practice. Yet overall, the harmony between Japanese management and labor is impressive and enviable. And according to Hodgson, the U.S. could easily "import" some of Japan's most productive attitudes toward labor.

First and foremost is simple respect for work. Second is a respect for the worker. "In the Japanese industrial hierarchy of values," Hodgson says, "the number one resource is the human resource." Third is a realization, by both management and labor, that a strong economy is beneficial to all, and that fair management-labor relations make that strong economy possible.

We have carried the adversary relationship of management and labor too far in this country. We appear to be losing sight of the principle that jobs and industry, management and labor should be naturally aimed at providing all with a high standard of living and an inflation-free economy. Work and workers have dignity. Management commands respect. These are the workable principles Hodgson brings us. He should be heard.

DESERET NEWS

Salt Lake City, Utah, October 22, 1980

For decades, the United States was the acknowledged leader in mass production engineering and technology. The world marveled at the precision-made interchangeable parts that poured out of American factories in huge volumes.

Managers and technicians from abroad came to study the speed and efficiency of Detroit's automobile assembly lines. The Japanese, before and after World War II, were particularly interested. They were busy note-takers and picture snappers.

In those days, the Japanese were known in this country as manufacturers of cheap imitations of American products. How that has changed! This week, some 500 members of the American Production and Inventory Control Society gathered in Detroit to learn manufacturing methods from Japanese experts. They represent such organizations as Toyota Motor Co., Yamaha Motors, and Waseda University.

The Americans are not expecting to learn how to make cheap imitations, but are hoping to gain some insights that will enable them to do as well as the Japanese have done in turning out high-quality products that are competing successfully with U.S.-made goods right on our own ground.

In steel, automobiles, and electronics, the Japanese plants built since World War II — with U.S. aid, to be sure — are highly competitive. Add to this advantage the dedicated Japanese work force, the cooperative government, accelerated depreciation provisions in tax laws, high savings and investment, rapid automation, and low labor costs in some areas, and you have an almost unbeatable combination.

That we can be humble enough and wise enough to realize that we can learn from our former students just as they hope to continue learning from us is a heartening sign. Trade rivalry should not prevent our continuing to progress together.

The San Diego Union

San Diego, Calif., June 10, 1979

What nation leads the world in per capita production of manufactured goods?

Of the major industrial countries, which boasts the fastest growing gross national product?

What nation possesses the most technologically advanced industrial base?

The correct answer to each of these questions is not the United States, but Japan.

Japan's per capita production of manufactured goods exceeds that of the United States by 50 percent. Its rate of investment and GNP growth is more than double that of the United States.

Japan's workers, contrary to the cheap-labor stereotypes of a decade or so ago, now have more disposable income than most Americans.

And, although the aggregate wealth of the Japanese economy remains a distant second to that of the United States, the comparative growth curves leave little doubt that the gap will continue to narrow.

That the Japanese have registered such spectacular gains in just over three decades since their crushing defeat in World War II is truly remarkable. That they have done so on an overcrowded island nation barely the size of Montana with almost no natural resources should command the admiration of the world.

It should also, as Harvard Prof. Ezra Vogel notes in a thought-provoking article in Saturday Review, stir Americans to a closer examination of what might be learned from Japan's success.

These lessons are particularly appropriate at a time when the United States faces a growing requirement for energy and raw material imports and a concomitant need to sell more American products abroad. After all, Japan's economy is built around an even greater dependence on imported energy and foreign trade.

Japan's $20.6-billion trade surplus last year and the United States' $29-billion trade deficit for the same period suggests just how well each country has met its respective economic challenge.

Chief among the applicable lessons cited by Prof. Vogel is the need to stress cooperative rather than adversarial relations between government and the private-sector economy.

The Japanese government's willingness to use the tools of national economic policy to stimulate exports and help keep key industries competitive on world markets is hardly mirrored in Washington.

By way of contrast, whether on taxes, antitrust policy, or regulation, government and corporations on this side of the Pacific are more often at odds than cooperating in the interests of a coordinated economic strategy.

Few in Washington have paid sufficient attention to the critical shortage of capital available for investment and the modernization of this country's aging industrial plant.

Research and development lag partly because of apathy in Washington, partly because of shortsightedness in corporate board rooms, and mostly because the nation has yet to grasp the extent to which foreigners have stolen a march on our traditional technological lead.

Japan's enviable record of labor-management peace contributes significantly to the competitive advantage Japanese products enjoy in world markets. American labor and management could learn much from this example, but few seem interested.

The Japanese have always been renowned for their willingness to copy the West and improve on what they borrow. They are now in a position to return the favor and Americans would be foolish indeed to ignore the lessons of Japan's dazzling economic successes.

The Dallas Morning News

Dallas, Texas, November 13, 1980

THE BIG push is on to get America's economy working again. On both sides of the aisle, on both sides of the political fence, it is now clear that the economic machinery that once made our standard of living the highest in the world has deteriorated. We must restore it.

How to do it? Looking around the world, we see that many other industrial nations regard Japan as the model, the leader. It is the Japanese team that the other competitors look up to.

French businessman Alan Chevalier, at a recent meeting in Strasbourg, explained why:

"Japan has become the prop of the liberal world economy. It practices capitalism without complexes."

In Washington, a town preoccupied with "supply-side economics," "reindustrialization" and the like, that phrase ought to shine with refreshing clarity: "capitalism without complexes."

That says it all. The Japanese realize that a producer who produces something that the consumers in the market want or need should be allowed to earn a real reward for his effort. A real reward, that is, not a scam in which the market gives with one hand what the government takes back, through taxes and inflation, with the other.

But in addition to the cash, the productive system deserves to get credit, too, from the public it serves. While some Americans have tried to make our economic system a villain, Chevalier rightly points out that "the great strength of the Japanese today is their belief in their industrial strength."

The Japanese know something that we used to know and need to relearn.

Newsday

Long Island, N. Y., July 14, 1980

There was flurry of talk, much of it encouraging, last week about the future of the American automobile industry.

General Motors, which had been among the leading complainers about government fuel-efficiency standards, announced that by 1985, 50 per cent of its cars would get more than 30 miles on a gallon of gas. That's a better performance record than the federal standards have mandated.

General Motors also said it would begin marketing an electric car in 1984.

At the same time, Ford disclosed that it was talking to Toyota, the huge Japanese auto maker, about a deal to produce cars jointly in the United States.

And President Carter, on his way to Japan, stopped off in Detroit and outlined a plan to help the U.S. auto industry by easing federal emission standards, providing federal credit to auto dealers and seeking ways to reduce the number of Japanese imports.

Those are reasonably hopeful signs for the future. For the present, the news is not so good: For the first time in history, the United States is now the second largest auto producer in the world. The largest is Japan.

There are a number of reasons Japan—a nation the size of California with half this country's population—has outstripped the United States in auto production. The most obvious one is that the Japanese are manufacturing cars that sell, not only in this country but around the world, because they get good mileage, are well built and are a good buy for the price.

But surely the United States, with its vast technological capacity, could meet all those requirements. So what's wrong?

Part of the answer, at least, lies not just in the auto industry but in industry generally and how American industry operates as compared to Japanese industry.

To begin with, forget the old myth that the reason Japan can do better in the marketplaces of the world is that it's products are made by cheap labor. Japanese workers are now among the very highest paid in the world.

What's more, they work in plants that are in many cases more modern than ours—partly because Japan was devastated in World War II and was forced to rebuild its industrial base almost from scratch.

Once Japanese workers go to work for a company, they virtually never leave until retirement—and the company virtually never lays them off or fires them. There is no unemployment insurance program in Japan, which means there's no expensive bureaucracy to administer it. If a company must cut back because of sagging business, it gets government tax breaks or other assistance to enable it to continue to pay its employees. Strikes as we understand them almost never occur.

Government and business work closely together to make sure that tax laws and other regulations serve not to limit an industry's growth and prosperity but to stimulate it.

Most important, perhaps, the Japanese worker and the Japanese entrepreneur seem to have escaped the old cliches of western industrialized societies: worker and manager alike understand that rather than being combatants in a class struggle, they are participants in a joint enterprise. If it does well, they all benefit.

The result is a remarkably high level of productivity and an astonishing pride at every level in one's company, in what it produces and in oneself.

It's unlikely that all or even most of Japan's innovative industrial experience can be applied in this country.

But in the search for formulas for the so-called re-industrialization of America, it would make a good deal of sense for American entrepreneurs to examine the Japanese approach very carefully indeed. There might be a few lessons they can learn.

THE SACRAMENTO BEE

Sacramento, Calif., December 6, 1980

With Japan's enormous industrial growth, its productivity and marketing successes, its conflictless labor-management relations and superior products, its high personal savings rates and smooth government-industry cooperation, it is hardly surprising that Americans are looking there for clues to solving our own economic problems. In fact, American government, business and academic researchers have been visiting Japan in droves lately, and coming away with ideas for all sorts of potential panaceas — from new quality-control procedures to new tax write-offs — that they would like to adopt here immediately.

Many of their discoveries are fascinating — for instance, the fact that Japanese workers, because their jobs are guaranteed for life, have been enthusiastic supporters of automation. Some of the Japanese lessons may even be useful here, although much of the Japanese experience does not seem to be adaptable to American economic conditions. But probably the most interesting comment on Japanese industry that we have read lately was about the drawbacks of Japan's business success — drawbacks that suggest maybe the United States would be better off not trying to follow the Japanese model too closely.

According to an article in Science magazine, Japanese industrial success, to a great extent, has been a matter of brilliantly applying technologies and management theories that have been developed outside Japan — mostly in the United States, in fact. But the very conditions that have helped the Japanese economy take such successful advantage of already-developed technologies have also left it highly unprepared to develop new ideas of its own.

Japanese research and development money, for instance, is tied up almost exclusively in "market-centered" rather than pure research. So is the Japanese university system. This makes for swift development of industrial applications but few technological breakthroughs. Similarly, the flip side of Japan's famous teamwork and cooperation is a near taboo on individual deviance or creativity. "Japan has proved herself remarkably capable of absorbing foreign technologies," the Science article reported, "but now that she has reached full technological maturity, the question that is currently fashionable (in Japan) is: Are the Japanese capable of genuine innovation?"

More than likely, the answer is Yes. The virtually resourceless island that has adapted so remarkably to industrialization probably will figure out a way to adapt to innovation, too. But the notion that strengths also imply weaknesses is worth pondering for a moment.

The House Ways and Means Committee, after its own study of Japan, concluded that "Japan's rate of industrial progress and stated economic goals should be as shocking to Americans as was Sputnik." This call to action may be useful politically, but if it leads us to imitate those Japanese policies that would undermine the strengths of American social and economic organization, it could be a terrible mistake.

The State

Columbia, S. C., August 22, 1981

IT'S TIME for Americans to copy the Japanese, for a change, in achieving better quality, better engineering and management, and better instruction for workers in broad technical skills.

We agree with this opinion, which comes not from a union leader but from a highly respected manager, Robert Lynas, of Chassis Components, Automotive Worldwide, TRW Inc.

Mr. Lynas' arguments, which defend the American worker, merit particular attention since they were aired in *Enterprise*, a publication of the National Association of Manufacturers.

Mr. Lynas believes quality is not stressed enough by management and is a primary cause of warranty problems and product recalls, while the Japanese truly believe in the "design it right, make it right" concept.

He adds that U.S. management systems, being cost-driven, sacrifice reliability and repeatability of machines and tools for pieces-per-hour emphasis.

The industrialist suggests new organizations and simpler management structures, reducing the number of levels so managers get closer to the work, the people and the decisions that actually control quality and affect productivity.

Among other sacred cows, the executive zooms in on time and performance standards, which he calls "the root of the demand philosophy (pieces-per-hour) of managing . . . . Time and performance standards should be used as part of the evaluation of costs and as criteria for designing appropriate technology centers. We must take time standards from the plant floor and substitute personal responsibility for them."

Mr. Lynas also blasts an undemocratic industry practice, where "second and third class citizens" are created in the work force through practices such as using the terms "office worker, skilled tradesmen, production laborers." Instead, he suggests using terms like "business citizens" and eliminating time cards and time clocks.

More technical training is urgently needed, too, he says, to match Japanese workers' technical skills.

Basically, Mr. Lynas urges American managers to "have greater confidence in the intelligence of employees and take advantage of their ingenuity and individualism." Making such a concept work in a confrontative environment, such as a union situation, is not easy, but the thrust of his analysis should stimulate provocative thought among America's managers.

THE SAGINAW NEWS

Saginaw, Mich., September 14, 1980

The catchword of the 1980s in these United States may well be a polysyllabic jawbreaker called "reindustrialization."

In some economic and industrial circles, it is being heralded as a brave banner under which America can reshape part of its way of life and march forward to labor peace, substantive gains in national productivity, and world industrial leadership.

In other circles, it is being translated into: Buy American, but think, work, plan, and invest the Japanese way.

For it envisions nothing else than a massive change in the way we have been managing our industrial manpower, resources and technology — before we came face to face, belatedly, with the fact that global competition was no longer something to joke about in the top executive echelons.

The industry in question now getting all the media attention just happens to be the auto industry, the backbone of Michigan's productive force. But before it, if you will look back, there were major segments of the electronics industry, and textiles, and optics. And, in the future, rubber, petroleum, chemicals, steel....

In the Metro section of today's News is a package of information about one of the top Japanese auto plants. Written by News Metropolitan Editor Alfred Peloquin and photographed by News Director of Photography Bill Gustafson, this is important reading for anyone concerned about Saginaw's economic future.

Reindustrialization would work to stamp a Made-in-America label on the harmonious relationships between government, labor, industry, and even education that have evolved in Japan, under American auspices, since it sank to its knees after World War II.

The trend toward quality control circles, conceived by Americans but developed by the Japanese, is one encouraging sign that we may yet learn — to learn from the efficiencies of other industrial nations.

The move to expand the National Productivity Council to include labor and management, along with government, is another sign that official attention is more and more centering on innovative solutions to old problems.

Media publicity itself is another welcome signal. Popular education is more and more dependent on mass communications. Newspapers, magazines, radio, TV are, perhaps later than they should, concentrating more and more space and time on informing Americans not just that we have lost our national way, but WHY we appear to be staggering through increasingly dangerous quagmires.

It will be no easy task, despite what some of our intellectuals seem to think. Grafting the group consensus policies of a nation as homogeneous as Japan on the ethnic melting pot of the United States may prove an operation beyond the tradition-bound skills of our national leaders, in whatever sphere.

Winning popular support for policies that threaten to subordinate individual rights in favor of national survival, no matter how logical that might seem, comes at a time when a good share of the nation has narrowed its goals to "me first, and the devil with the rest of you" special interest programs.

Avoiding the game of pinning the blame for past failures on labor alone, or management alone, or government alone also poses problems of face-saving not indigenous to the Far East. Auto management has yet to beat its breast in public. Labor leaders are not prone to push their members toward management goals. Government is elected by those thousands of votes from hundreds of special interest groups.

So it will take more than what one economist has called the "grandiloquent slogans" of reindustrialization to turn the nation around, and to face up to wrenching, painful, even traumatic changes in thinking and outlook and lifestyle.

It will take a deep-seated realization by a majority of Americans that our freedoms and our way of life can be wiped out as easily by the silent guns of economic global competition as by the nuclear warheads of a belligerent dictatorship.

It will take a national leadership effort that no one foresees before Nov. 4th is over and done with — and that many cannot forecast in the short term beyond that.

The peril is not the shock of a Pearl Harbor, setting off a massive national rally superseding all other interests. The peril is the almost glacial erosion of our economic strengths and the lack of any national campaign to fire our enthusiasms and our energies.

Empty, fanciful slogans are not enough.

the Charleston Gazette

Charleston, W. Va., August 16, 1981

Japan's auto industry today manufactures and sells more motor vehicles than Detroit.

Is there a lesson in America for Japan's success?

Doubtless there is, but Americans who think that Detroit's Big Four can abruptly shift into some mysterious Japanese gear and regain the lead that has been lost are badly deceiving themselves. So says Isaac Shapiro, corporate lawyer, longtime resident of Japan, president of the Japan Society from 1970 to 1977 and for the last few years West Virginia's business and industrial sales representative in Tokyo.

Japan's management and labor models aren't transferable to the U.S. Japan is "a closed, homogeneous, disciplined society," Shapiro writes in *The Wall Street Journal*, "and the United States (is) open, heterogeneous and nonconformist."

Japan's economic success was discussed at length in a late March issue of *Time*. Five cultural traits supposedly are basic to Japan's prosperity: emulation, consensus, futurism, quality and competition. None of these traits is a stranger to Americans.

But Japan's economic strength depends upon factors other than *Time's* five magical qualities, says Shapiro. "The basic characteristics of Japanese society that account for its relative success are a high level of universal and uniform education, homogeneity, sense of duty, conformity and acceptance of authority and a survival mentality."

Education largely makes possible high Japanese productivity, believe many observers, including former U.S. Ambassador Edwin Reischauer.

Is this country prepared to adopt Japan's educational system? Curriculum content and administration of all public schools are controlled by Japan's Ministry of Education. Moreover, Japanese education, says Shapiro, 'relies heavily on the lecture method of teaching, memorization and academic achievement.

"In addition to mastering a very difficult language, every Japanese child must take courses in ethics, which stress moral values, proper social behavior, work and community obligations. ... The Japanese find no inconsistency between teaching morality in school and the constitutional separation of religion and state."

This system is diametrically opposite America's. What are the chances of Americans surrendering the local sovereignty of their thousands of different public school systems to Washington? What would be the response from American parents were they to learn that a course in correct social behavior, formulated in Washington without any imput from them, was being taught their offspring?

In this nation, ethical values that are taught "stress self-reliance, self-development, tolerance of a multiplicity of opinions and lifestyles rather than conformity to group mores.

Japan's cultural model is good for Japan. It wouldn't necessarily be good for the United States, and the test in any event will never be made. This society, as is true of all societies, has developed its own culture in its own way. Even if it wished to, it couldn't suddenly graft onto itself a foreign culture.

The Cincinnati Post

Cincinnati, Ohio, March 2, 1981

We're beginning to appreciate how Gen. Cornwallis must have felt at Yorktown as his army surrendered to the upstart Americans and the band played "The World Turned Upside-Down."

We're told that the words "Made in U.S.A." are now equated with poor workmanship. In Japan.

For Americans whose memories go back to before World War II, this is indeed a revolution—a turning over. For back then, "Made in Japan" was another way of saying shoddy.

That has not been true for a good many years now, as millions of proud owners of Japanese-made automobiles, television sets, cameras, watches, etc., can attest. With all due credit to a benevolent conqueror, the Japanese pulled themselves up out of the ashes of World War II and have fully earned their economic success.

And with all due discredit to Japan's trade restrictions, Americans have earned their present economic difficulties, not least of which are flagging productivity and increasing inability to compete.

Japanese consumers not only reject American products because of doubts about their quality, but also Japanese businessmen hesitate to open plants in this country because of doubts about the "quality of labor" and because of the adversarial nature of labor-management relations here.

It may be time for Americans to reverse history again and begin copying the Japanese.

We don't mean American workers should exchange their rugged individualism and mobility and independence of spirit for lifetime employment with one company. But many need to start relearning the meaning of words like loyalty and pride.

For their part, American managers must turn their focus from short-term profit to long-term improvement and infuse both themselves and their employees with dedication to quality.

The process couldn't start too soon.

THE ARIZONA REPUBLIC

Phoenix, Ariz., October 24, 1981

THE United States is expected to suffer a record trade deficit of up to $15 billion with Japan this year.

The Nikko Research Center recommends that Japan offer the U.S. foreign aid to avoid a growing resurgence of protectionism in America.

In a long report, the research group suggests that Japan help the U.S. develop new growth industries — such as semiconductors and computers.

The researchers say the U.S. lead in these fields has been surpassed by Japanese firms which should now assist the Americans.

The Nikko study says U.S. industries have lost their competitive edge to the Japanese.

Among the reasons, as the Japanese researchers see them, are a slowdown in U.S. investment in new equipment, policies seeking primarily short-term returns on investment, less competition among companies, declining labor quality and inadequate labor relations, as well as inattention to export opportunities.

The report also criticized what it calls the "declining work ethic" and blamed it on loss of personal values and "excessive social welfare."

The Japanese also blamed decreasing quality in U.S. education, increasing crime, and declining social discipline.

The institute suggested both increased Japanese investment in American-based production and the transfer of their technology to U.S. companies under licensing agreements. It also recommends providing Japanese expertise so Americans can modernize and improve productivity and quality.

Calling its proposals a "helping hand" program, the institute views its recommendations as a private initiative — not government intervention — to help American industry and the U.S. economy.

The institute's proposals come at a time of increasing concern in both Washington and Tokyo that the two nations are headed toward a trade battle.

Japan's role as an industrial superpower is a curious reversal of post-World War II years, when the Japanese were noted more for imitation than innovation.

Now they have become the world's largest auto producer, the largest shipbuilder, and a major threat to U.S. electronics and computer industries.

Their dominance in the market has come from investment and management principles they now are willing to share with U.S. industry.

Is this an offer that's too good to refuse?

CHICAGO Sun-Times

Chicago, Ill., December 28, 1981

To a great extent, the U.S.-Japan clash over auto markets is yesterday's conflict. Economic competition between these powerful free-world allies in the 1980s and beyond will be waged on high-technology battlegrounds.

But at the moment, Japanese car sales are a serious issue in Detroit's eyes.

Second of a series

Japan's penetration of our auto market has risen to more than 20 percent. Detroit's output this year sputtered to its lowest point in a decade. The industry's year-end losses could total $1.6 billion. More than 211,000 laid-off than 211,000 laid-off autoworkers—and thousands in related jobs—want to know what can be done *now.*

The painful answer: not much right away.

The problem, having developed over many years, will take time to solve.

Japan has modern plants. Detroit is getting those, too, with an $80 billion investment by the mid-'80s to overhaul factories and equipment (although some plans are being delayed by the industry's current cash crunch).

Japan has assembly-line robots. Detroit does, too. Japan makes computer-operated engines for better fuel economy. Detroit does, too. So where is the Japanese edge?

First, Japan now has a reputation for higher quality. Second, it was already producing small, fuel-efficient cars when U.S. motorists suddenly rejected their heavy gas-guzzlers as gasoline prices took off. One analyst thinks each 10-cent increase per gallon cuts U.S. car sales by about 1 million.

Also important, Big 3 auto workers were paid about 70 percent more than the average U.S. manufacturing wage—*despite lower productivity.* With fringes, they get about $20 an hour—66 percent more than their Japanese counterparts. Add fatter pay and bonuses for U.S. management, and Japan can sell its cars here for less (up to $1,500 less) than Detroit.

There's more. Japan's better inventory controls cut costs. "Quality circles," small groups of workers, emphasize teamwork and product improvements. Japanese productivity rose at a 9.3 percent annual rate in 1975-80, compared with 1.6 percent here. Bonuses for workers and management hinge primarily on profitability. Workers belong to company unions, not industry-wide organizations like the United Auto Workers—a difference that helps explain why Japanese labor-management relations are among the best in the world.

In addition, there's a uniquely Japanese sense of trust and mutual responsibility between labor and management. Lifetime employment is a real goal. Layoffs are a last resort. No wonder Japan's last auto strike was in the early 1950s.

Our diverse American workforce will undoubtedly reject many ideas for harmony and cooperation that homogeneous Japanese workers endorse as perfectly natural. But beyond dispute, Japanese *attitudes* toward their companies, their jobs and their employees have contributed to economic success.

The Detroit News

Detroit, Mich., June 30, 1981

The UAW recently sent a study group to Japan to find out how the Japanese are able to maintain a production rate far surpassing America's.

The UAW, we suspect, found that labor's attitude is a major factor. But perhaps the union also discovered that Japanese managers, unlike U.S. executives who are biased toward short-term results, are much more oriented toward long-term objectives.

Critics of American industry's emphasis on the short-term inevitably point to Japan, but the comparison is not entirely useful. Unlike their American counterparts, Japanese companies aren't subjected to the pressures of securities analysts and shareholders, because in Japan the financing comes from banks rather than from the sale of stock.

Consequently, Japanese businessmen need not answer to impatient stockholders — only to bankers who are more sympathetic to their strategies.

The result of the struggle between short and long-term considerations is that too many executive compensation plans favor immediate results at the expense of long-term growth.

With his tenure in the top job likely to be just a few years, the chief executive officer (CEO) in this country tends to stress short-term profits, so his leadership will be judged successful, so his stock options will increase in value, and so he merits his annual performance bonus.

Basically, the CEO must balance immediate and future considerations in order to position his company for the long haul in matters of investment, research, development, and structuring that will affect his company's prospects. Some of these factors must occasionally be favored at the expense of current results. But the problem is that too often the short-term wins out with deleterious effects on the future of American industry and the entire economy.

While some critics point to the demands of the investment community, others blame inflation and business schools for much of the problem.

Economist Alan Greenspan says inflation shortens the investment horizon, and experts believe the analytic techniques most business schools use emphasize the short-term.

"The long-term problems are very difficult to model . . . very difficult to analyze," asserts Robert H. Hayes, a professor of business administration at Harvard. "We tend unconsciously to develop an inclination to take a short-term perspective."

In order for the United States to compete more effectively in world markets, American companies must change their priorities.

Toward that end, we would suggest a restructuring of executive compensation plans. Boards of directors should consider paying out bonuses only in stock certificates, which could be redeemed only after retirement in, say, 10 installments. The same restrictions should be imposed on stock options.

This plan would serve to alter priorities in favor of the long-term health of companies, industries, and the economy. And the plan should appeal to executives because of its impact on taxes. The taxes imposed on gains arising from the bonus stock certificates and stock options would be reduced because they would accrue to the managers in post-retirement years.

Harvard's Professor Hayes adds: "Management has got to get its house in order."

We like to think our modest plan would be a step in that direction.

Houston Chronicle

Houston, Texas, November 11, 1981

One of the latest major projects of Japan's legendary Ministry of International Trade and Industry is a coordination of efforts between large, private Japanese electronics manufacturers to find and create new markets for their newest high technology. Under MITI's supervision representatives of the giants — Toshiba, Matsushita, etc. — regularly sit down to discuss market shares, techniques and strategies to avoid destructive competition among themselves, and to compete more effectively against foreign, principally U.S., manufacturers.

A government role in guiding industry, such as this one lately undertaken by MITI, long has been accepted practice in Japan. Even before World War II, MITI played an active role in Japanese industrial life, and in the late 1960s and early 1970s its leadership navigated Japanese industry through such straits as South Korean challenges to Japanese pre-eminence in textile and steel production. Government is part and parcel of the "Japanese miracle," and that success has led many Americans returning from Japan to raise questions about what the American government's role in our own "new industrial revolution" might properly be.

Faced with mounting trade deficits, layoffs in key industries and productivity problems, some American experts now suggest remodeling the American industrial infrastructure after the Japanese, including giving the government a more direct role.

While it is true that innovations such as quality circles, which demonstrably improve worker productivity, are finding their way into American industry with good result, government "coordination" of industry such as the Japanese practice is another matter. It may work well in a highly-ordered, disciplined society such as Japan's, but it portends only trouble in a society as diverse as ours. The debacle of the Energy Department is evidence of the sort of disastrous result such a system could produce.

That does not mean government should not have a role in ongoing American reindustrialization. Necessarily, it has a vital one. One useful step, already under way, is for those in government to replace ingrained attitudes of hostility toward those in the private sector with a spirit of cooperation on matters such as the design and administration of regulations. Others, also being taken, include the easing of requirements for capital formation and for depreciation of plant and equipment.

There is much to admire about Japan and the way Japanese society works. But Japan is not the United States. Many good ideas still bear the label "Made in U.S.A." In the long term, these will serve us better than mimicry of the Japanese.

THE PLAIN DEALER

Cleveland, Ohio, January 12, 1981

Japan is exporting to the United States more than just its fuel-efficient autos and quality electronics goods. Suddenly, America's corporate managers are beginning to take notice and move toward acceptance of a form of Japan's paternalistic style of management-labor relations.

Already an official of Westinghouse Corp. says that when workers are asked "what to do, instead of being told what to do, they respond with more than good ideas."

Worker alienation goes down. Productivity goes up. Quality and reliability of goods are increased. Workers even tend to welcome technological change when they feel they are in charge of change.

Westinghouse is using Japan's practice of involving management and labor together in discussions of issues in the workplace. Ideas are exchanged freely between management and labor on such critical problems as productivity and the quality and reliability of what is produced.

The idea of joint management-labor councils is not new. Japan's corporate managers actually learned of their use and capabilities while being educated in America during the 1950s. They have applied diligently the lessons learned here but neglected for far too long by America's corporate management.

Once hailed overseas as aggressive and ingenious, America's corporate managers now no longer are held in such great esteem. In fact, executives of foreign corporations tend to find their counterparts in the United States to be wanting in this critical area of management-labor relations.

At Japanese-owned television and motorcycle manufacturing plants in the United States, workers are presenting a clear case that the Japanese style of management-labor relations contributes to higher productivity and quality. Productivity and quality are, of course, strongly related in an era of scarce resources and soaring costs of materials. The attitudes of workers are more important than ever before.

There is no doubt that America's industry is in need of tax help from Washington and relief from some costly and cumbersome government regulations. But any corporate manager who puts all the blame for declining productivity and a lowered standard of living on government is overlooking the failures of management.

We believe that Westinghouse and some other American corporations which are drawing on Japan's experience are on the right path. The traditional adversarial relationship between management and labor has grown too counterproductive.

The two sides have an equal stake in improving productivity and quality. Accomplishing both these aims does not rest entirely on dollars invested in rebuilding old factories or constructing new ones. There is still the intangible human factor, which America's corporate managers have neglected.

TULSA WORLD

Tulsa, Okla., October 4, 1981

JAPAN'S Honda Motor Co. is in the midst of a daring experiment at Marysville, Ohio, to see whether American workers can build motorcycles and autos to the same standards as Japanese workers.

To any American older than 40, the thought that Japanese would consider American products inferior their own is hard to take. After all, for decades the Japanese flooded this country with a wide variety of cheap products.

The worldwide recognition of the ability and productivity of the American worker has been a source of pride in this country for generations.

Yet the Japanese can prove a lot of what they claim; the statistics on absenteeism, education and dedication are in their favor and the quality of Japanese autos, electronics and optics is legendary.

Then how are things coming along at Marysville? According to a report in the Wall Street Journal, just fine. The quality of the motorcycles is good. Better, claim the managers of the Maryville plant, than the same cycles made in Japan.

Encouraged, Honda will soon start building autos at the Maryville plant, betting the buyers won't be able to tell them from those made in Japan.

In fact, there is considerable evidence to suggest that the American worker has gotten a bum rap. Japanese managers at Maryville are successfully installing some of the policies that in general keep Japanese workers the happiest in the world.

They have gone to great lengths to eliminate the hierarchy that exists in U. S. plants. The manager of the plant dresses in the same white uniform with the same first-name tag as his workers. He eats in the same company cafeteria. He has no reserved spot for his auto in the parking lot.

Management seeks to make employees a part of the company family. They are consulted on how to cope with problems: how to best get the job done. (The Americans, by the way, respond at first with a shrug and a 'whatever you say' attitude.)

Employees are not guaranteed lifelong jobs, as they are in Japan, but they are made to understand that their jobs depend on the success of the company.

"As long as we put in good quality, people will buy our products and I'll have a job," said one American employee.

Working for the Japanese managers makes a quest for quality contagious, says another. "I want to do my best, because that's the way they do it."

Perhaps there's a lesson here for Americans, whether they be managers or employees.

The Boston Herald American

Boston, Mass., August 28, 1981

The lower wages and higher productivity of Japanese autoworkers accounts for a major share of the average $600 differential between comparable Japanese and American cars.

So, with Japanese imports continuing to command a full 25 percent of the American market while domestic automakers struggle to earn even modest profits, moves to produce American cars overseas were probably inevitable.

Recently, General Motors announced plans to design and build a subcompact mini-car in a joint venture with Japan's Suzuki Motor Co. The new subcompact, which may get up to 60 miles per gallon, is likely to be built in Japan for sale there and in the United States.

If the venture succeeds, it will offer American consumers unsurpassed value and provide General Motors with a needed infusion of black ink.

But where would this leave domestic autoworkers in an industry that has already furloughed nearly a quarter of its work force?

The answer is as obvious as the solution to the export of American jobs. Labor, in this case the United Auto Workers, must be willing to share in the cost-cutting essential to meet foreign competition or face the certainty of fewer jobs in shrinking domestic industries.

That won't require American autoworkers — currently the best paid of any in the world — to settle for the wage rates prevailing in Tokyo or in Hamamatsu, where the GM-Suzuki mini-car will probably be built. But it will mean that the UAW must persuade its members of some sobering economic realities.

As Chrysler Chairman Lee Iacocca put it last year in urging the UAW to accept a renegotiated contract, "We have thousands of jobs at $18 an hour. We have no jobs at $20 an hour."

Chronic absenteeism and shoddy workmanship have plagued domestic automakers for years. Labor's help in reducing both would represent the kind of cooperation that might keep the GM-Suzuki venture from triggering a trend that no American should welcome.

The Chattanooga Times

Chattanooga, Tenn., December 31, 1981

Unfavorable comparisons of the rates of productivity in today's U.S. and Japanese manufacturing complexes are often cited as evidence of a rapidly declining American dominance in the fields of industrial know-how and technological advancement.

It may be a valid point but the consideration should not stop short of searching for causes while reading the symptoms.

The views of a Japanese industrial analyst, familiar with the strengths and the weaknesses of both nations in this regard, offer a valuable commentary on how American companies can turn the situation to their advantage by learning from the demonstrated successes of their competitors.

Yoshi Tsurumi, writing for the *Global Review* of the Southern Center for International Studies, points up some of the misconceptions prevalent in this country concerning the growing importance of Japanese products in world markets as well as some of the ways in which U.S. industry can improve its standing. Mr. Tsurumi presently is professor of international management in the City University of New York.

He begins by dispelling the myth that Japanese firms are excelling their counterparts in all areas of industrial endeavor. U.S. and Canadian companies, for example, out-perform the Japanese in such fields as aircraft, advanced telecommunications equipment, pharmaceuticals and heavy earth-moving machinery. It is in the mass production of such goods ranging from automobiles to household appliances and computers where quality is so vital that Japan's manufacturers have moved toward top place.

Nor. he writes, is the "Japanese work ethic" the point of difference. He cites two kinds of examples in support. The first is the failure of U.S. firms to increase production over their domestic level when they build plants in foreign countries and man them with native labor. The second, in contrast, is the success of Japanese managers in rapidly increasing productivity after buying existing American factories and continuing operations with the same American workers and the same American machines but with Japanese management methods.

The deciding element, he concludes, is the attitude from the top down toward assuring built in quality in the goods produced. The Japanese approach is to bring workers more into the decision process and to give them the time and equipment needed to turn out quality goods. Elaborate quality controls to check the units which have been built are fine, he says, but why not do it right on the assembly line? Too often, he finds, top management in this country "sees productivity and quality as technical and operational problems which should be delegated to some lower-rung supervisor...while they concern themselves with quarterly share earnings."

Here as elsewhere, over generalization can lead to false conclusions, but it is pertinent to wonder if there cannot be profit in considering quality in finished products as an achievable goal rather than dismissing it as a lost art in the changing ethics of America's workers. The answer should be sought in the nation's boardrooms as well as in hiring halls.

THE CHRISTIAN SCIENCE MONITOR

Boston, Mass., February 2, 1981

Why is it that most new jobs – and much of the genuine innovation – that occurred within the US business community the past decade took place among smaller firms instead of the nation's largest corporations? And what lessons from these twin factors are there for the Reagan administration as it works to boost American productivity in the face of tough challenges from such major overseas competitors as the Japanese auto industry and European and Japanese steel firms?

The answers are as varied as American industry itself. But when one talks with business experts, it becomes increasingly clear that US industry must be as bold in pruning away its own cumbersome management practices that inhibit productivity as is the new administration in going after outmoded regulations within the federal establishment. As one very astute observer of the giant multibillion dollar US steel industry reminds us, "bureaucracy is not just to be found in government."

Steel is a case in point. Very few top leaders of that industry, our management friend reminds us, come "from outside the industry." There is little mobility within firms, and the "path to the top is very slow." The most aggressive young MBA graduates that enter steel firms all too often eventually give up and seek their challenges in more lively corporate environments.

More "productivity" is as much needed in "front offices" as on the "production line."

The US auto industry offers both some good – and some unpleasant – insights into management innovativeness. The Detroit firms, it must be noted, are truly huge enterprises, with assets larger than the gross national product of most nations. Given the very size of these concerns, "turnabout" time – the time it takes management to switch direction from one product line to another – has become very slow, usually five years or more. Economies of scale, (producing as many units of one product as possible to hold down costs) work as an incentive for retention of specific model lines over long periods of time and, in part, past their usefulness.

The US auto industry, more than the steel industry, has welcomed and rewarded bright young business school graduates.

On the other hand, if you saw the CBS documentary "The Toyota Invasion" last week, it quickly becomes apparent that the US auto industry has some catching up to do in management practices. Toyota is now the world's third largest automaker, and its Corolla the world's best selling car. Beyond modernized plants and a dedicated, zealous work force, Toyota's operational structure is as sleek and no-nonsense as its cars. Toyota executives, from the chairman of the board and chief operating officer down, share offices to encourage greater communication. The corporate emphasis is geared to long-range development and market research. Quality of product remains paramount. If a product fails to sell, it – and the market in question – are analyzed again and again and again with a patient doggedness until success is achieved.

While obviously US auto firms cannot – and should not – be carbons of their very hierarchical Japanese counterparts (which, after all, to a great extent based their techniques in the past several decades on Detroit), it can be argued that US corporate officials need to be far more zealous and enterprising than is currently the case.

That is true for US industry as a whole, not just autos. Like the administration in Washington, many a corporation and business firm needs to "hit the ground running." This is particularly important now that the Reagan team is moving against many of those federal regulations that the business community has long pointed to as time-consuming or injurious.

We are certainly not suggesting that US firms "retire" their older officials just to favor the young. What is important is that American firms be more innovative, more insistent on streamlining, cutting irrelevant business practices, curbing the hours spent in unnecessary meetings, discarding the insularity that has often prevented the very flow of ideas that leads to inventiveness in the first place.

Americans are a pragmatic, enterprising people. The greatest geniuses of corporate America – Andrew Carnegie, Henry Ford, Alfred Sloan, Thomas Edison, to name just a few – have been highly motivated individuals who cut right to the central nub of an idea. US corporate industry must again welcome that type of enthusiasm and persistence – and individual.

The Houston Post

Houston, Texas, April 18, 1981

The Japanese are attempting to explode the myth that their success in industrial productivity is a result of some secret formula. American industrialists might do well to listen. A consensus is forming among both Japanese and American experts that the United States already has all the ingredients to establish a productivity level equal or superior to Japan's if those methods would be brought together. The Japanese Labor Ministry has issued a report listing the three keys to productivity success:

1) Automation and technological development, **2)** application of certain well known management techniques, and **3)** a combination of stable labor-management relations, low absenteeism and cooperation in the introduction of new technology.

If that fails to sound like inscrutable Oriental philosophy, there is good reason. The Japanese learned their lessons from Americans. Tozo I[illegible]kichi, who heads systems development for Japan's industrial giant, Hitachi Ltd., recently told U.S. executives at a seminar in Chicago, "Twenty years ago, I visited your country for the first time and was taught the American way of management." He said that over those 20 years Japan faced many crises but overcame them by applying and adapting American management techniques.

Japanese terms add to the mystique surrounding management and production techniques which American executives are flocking to Japan to learn. Those techniques have such exotic names as *Ringi*, *Ukezara*, *Kanban* and *Jidoka*. But, when translated, they stand for consensus management decision-making, careful preparation of the factory for new management techniques, reducing inventories to cut costs and quality control through giving workers the right to bring a production line to a halt. Hardly anything mysterious or even new about any of that. Upon examination, many American industrial firms considered as models in their fields are run largely the same as their Japanese counterparts. What we need in this country is a wider application of the productivity-improving techniques that we already know about.

DESERET NEWS

Salt Lake City, Utah, February 11, 1982

How did Japan turn from a manufacturer of cheap, shoddy merchandise before World War II into a competitive giant building technologically-advanced, high-quality products that are still reasonably priced?

That's a question American businessmen are asking more and more as they seek to compete in world markets with Japanese products. U.S. business executives have flocked to Japan in recent years to discover those answers. Another group departs from San Francisco Feb. 12 for a two-week study mission, sponsored by the U.S. Chamber of Commerce and the Japan Productivity Center.

They shouldn't have to look far for answers. The Japanese, who were so clever at imitating foreign products before World War II, seized upon a U.S. idea in the post-war years: quality control. By carefully prescribing tolerance limits and rejecting work that didn't measure up to specifications, U.S. manufacturers found, a product of uniform quality could be produced. As Andrew Carnegie noted in the early 1900s: "The surest foundation of a manufacturing concern is quality. After that — and a long way after — comes cost."

The Japanese also have borrowed another idea developed in the United States by such companies as Du Pont: Develop a new product, stay with it until competition starts cutting the price, then go on to other technologically-advanced products. The Japanese are quick to abandon dying products or industries.

Some, but not all, of those lessons can be re-adapted for the United States. More likely, U.S. businessmen will have to cultivate American strengths — high technology, space spin-offs, medical advances, for example — to make it in this competitive world.

The Dispatch

Columbus, Ohio, October 27, 1981

IT IS NO fluke that the Japanese are manufacturing good products and selling those products in the free markets around the world. The Japanese learned much about business and manufacturing from the other industrialized nations, did some streamlining of their own and put it all to work.

Now U.S. industry could learn some things from the Japanese. That applies to U.S. management and U.S. labor unions.

In this country, management and unions seemingly have gone to war against one another annually, biennially or triennially, and they have fought many costly skirmishes in between. Management was out for management and stockholders, and unions were out for unions and their members.

These wars and these skirmishes were fought with narrow vision and foresight while the markets of this country became flooded with products from abroad that were often cheaper and often better than those "made in the U.S.A."

And the situation for U.S. industry could get worse.

In the rapid trend toward automation and labor-saving devices through the use of integrated circuits, computers and robots, many industries in this country face serious labor problems trying to make use of such devices.

Kimpei Shiba, a columnist for the *Asahi Evening News*, wrote that Japan's firms and labor unions are far better equipped to deal with modernization problems than are firms and unions in the United States.

Shiba said this is because Japanese unions are structured vertically as company unions in contrast with horizontal American unions. In Japanese firms, management and unions work together to shift employees to new jobs in a framework of lifetime employment policies, as opposed to employees being fired when displaced by new equipment.

The Japanese system lets Japanese firms take advantage of technological advances in producing competitive products, and this in turn benefits both the firms and the employees.

Shiba pointed out, as an example, that in the Japanese auto industry, Nissan's conversion to welding by robots in an ultra-modern plant presented relatively few labor problems. Welders were trained for other jobs.

As we said, U.S. management and unions could learn some things. This could be to the advantage of both — and to all of us.

Part II: Research

When Japanese goods began to encroach upon U.S. markets, Americans reassured themselves that the challenge was only in manufacturing; in the area of invention, the U.S. was still number one. The Japanese, went the common wisdom, were good at adapting items for the mass market but not at innovating; they were imitators, not pioneers. That myth is rapidly being put to rest, as Japan creeps up on the U.S. in new computer parts, drugs and energy technology.

The outlook for U.S. research is not yet at the crisis point. U.S. scientists still capture a healthy proportion of Nobel Prizes. From 1901 to 1980, Americans were awarded 121 Nobel Prizes, while Japanese scientists received only three. The transistor, which was responsible for Japan's success, was not invented in Japan; the Japanese merely turned it into an item for mass-market use. The Japanese themselves recognize their weakness in innovation, and they have embarked on a drive to invent with the same determination they used to gain dominance in manufacturing. Japanese industry supports basic research to a greater degree than in the U.S. Industrial organizations and the government help by providing low-interest loans. The U.S., in contrast, leaves much more of its research to the hazards of the marketplace. The government stake in research and development has narrowed considerably in recent years. In 1970, about 43% of all R & D was government-funded. The proportion fell to 31% in 1981.

Since private industry controls most of U.S. R & D, it is not surprising to find that more research is directed toward developing products than toward searching for new ones. Development took up 78% of industry R & D spending in 1979, while basic research received 3% and applied research 19%. Critics accuse American industry of concentrating on marketing already existing products rather than inventing new ones. As Sen. Harrison Schmitt, the former astronaut, observed at a National Academy of Sciences conference in 1978, "There is a very direct link, although a long-term one, between the health of our basic research community and what will eventually become invention or innovation."

Carter Sets '79 Research Outlay: Congress Increases the Total

President Jimmy Carter called for $5.1 billion in spending on science, technology and space exploration for the 1979 fiscal year (which began in October 1978). Aside from paring the budget of the National Aeronautics and Space Administration (see Chapter IV), Carter's proposals did not differ from the plans of the previous administration. Current R & D spending (for the 1978 fiscal year) including NASA, would reach $4.8 billion, according to estimates. It was an increase of $100 million from the spending level of the previous fiscal year. In reviewing Carter's budget, Congress modified the results slightly, authorizing an extra $100 million for R & D in 1979, which brought the total to $5.2 billion, but the legislators scheduled actual outlays of $5 billion, less than Carter had called for.

Much of R & D funds is "hidden;" that is, it is allocated to specific departments: agriculture, education, energy, health and welfare. The general science, technology and NASA budget line does not tell the whole story. Actually, government spending on research is a mere fraction of total national R & D spending; private industry is responsible for more research and technological innovation than government. Federal policies, however, set the tone.

THE SUN
Baltimore, Md., January 25, 1978

After years of relatively thin gruel for pure scientists at the federal table, the President's fiscal 1979 budget should delight most of them. Major increases for basic research are proposed for almost all federal agencies that deal with science and technology. The overall increase is about 11 per cent, with increases for some specific agencies—NASA, the Department of Defense and the Department of Health, Education and Welfare—rising above the average. The 11 per cent increase represents an increase of about 5 per cent in real dollars after inflation.

But while President Carter, a former engineer, asks for major increases for basic research, he proposes much smaller increases in outlays for what is the work largely of engineers: *applied* research and development. Proposed is a 5 per cent increase in applied R&D, probably not enough to keep pace with inflation. This de-emphasis can be attributed to the administration's view that most of the costs for developing promising new technologies ought to be paid for by private industry. Critics contend that the best payoff on federal investments comes in the form of exportable new technologies that can reduce the nation's balance of trade deficit. In recent years, Japan has made vast inroads into the world electronics market and threatens the all-important U.S. lead in computers. Sweden in telephones, Germany in nuclear power, France in solar energy—these are areas where the loss of U.S. technological dominance is showing up in adverse trade figures. Renewed emphasis on both basic and applied research is needed.

But the new empnasis on basic research is not necessarily a return to the era of the 1960s when almost any scientist capable of writing a grant application could get federal dollars for almost any kind of research, no matter how impractical. Basic research to Mr. Carter and his advisers clearly seems to mean something different. For instance, the administration is asking for a very practical 39 per cent increase in funds for research aimed primarily at determining whether man is altering the quantities of ozone and carbon dioxide in the atmosphere in ways that could be disastrous for civilization. Moreover, basic research can very quickly become applied research nowadays. The prime current example is the success of California researchers in using "genetic engineering" to coax bacteria into making a humane hormone; that bit of basic research almost instantly became commercially valuable—and probably soon will be producing biochemicals to sell overseas.

SYRACUSE
HERALD-JOURNAL
Syracuse, N. Y.,
January 28, 1978

Until World War II, 65 to 70 percent of the basic scientific inquiry in this country was paid for by foundations, government and universities.

Industry couldn't justify such research because immediate payoffs could not be charted.

Still, industry shared with government and the universities the cost of applied and general research such as adaptation of a vaccine to general use or development of a product.

World War II shook up this division of labor and for years, government funding has inspired a big share of all kinds of research in this country.

Twenty years ago, shocked by the Soviets' Sputnik, we poured tax money into basic research as if it were water filling a dry hole in the Arizona desert. But, federal interest tapered off after the 1969 moon landings and, consequently, basic research has suffered.

President Carter is trying to return to the pre-World War II division of research labor with an 11 percent increase, some $319 million, for basic research.

What's recognized as applied research is up 9 percent, and development 6 percent, with the intention that industry take over most programs closest to the payoff stage.

Whether we're on the right governmental track won't be determined for years, but the administration, significantly, is directing more and more research dollars toward resources and energy.

This is seen in the 40 percent increase in research funds allotted the Environmental Protection Administration and another 40 percent increase among several agencies under the Commerce Department for climate research.

Elsewhere — nutrition, health, armaments, energy — the budget trails the rate of inflation by a point or two.

But the big news — of concern particularly to young scientists and their colleges and universities — is the reversal of the downtrend in basic research funding.

For the country, that's recognizing a basic need.

The Houston Post

Houston, Texas, February 17, 1978

Our grand sweep of Nobel Prizes in the past two years gives us the false notion that we are leading the world in scientific research. Actually, it means that we *were*. Scientific research requires time. The 1977 and 1978 prizes stand for research done 10 to 20 years ago and drawing on research done in even earlier decades. They conceal a waning research effort today throughout the United States.

The American share of scientific papers published in the world dropped in many fields between 1965 and 1973: From 25.9 percent to 21.2 percent in chemistry; 49.9 to 43.7 percent in engineering; 41.3 to 38.4 percent in physics; 35.8 to 30.8 percent in systemic biology. The number of patents awarded American individuals or companies rose from 1960 to 1971, and has been declining ever since. And our government has spent less and less on basic research. Between 1968 and 1976, our federal investment in basic research declined 15 percent. Federal investment in research and development facilities dropped 77 percent from a peak of $126 million in 1965 to a low of $29 million in 1974.

Economists report on our slowing rate of gains in productivity We have fallen behind every major industrial country in the rate of gain. We need research and development to revitalize our production and enhance our ability to compete with other nations. It was therefore disappointing to see that of the large amount President Carter budgeted for research and development, the bulk is earmarked for military use. If the national structure rusts, the military will have little support for any technological breakthroughs it may develop.

A gardener cannot wait until he sees tulips blooming to plant bulbs and get spring flowers. A farmer has to have long-range plans for the enrichment, planting and harvesting of his land. Both gardener and farmer think in terms of years, not seasons. Basic research — disinterested, pure research — is an investment in our future. Whatever decisions we make today on scientific research will effect the nation in the 1980s and 1990s. No one can predict specific results. But we can predict that if we continue to skimp on our basic research now, year after year, we will pay in the 1980s by loss of achievement. We stand ready to lose our world leadership in science.

Detroit Free Press

Detroit, Mich., July 4, 1978

A FEDERAL budgeter has warned America's scientists that they must "accept a pause" in the funding of basic research. That is a mild phrase to describe what threatens to become an economic and cultural catastrophe.

Speaking to a gathering of scientists in Washington, W Bowman Cutter, of the Office of Management and Budget, blamed inflation, the federal deficit, the tax limitation mania and the skepticism of conservatives in Congress for the impending cutbacks. But federal support for basic research has been sliding downhill since 1968.

In the last decade, federally financed doctoral fellowships have fallen from 50,000 to 15,000 Laboratory facilities at major research universities are so outmoded or deteriorated that the quality of work is being affected Support for astronomy research has fallen by 40 percent, for physics and engineering 30 percent. Only medicine and the environmental sciences have seen an increase in funding levels.

Congress has become increasingly reluctant to fund any projects that do not promise a quick payoff But every major scientific and technological advance rests on a mosaic of studies that once seemed arcane, or obscure, or even laughable. Science cannot tell in advance which are the blind alleys until someone has groped down them; it cannot pay off, in the way that businessmen and politicians like to see it pay off, until someone has gambled. Today that groping and gambling most often must be underwritten by the federal government, or not at all.

As Congress reviews the budget for fiscal 1979, the wiser men in that body should fight to increase the funds earmarked for research and the replacement of obsolete laboratory equipment. Cutting off funds for basic research is like eating the seed corn during a famine: It robs us of the future.

Sunday Journal and Star

Lincoln, Neb., September 10, 1978

The Carter administration is not convinced all possible good ideas related to food and fiber production must come from traditional agricultural research centers.

This really doesn't seem to be such a revolutionary view.

Nevertheless, it appears to have mightily upset the House Agricultural Appropriations Committee and, especially, its chairman, Rep. Jamie Whitten. As a result, the committee is blocking the administration's attempt to increase the Department of Agriculture's program of competitive research grants from $14.4 million this fiscal year to $30 million next year.

Whitten and associates object to any deviation from the USDA's established pattern of farming out research projects to land grant institutions, such as the University of Nebraska, and agricultural experiment stations. As we understand it, the allocation of federal ag research moneys has been essentially on a formula basis, with competition accorded a lesser role.

Ag Secretary Bob Bergland does not fault the results of the customary system. They have assisted in bringing enormous increases in crop production. Yet, as Bergland says, the country could benefit by "broadening the base of scientific effort applied to agricultural problems."

Even though the most recent group of 21 USDA grants went largely to traditional ag research powerhouses — Iowa State, Kansas State, Purdue, Wisconsin — Whitten probably was outraged that Harvard got $120,000 for work on corn photosynthesis. And other awards, to Johns Hopkins University and the Carnegie Institute, couldn't have contributed to Whitten's composure, either.

Utter parochialism and political clout ought not absolutely control the matter of investing federal research dollars, agriculturally targeted or otherwise. There is a place for competition among different researchers and the Carter administration, in this circumstance, appears to be moving carefully to strike a proper balance among all interests.

Surely there can be no disagreement with Bergland's belief the United States will have to have more "technological breakthoughs to continue to feed and clothe the nation in the face of dwindling supplies of oil, water and prime agricultural land."

Let's use the best brains the country has, regardless of where they are located.

Substantial Spending Boost Set for '81 Research Budget

In the last year of his administration, President Jimmy Carter was more inclined than before to emphasize R & D in the federal budget. While holding down spending in many key areas, his proposed 1981 budget called for a 13% increase in R & D spending. Only defense, which received a 12% increase, came close to receiving the same attention in the $615.8-billion budget. Carter called for a total of $36.1 billion for R & D, which would be spread among 31 separate federal agencies. The National Science Foundation received $1.1 billion of this total, to assist it in funding basic research projects around the country. Carter echoed the growing national awareness of the importance of research when he observed that "in the long run, economic growth depends critically on technological development."

The Oregonian

Portland, Ore., July 10, 1980

Budget-cutting can go too far.

A report by the National Commission on Research says that lack of funds to replace deteriorating or obsolete research equipment in state universities threatens a loss of research efficiency and ability to compete in the huge international research market.

Researchers trained by our state schools who are unfamiliar with the latest equipment will lose jobs to more highly skilled technicians from other countries, the report warned. The quality of their research also will be reduced by lack of top-quality equipment.

"Research is like a bean plant," said Clarethel Kahananui, assistant chancellor for academic affairs for the Oregon State Board of Higher Education. "It may appear healthy one day and wilt the next.' '

Mrs. Kahananui warns that the bean plant already has begun to wilt in Oregon. The Oregon State System of Higher Education has only enough funds to replace equipment every 50 years — much longer than the useful life of most scientific instruments. The 1981 Legislature will be asked next January to appropriate $2.7 million to stem the scientific decay.

Even at times of desperate budget-cutting, the importance of research at our institutions of higher education must not be undervalued. Reducing already low budgets for research equipment replacement would be a false economy. It would steal from our future. Oregon needs more of the high-quality scientists our state schools are now turning out if it is to solve the problems of an increasingly complex society. But without proper equipment, state schools will produce fewer and less well-trained researchers.

The Chattanooga Times

Chattanooga, Tenn., February 3, 1980

Tucked away in President Carter's mammoth $616 billion budget for fiscal 1981 is a category for which the president has proposed a significant increase in spending, reversing a trend that, if continued, could have serious implications for the United States' ability to retain its lead in a vital area. We refer to scientific research and technological development.

Mr. Carter has suggested a 21 percent increase, to more than $500,000, for basic research by the Defense Department. And financing for the National Science Foundation will jump by 17 percent, more than the rate of inflation for 1979 to well above $1 billion.

Considering the challenges facing the United States in the next decade — particularly in the area of meeting incredible demands for energy — the proposal for increased appropriations is timely and well conceived. If Congress goes along, the greater emphasis on basic research should prevent the United States from lagging behind other nations in this vital area.

President Carter is known to believe that scientific research, which previous administrations have downplayed in importance, is essential to the United States' economic growth and national defense. The former will benefit from research into such areas as energy, space, physics, industrial innovation, computer sciences and deep-sea drilling because more jobs will be created, strengthening the national economy. An increased commitment to research and technological innovation is evidence that the American economy cannot survive and prosper by a deadening allegience to the status quo.

The administration has requested $36.1 billion for fiscal 1981, a $4.2 billion or 13 percent increase over the current budget year. If the administration's proposal is accepted as is, basic research alone would benefit by a $543 million increase over fiscal 1980. Federal money supports two-thirds of the basic research done in the U. S.

The government has done much in recent years to support medical and biological research, and the results have been extremely valuable. An increased emphasis on the more basic types of technological research, however, is long overdue. If Congress expects the United States to maintain its record of the type of technological innovation that has made the country the envy of the industrialized — and non-industrialized world, it will support the president's request for increased expenditures for essential research.

THE ANN ARBOR NEWS

Ann Arbor, Mich., November 26, 1980

ONE OF the most important holdovers from the Carter administration will be the new National Science Foundation Director, John B. Slaughter. He will take office in January.

Slaughter will need close and skeptical watching from all who worry about the need to do more than keep up with inflation in this country's financial support for scientific research, both applied and basic. He is talking as if he might be willing to see the National Science Foundation and other federal agencies direct their research grants in ways that could downgrade basic research, meaning research that uncovers new information and raises new questions without being obviously useful.

"The engineering societies," Slaughter said a few days ago, "have strongly proclaimed the need for improved support to engineering research and the NSF unquestionably has to deal with that-...What we need to be doing is supporting good science, and identifying processes by which we can identify and support good science. We should not worry so much about how far it is to the basic side in making the decision that it is good science."

THERE'S nothing wrong with Slaughter's generalization in favor of "good science." And there's assuredly nothing wrong with the idea that federal agencies should be doing more to stimulate innovation in engineering and industry generally.

Apprehension about the level and direction of federal support for research is raised by Slaughter's seeming casualness in blurring the difference between basic and applied research.

It's true that there isn't always an obvious distinction. Scientists paid to test certain materials in a lab, and find an answer to a fairly specific problem, always have a chance to spot unexpected new information or something utterly puzzling, and this would amount to basic research. But applied research, essential as it is in fields ranging from auto safety to evaluation of possible carcinogens, can never substitute for relatively undirected research by scientists free to observe and theorize without pressure to produce useful results.

Slaughter is surely well aware of this distinction. He's an engineer, but he's had plenty of exposure to basic research viewpoints as provost of Washington State University and as an assistant NSF director.

President Carter's own record in seeking to advance federal support of basic research is excellent. He would not have made this appointment if he thought Slaughter would be willing to increase grant support for applied research by taking money from basic research.

Slaughter himself says he hopes federal support for applied engineering research can be increased with "new money" rather than with a cut in basic research grants. This isn't very reassuring, though, considering that President-elect Reagan's goals and priorities in national science policy are considerably less clear than Reagan's determination to make significant cuts in the national budget.

EVEN with President Carter's clear commitment to federal support of all scientific research, inflation is barely being offset. At the University of Michigan, the annual report this fall from Charles G. Overberger, vice president for research, includes this comment: "From fiscal 1979 to fiscal '80 (U-M) research expenditures increased over 10 percent, but when discounted for inflation the increase is a little over 1 percent. If we examine the 10 years since the base year of 1970, (research) expenditures increased 98 percent; however, when discounted for inflation, the increase is 2.4 percent...Our largest source of funds is the federal government..."

Scientific research, basic and applied, really needs no justification beyond the one it gives by producing new information and by constantly showing how little humans yet know. President-elect Reagan can, if he wishes, find further justification by bearing in mind that his intention to present the USSR with stepped-up competition should include strong support for basic research, not simply for building weapons and solving known problems.

NEW YORK POST

New York, N. Y., January 29, 1980

President Carter deserves credit for boosting basic research in his new budget. For far too long we have all been indifferent to the decline in high technology in this country — and yet it was the development of new ideas, the exploitation of advanced technology which gave the U. S. its leading role as an industrial power.

Basic research by NASA, the Energy Dept. and the national science foundations is the key to the future. Tax credits to industry can provide only so much incentive. Basic research on specifically government-funded projects is the route to the creation of new industries, new jobs, and, it is to be hoped, a booming economy.

The Union Leader

Manchester, N. H., April 13, 1980

If you think the United States has been doing poorly in world economic competition of late, brace yourself. The situation is liable to get worse before it gets better — if ever.

Or so suggest the results of a Conference Board survey of comparative research and development efforts in this and other major industrial countries.

The signs of continuing U.S. decline in an area in which the country was once pre-eminent are in a set of dismal statistics.

American R & D expenditures as a share of gross national product, a primary means of measuring scientific progress, have dropped from 2.64 percent to 2.22 percent in the last decade. During the same period, Soviet outlays have risen from 3.23 percent to 3.4 percent and spending by our friendly competitors in the Free World shows a similar upward trend.

In an increasingly technological world, stinting on the basic research from which new industries must grow all too easily and quickly translates into a stunted economic future.

Don Graff/NEA

Houston Chronicle

Houston, Texas, June 13, 1980

R and D" — shorthand for research and development — is the engine of military progress. Without the technological tools tested out in a vast network of laboratories nationwide, there would be no new missiles, satellites or tanks.

As with political and economic systems, the "R and D" styles of the United States and the Soviet Union are vastly different. In fact, to a large degree, the "R and D" styles of the two countries reflect the differences of the societies.

A study of those differences, commissioned by the Rand Corp., the California think tank, is instructive. It depicts the Soviet system as highly organized, bureaucratized and amply funded. In contrast, according to the Rand report, the American system is widely considered to be less well-organized, often less cost-efficient and, in recent times, less well-funded.

The surprising part of the study is its conclusion that the American system is, overall, the superior of the two. For all its imperfections, U.S. research and development is more innovative and more flexible. Despite its funding and organizational advantages, the Soviet system is resistant to change and the quality of its products is less certain.

The study concludes with an apt paraphrase of Winston Churchill's often-quoted observation about democracy. The U.S. system, it observes, is the worst possible way of doing "R and D" — except for all the rest. The thought is reassuring.

Reagan's Budget-Slashing Hits Research Sector

President Carter's outgoing budget proposals for the 1982 fiscal year (beginning October 1981) included $1.2 billion for the National Science Foundation. The incoming administration of President Ronald Reagan promptly reduced the NSF budget by $155 million, including slashing $30 million from the proposed $40-million NSF budget for sociology, psychology and economics research and $75 million from the NSF's fund to help modernize college and university laboratories. The NSF's education programs were slated for an initial $47-million cut, which was almost doubled in March to $87 million. The National Academy of Sciences warned that the administration's policies would do "irreversible damage" to American science. The NAS concluded a two-day conference in October with a call for increased U.S. investment in research. The concluding statement noted that "because of the important relationship between research, technology and increased productivity, the expressed goals of this administration for a strong economy and improved national security demand more rather than less investment in basic research." One conference member warned that the U.S. was entering the "pre-Sputnik" era in terms of the number of physics trainees, a reference to the first Soviet space satellite, which had jolted Americans into a massive effort of improving science education and research.

TULSA WORLD

Tulsa, Okla., July 4, 1981

AMERICAN industry continues to pump money into research and development at a pace outstripping inflation.

A survey of 744 companies by Business Week magazine shows an average increase in research and development expenditures of 16.4 percent over 1979. That figure is 4 points higher than the rate of inflation.

Within some segments of the economy, the rate is even higher. Among oil service and supply companies, for example, R&D increased 37.5 percent.

This is encouraging news for our country. If America is to regain her role as the world's leading industrial producer, it will come as a result of investment in better equipment to do the job more efficiently. The Reagan Administration's supply side economic theory rests on the belief that money returned to business through tax cuts will be used to develop new equipment which will spur American productivity.

But the picture is not entirely bright. R&D expenditures as a percentage of Gross National Product continue to decline. In addition, the cost of industrial research is increasing faster than the rate of inflation. Finally, Federal budget cuts may mean cuts in Government-sponsored research by American colleges and universities. Academic research has long been a major source of information for business and industry.

Research and development is the backbone of any nation's industrial might. Business Week's figures describe the nation's good performance, and warn that good is no longer good enough.

The Boston Globe

Boston, Mass., March 4, 1981

Preliminary reports indicate that the Reagan Administration plans additional reductions in funding for research and development projects as part of the President's overall drive to hold the federal budget to a total of $695.5 billion next year. It should move slowly on this question.

While an argument can be made that virtually any federal program is cuttable, the Administration ought to differentiate carefully between expenditures with no hope of future pay out and those with real potential benefits. Research and development, in this context, ought to be viewed as an investment rather than an expenditure.

The Administration is approaching the problem under unfortunate circumstances. Reagan has yet to name a science adviser, and is presumably getting little or no counsel on the complex issues involved in R & D programs.

The real hazard in approaching R & D expenditures in the mood of slash, slash, slash that permeates Washington today is that the scientific community will lose at least some of its sense of momentum. As was pointed out in a recent issue of Science magazine, the publication of the American Asssociation of Arts and Sciences, "science is a long-term creative process, and a multi-year retrenchment would damage seriously the nation's scientifc and technological base. Investment for R & D has its place in supply-side economics."

Furthermore, there are indications that the Administration indends to reduce support of training of engineers and computer specialists at just the time the nation is experiencing shortages of both. It is significant that Japan, whose economic success is apparent to all the world, has a much higher ratio of engineers to blue-collar workers than is the case in this country. Japanese productivity gains continue to soar, to the disadvantage of American industry.

The Administration understandably wants to make its fiscal point as dramatically as possible. Ronald Reagan was elected in part by a majority of people who are skeptical about the utility of many government projects, rightly in some cases. The Administration is also firmly dedicated to strengthening the American economy for both the immediate and the longer-term future.

That future is going to be profoundly influenced by science and technology, much of it lying in universities and academic laboratories in need of public funding. Carving indiscriminately at such funding could betray the Administration's own laudable objectives for the future of the American economy.

The Wichita Eagle-Beacon

Wichita, Kans., March 3, 1981

American scientists rightly are concerned about plans to cut further into the important federal funding that helps sustain the National Science Foundation. But it shouldn't be only scientists who are worried about the impact of such a decision: All Americans have a stake in the scientific future, and could pay a heavy price if the philosophy of continually expanding humankind's horizons of knowledge and understanding is sacrificed or compromised for temporal budgetary considerations.

The science foundation already has had its federal support trimmed by a whopping $47 million, and another $40 million cut now may be in the offing.

Even if the matter were to be decided strictly on the basis of investment versus return, continued support of all sorts of programs of scientific inquiry not only is justifiable, but imperative. Who could put a value on the worth of the Salk polio vaccine, or an eventual breakthrough in cancer treatment? America's commanding position on the leading edge of semiconductor and computer technology probably would never have been, had it not been for the healthy scientific climate of this country's space exploration program.

Not all scientific research and experimentation produce such dramatic, tangible results, but nearly all of it, through its cumulative effect, can be used in some way to enhance the human condition. Clearly, scientific endeavor has tipped the scales far to the positive balance.

Giving science short shrift isn't likely to help the United States out of its economic doldrums, either. Innovative scientific thinking and questioning helped put America in the forefront of modern industry. That position now is threatened, mainly by nations willing to make the commitment to science and technology. Abandoning American know-how isn't the answer to such a challenge.

This country should not abrogate its leadership role in the realm of science to become a follower, because that's something it doesn't do well: follow. It is far too early to end, or even restrict, the quest for knowledge, and those willing to do so strictly as part of a budget balancing act ought to rethink that position.

The Dallas Morning News

Dallas, Texas, November 4, 1981

Throughout the '60s and into the '70s, science rode high in the United States. Enough money could not be invested in research. Today we're reaping the benefits of much of that work.

But Dr. Henry McGill Jr., scientific director at the Southwest Foundation for Research and Education, is sounding a warning for the future: The United States is in danger of losing its lead in biomedical research. He could have included many other facets of research as well.

Washington is grudgingly accepting the need to trim federal expenditures. But congressmen want to take the easy way out — to cut the spending that causes the least political pain. Unlike income transfer programs, research funding, especially for pure science, doesn't have a loud constituency. The budget knife, therefore, can be applied deeply.

That's false economy, however. Cuts in research funds might not be painful today; tomorrow they might be disastrous. Research spending and results don't always walk in lockstep. For example, today's breakthroughs in biotechnology are based on a solid foundation of research that has taken place for the past 20 years. Without that solid base, there would be no weekly announcements of advances, and without the funding that took place two decades ago, there would be no base to build upon.

Great care must be made to maintain a healthy research atmosphere today and that means continued investment by both the federal government and private industry.

The San Diego Union

San Diego, Calif., March 18, 1981

The question arises: What is the outlook for scientific research in this country in light of the economies being projected by the Reagan administration?

A broad fear is that scientific inquiry might suffer inordinate sacrifices because it does not enjoy a noisy constituency and a powerful lobby to compare with most other threatened programs. Would the Reagan administration commit the scientific equivalent of farmers eating their seed corn?

Fortunately, the President and his budget people seem to be more selective in their fiscal chopping than the headlines would suggest. The short-term outlook is heartening, all things considered.

Although general education will be feeling the effects of the proposed cutbacks, basic research seems to have been protected fairly well. Moreover, substantial portions of the Pentagon's budget increase are earmarked for research and development that will have a beneficial ripple consequence in the scientific community.

About 60 percent of the basic research being carried out in this country is supported by the federal government through the work of universities of which the University of California, San Diego, is a splendid example. And because basic research seeks knowledge for its own sake without regard for immediate application or economic payoff, it has suffered curtailment during recent years in both private and federal support.

Most of the applied research, together with some basic, is being carried out by private industry in an informal and natural division of research labor that is working well.

Unfortunately, the long-term trends for research in this country are discouraging. Dr. Richard C. Atkinson, UCSD chancellor and director of the National Science Foundation from 1975-80, warns that while research and development as a fraction of our gross national product dropped 19 percent from 1968 to 1980, it went up 14 percent in the Soviet Union, 16 percent in West Germany, and 19 percent in Japan. Even though the United States is still the world's leader in science, other countries are catching up rapidly.

Research is indeed the seed corn of the future. The laser is a classic example of this. Little more than a scientific toy in the 1960s, this newly discovered phenomenon now plays an essential part in eye surgery, videodisc players, satellite communications, and the production of energy by nuclear fusion. So, even though the Reagan administration is sparing research from its pruning, the known relationship between research and economic growth suggests the status quo that is not altogether satisfactory ought to be improved upon. Unlike money appropriated for many government programs, tax funds spent for research represent an investment and not an expenditure.

The Knickerbocker News

Albany, N. Y., December 4, 1981

The nation's university research laboratories are about to learn a practical and painful lesson in Reaganomics: Just as some of the most promising new fields of scientific research are holding out hope for dramatic results, Washington is ready to curtail grants to researchers whose work might lead to critical breakthroughs.

The fields of study could have enormous impact on everyday life, not just academic stature — interferon and its capacity to inhibit tumors; DNA techniques that could substantially enrich farmers' harvests, gene splicing to produce human insulin.

Work in all of these fields is being conducted nationwide and in the biology department at State University of New York at Albany, where PhDs are warning of the grave consequences ahead. In some cases, Reaganomics will mean the funding levels of research grants will be cut 15 to 30 percent, while the number of grants could be slashed in half.

In the long run, everyone will pay for these cutbacks. As Dr. Leonard Lerman, chairman of the SUNYA biology department, notes, research dollars that led to the development of the polio vaccine spared taxpayers billions of dollars by eliminating all future need for costly iron lung support systems. Similarly, research monies that supported gene splicing studies have spawned a thriving new industry that will pump billions of new dollars into the economy in years to come.

But now Washington's tight fist could wreak long-term damage to the ranks of dedicated academic researchers. Without necessary funds to support their work on campus, they may well be siphoned off to industry, which can afford to pay them as much as five times their faculty salaries. And new graduates are unlikely even to consider an academic career under such circumstances. Without young replacements, universities could face the prospect of having faculties that have "skipped a generation" and United States research could run a poor second to competing nations, particularly Japan.

In assessing the Reagan cuts, Prof. Lerman warns, "We are sacrificing the future." He is right. And it is too high a price to pay for short-sighted savings.

The Providence Journal

Providence, R. I., November 19, 1981

The channeling of millions of dollars in scientific research grants in the United States is a touchy business. It has long been assumed that scientists could agree, with a high degree of objectivity, on the merits of proposed research projects and the ability of other scientists to carry them through. Such evaluations, it was thought, could achieve an almost scientific accuracy.

The assumptions, it seems, were faulty. Last week the results of a study announced by the National Academy of Sciences (NAS) caused some embarrassment concerning the peer review system of selecting grant recipients. The experts often don't agree. Indeed, they must share their role in the process with a lady called Luck. The red faces that caused also led the NAS to withhold the study for the last two years.

The nation should have been spared the academy's timidity. The National Science Foundation (NSF), a federal agency with about $1 billion a year to distribute, doesn't quarrel with the findings. Its assistant director, Jack T. Sanderson, said the study was "very good from an analytical point of view. They did a good study, and it's right. But I don't think you're going to find any better method."

The problem feared most, apparently, is basically political. Will the administration and/or Congress seize upon the element of chance involved in awarding grants as justification for cutting back funds allocated for scientific research? That shouldn't happen. If the United States is to maintain its pre-eminence in scientific affairs (witness the preponderance of Nobel prizes for science won by Americans, for example) a pound foolish policy toward research would be counter-productive.

The authors of the study were emphatic on this point. Their findings, they said, make it "absolutely essential" to finance a "broad range of research".

It may be possible to improve the peer review process, but it is hard to see any drastic departure emerging as a substitute for the current professional assessment of project proposals and individual scientists. To "punish" the scientific community and the nation because the best, or near best, is not good enough would be posturing by public officials at the nation's expense.

The Indianapolis Star

Indianapolis, Ind., May 24, 1981

Most Americans earn their livelihood in industries that didn't even exist not many years ago — automotive, aviation, radio, television, computers, aerospace, the modern chemical, electronic and pharmaceutical industries, data processing and copying and many others.

Research and development were the key to today's complex and interrelated, technologically sophisticated industries. They can be the key to tomorrow's.

That is, they can if they get the proper push.

In R&D current U.S. spending as a percentage of gross national product is well below that of 1964 and also far below that of West Germany and Japan.

R&D suffers quickly under inflation. Outlays for research are among the first to be cut back. Since research usually takes 10 to 15 years to bear fruit, the harvests that could mean bumper crops of prosperity are scanty.

"The result of our decreasing emphasis on R&D has been a decline in our rate of economic growth and in the competitiveness of American products in international markets," Sen. John Danforth of Missouri, fourth ranking Republican on the Senate Finance Committee, said recently.

Who is not familiar with the specifics of that economic slump?

"The revitalization of American industrial innovation is the key to solving problems of persistent inflation, the inability of American producers to compete in international markets and, most importantly, our failing domestic productivity," Danforth elaborated.

"The ability of our economy to carry out technological innovation, to introduce successful new products, services and processes, is the foundation of both our domestic prosperity and our international competitiveness," he pointed out.

Productivity "supports much of our real economic growth which, in turn, permits a rising standard of living," Danforth said.

This may sound like primer-style economics. But that should not blind businessmen or labor union officials or rank and file, or anyone else interested in a healthy U.S. economy, to the fact that it also is emphatically true.

Where would the U.S. economy be without TV, jets, computers, photocopiers, radar, communications satellites, microwaves, transistors, miniaturization, plastics, self-developing film and the miracle drugs?

It might well be where it may well be in a decade or so unless R&D gets a new infusion of economic blood.

Danforth urges strong new tax incentives — which go beyond the 15 percent business tax reductions proposed — to stimulate economic growth.

Those who favor healthy economic growth should heartily endorse his recommendations, and make their feelings known — emphatically — in Congress.

Reagan Increases Research Outlay, but in Narrowly Defined Areas

President Ronald Reagan's budget plan for fiscal 1983 (to begin in October 1982) were clearer on the subject of R & D than his budget for the previous fiscal year. Total spending on R & D would rise to $44.4 billion, but only specific areas would receive federal funds. Other forms of research would have to rely on private industry or university sponsorship. Most of the federal R & D increase was defense-related, in keeping with Reagan's major military buildup program. According to an analysis of the budget by the American Association for the Advancement of Science and other science organizations, spending for non-defense R & D would actually decline in real (inflation-adjusted) terms. Especially hard-hit were projects in the social sciences. The analysis observed that federal research funding was divided into two major categories: one serving direct government needs (defense and environmental management) and the other serving national needs (agriculture, energy, health), which affected the economy as a whole.

San Francisco Chronicle

San Francisco, Calif., January 6, 1982

THE REAGAN ADMINISTRATION has next to nothing to say to the American scientific community beyond advising it to tighten its belt and get ready for ever sharper cuts in the financing of scientific research. This warning to get along with less was the opening greeting of the White House to the American Association for the Advancement of Science at its annual meeting in Washington this week and we think it is shocking. So do the scientists.

This country is in a crisis of support for scientific research. Says the president of the AAAS, D. Allan Bromley of Yale: "The fraction of our resources now being invested in long-term research and development is at a low ebb. and it is now less than at any time since World War II."

Just yesterday the prime minister's office in Tokyo proudly announced that Japan is spending $23.9 billion on research, up 14.5 percent from last year and amounting to 2.19 percent of gross national product. Though Japan is still second to the U. S. in total research spending, its GNP figure dwarfs ours. According to the House Ways and Means Subcommittee on Trade, ours is only 1.39 percent of American GNP.

WE CAN'T MEET the Japanese economic challenge unless the government leads the way to making the U. S. R&D effort more competitive, says the subcommittee.

Sometimes one gets the impression that the government's budgetmakers consider funding for science and technology an esoteric waste unless it's for weapons. American scientists are trying to tell the budgeters that this is no way to sharpen U. S. competitiveness in this high-tech world, and Ways and Means is doing its best to back them up.

The Washington Post

Washington, D. C., February 3, 1982

AS THE ANNUAL struggle over the federal budget gets under way, it is worth pondering what will happen to money for science. American science and technology are still the best in the world. But there are enough signs of strain to suggest that our accustomed preeminence—on which a large part of U.S. security and economic power depends—is fragile, even endangered.

The trouble starts with education. For more than a decade, secondary school curriculum requirements and achievement have fallen sharply in science and mathematics, while an opposite trend has been present in most other developed countries. The result is already evident. A NASA official reported recently, for example, that the space agency's cost overruns come in part from delays that are, in turn, the result of a lack of technically skilled workers.

Federal support for graduate education is in doubt for the first time in 30 years. In many fields, engineers are in short supply, but engineering schools cannot take in more students because they cannot find trained faculty to teach them. Shortage of faculty means heavier teaching loads and therefore less research. Schools do not have enough money to pay more professors even if these could be found, nor can they replace badly obsolete laboratories. Troubles that now afflict engineering are beginning to be seen in the sciences as well.

Money for basic research in this country has been essentially constant for 10 years. To the extent that scientific advance is linked to money—there is a close, but not rigid relationship—that means a decade without real growth. Meanwhile increases in research funds in Japan, West Germany, France and elsewhere have paid off with growth in both scientific and industrial productivity. And now federal research budgets face severe cuts.

Basic research, a long-range investment for the benefit of all of society, is properly and necessarily the responsibility of the federal government. Industry can be asked to expand its support of applied research and of development projects, but it is not industry's role, nor is the industrial setting the best environment for basic research. Yet less than 15 percent of federal research and development funds currently goes to basic research. Too much federal money supports development projects that are the proper province of industry. Allen Bromley, president of the American Association for the Advancement of Science, recently made the valuable suggestion that the ubiquitous but misleading term "R&D" be dropped, in order to separate the financing of these two very different activities.

If serious damage is to be avoided as the federal budget is cut, Congress and the administration should not only protect, but in some fields increase, basic research funds. Ways should also be explored to assure more continuity in the amount of support such research is given. It takes nine years to produce a PhD scientist, and years to assemble a research team and carry a project to fruition. When the money disappears for a few years, the people disappear too, and can seldom be brought back. Abrupt changes like those that took place during last year's budget cycle can therefore wipe out years of past investment and future productivity.

Newsday

Long Island, N. Y., January 25, 1982

"America's scientific and technological achievements have helped to make our country a leader among nations," President Reagan told the American Association for the Advancement of Science recently in a message read by his science adviser, George Keyworth. "These accomplishments contribute to human progress and assist us in addressing the complex problems facing the world in the years ahead."

But Keyworth also carried another message that was much less palatable to his audience. Scientists' decisions, he said, "must be carried out in the context of other national policies such as those concerning national security, international relations, energy, social sciences and the economy."

So in one breath the administration recognized the vital importance—to the country and the world—of continued advancement in scientific knowledge. And in another it took that recognition away.

Everyone at the AAAS meeting knew that in most of the areas Keyworth mentioned, notably excepting "national security," traumatic budget cuts had already been made. Since federal dollars are the necessary fuel of much of American science, his meaning was clear. Any lingering fog about administration attitudes was dispelled when he told the scientists they "cannot expect to be pre-eminent in all fields, nor is it necessarily desirable."

Even if that's true in the narrowest sense, lecturing scientists on limits is hardly in the spirit of a country that has done so much to unlock the mysteries of atoms and cells, to wipe disease from the planet, to explore interplanetary reaches, to plumb the universe of its secrets for the sake of human betterment.

Surely this is no time for American scientists and engineers to give up their quest for pre-eminence. That's too great a departure from a tradition of excellence, and false economy in the bargain.

The Oregonian

Portland, Ore., January 12, 1982

In dollars spent, the United States seems to be doing better at research and development than any nation but the Soviet Union, according to figures in the 13th annual report of the National Science Board released by President Reagan. But the sticker is the inflation that has reduced the $69 billion estimated 1980 expenditures to $35 billion in 1972 dollars.

In attempting to combat inflation by balancing the budget, the administration is cutting back federal R&D expenditures in the new budget so that there will be a further shrinkage of real effort. While 70 percent of R&D comes from the private sector, this spending also is slowing down due to the general recession, except in areas of military contract expansions.

While dollars don't always say what is being attempted or accomplished, they are reliable barometers of the national will. Our major competitors outside the Soviet Union, notably Japan and West Germany, are increasing R&D spending faster than their economies are growing, the report found. Both nations experienced rapid gross national product growth in the 1970s, making their percentage increases against their GNPs more remarkable.

In looking at capital investment as a percentage of output, the study found the United States has been lower both in manufacturing and the total economy than any other major industrial nation. This reflects the fact that the United States has experienced slower growth rates in manufacturing productivity than have most industrialized nations.

When all these figures are translated into human terms, they spell discouragement for graduate science and engineering students, deflecting some of the brightest into other fields. Scientific declines will cost future job opportunities for many other skills in the labor force and will directly threaten the nation's long-term economic health, if not its security.

This loss of technical leadership would prove far more damaging to the nation's security than would the failure to finance expensive arms programs, such as the B-1B bomber, super tanks or costly deployments of the MX missile.

While science efforts need some belt tightening and a fresh look at priorities in a time of federal fiscal cuts, the danger is that too much breath will be squeezed out of the very asset needed to conquer low productivity, inflation and unemployment.

The Arizona Republic

Phoenix, Ariz., January 17, 1982

AFTER years of neglect, the United States is beginning to make a larger financial investment in research and development.

Failure to modernize production equipment and facilities has allowed Japan and other nations to reach and, at times, surpass America's technological level.

Studies now indicate a dramatic turnaround. The United States is expected to commit between $750 billion and $1 trillion to R&D during the 1980s.

During the 1960s, Washington contributed most of the funds. But industry has steadily increased its share to where it now contributes about half the national total.

Most new R&D will, as in the past, relate to defense. But there are spinoffs affecting other important research.

But, still lacking is overall coordination of the direction of basic R&D.

Although decentralization of Washington power is a major goal of the Reagan administration, some cooperation among the nation's research communities is essential.

This would accelerate the ability to meet technological competition from abroad, and also help U.S. firms save research funds.

The Dallas Morning News

Dallas, Texas, January 26, 1982

We are well aware that what one person sees as fat in a government budget is what another regards as a treasured program. But the decline in funding of non-defense-related research and development in the federal budget has scientists worried.

Defense and space R&D rose this year. But all other federally backed R&D was reduced about 8 percent, according to the National Science Foundation.

There is merit to the argument that private industry should shoulder a larger share of R&D. And this sector does carry a considerable load in product-oriented research. But in our technological world, a solid base of pure research must be maintained, for it is often the base for project-oriented research.

Scientific research can't be turned on and off erratically. A pool of talented scientists must be maintained both for teaching and for directing projects. When constant funding is not available, promising students aren't attracted to research and dedicated teachers move to other fields.

Already there are indications that the Russians and our international economic competitors are moving in on our technological leadership. If we are not very careful in the way we cut federal R&D, the edge we have enjoyed will be in danger of disappearing.

THE CHRISTIAN SCIENCE MONITOR

Boston, Mass., January 5, 1982

Of all the money spent by private industry or the US government, perhaps none is more vital to the long-range prosperity of the United States than "R&D" dollars – investments in basic research and development. After years of neglect, American industry is now plowing more money into invention and product development than ever, an upswing that began in the late 1970s. At the same time, government funds for research in a broad range of scientific areas other than defense are being reduced. Perhaps most disturbing, there is as yet no firm strategy regarding basic research in the US.

Clearly, whether the US maintains its world economic leadership will in great measure be determined by the extent to which it stays abreast of the profound technological changes now transforming the industrial world. But to do this will require that business firms, major universities, national foundations, scientific laboratories, and the federal government itself set aside sufficient funds for new research. And this in turn is dependent on the degree to which the US develops an engineering and scientific community equal to the momentous challenges that will be faced in exploring the space-age technology of the future.

Unfortunately, it is not yet certain that the US will meet the R&D challenge as well as it should, though there are some promising indications of addressing it. Over the course of the decade, spending on research and development is expected to total something on the order of $700 billion to $1 trillion, no paltry sum. Industry now accounts for close to half the total annual R&D budget, compared to one third back in the 1960s when government was the major contributor.

But a significant share of the industry budget will flow into defense-related research. Surely there must be no scrimping on appropriate research for defense. But military spending must not be allowed to become such an enormous drag on the civilian sector of the economy as to divert professional workers and scarce resources away from promising scientific inquiry in nondefense fields.

Many of the federal cutbacks are dismaying. That is especially the case for "basic research" as well as projects related to small businesses, which tend to generate most new jobs within the American workforce. Thus funds for nonnuclear energy programs, including solar power, space research, and conservation research, are being slashed. At the same time federally supported research in the social sciences is being even more pointedly reduced.

It is also cause for concern that the government is cutting back on its support for engineering and scientific education research projects. There is now, for example, a scarcity of engineers receiving doctoral degrees, and many college-level teaching positions are unfilled. By contrast, Japan, America's major competitor in the electronics, high-technology area, is devoting special attention to boosting its professional workforce of engineers. Moreover, the US has not kept up with the modernization of equipment and facilities in its engineering and research laboratories.

Most serious of all, however, is the fact that there is currently no effort underway in the US for coordinated planning involving R&D expenditures. That is not to say that all such funds should fit neatly into some sort of federal "master plan." Such a centralized approach would garner little public support during a period of federal deregulation.

But at a time when budgets are lean, a modicum of government, industry, and university planning in the research area – if only by reaching a broad consensus that certain forms of research need more focused support than others – seems logical. Such an approach is taken to an extent by the Japanese. One need only think back to the 1950s and 1960s when America established its highly successful space program to realize that, when all parties to a scientific inquiry put their skills to a task, remarkable results can occur.

AKRON BEACON JOURNAL

Akron, Ohio, January 20, 1982

JOHN GLENN is right: It is penny wisdom and pound foolishness for the U. S. government to cut sharply back, as the Reagan administration plans to do, on its support of research and development.

As the Ohio senator points out, it is "basic R&D that has made this country what it is today," and without it "we will become a second-rate power in the world."

Sen. Glenn

Truly basic research — the area that yields the most fundamental and crucial breakthroughs in science and technology — cannot safely be left to private business and industry, as President Reagan would prefer to do with many things in which the government has in recent times played a large role.

Although some have done it anyway, private corporations have little economic incentive to fund the most basic kinds of research — because the yields from it are altogether unpredictable and often cannot be patented and turned to the exclusive profit of those whose support made them possible.

And because of the huge costs entailed in fundamental research in many areas of great promise, the universities, seedbed of the greatest advances in science and technology, cannot be expected to do what is necessary without outside support.

Thus there is a double necessary role for government in this field: To support basic research in Academia, despite such barbs as this occasionally draws from Sen. William Proxmire in his "Golden Fleece" awards, and to keep up research of its own in such agencies, for example, as the National Aeronautics and Space Administration and the National Institutes of Health.

In the long term, the benefits to be derived from this kind of activity are huge by comparison with their cost to the taxpayers — a tiny fraction of the federal budget, even when support has been the strongest.

And to be fruitful, research and development work must keep moving at a steady and secure pace. It cannot be momentarily "turned off" at a time of economic stress without disastrous loss of talented people and crippling interruptions.

It may be possible to trim waste in this work, though waste in it is peculiarly hard to define and identify because it is so hard to tell which avenues of research are likely to be fruitful and which are not.

But jolts like a reduction from the 1980 level of $712 million for energy conservation programs to the proposed 1983 level of $19 million, for example, do not merely trim and delay the work — they destroy it.

As Sen. Glenn points out, none of this means that "American scientific enterprise is in immediate threat of extinction."

Americans have long been dominant in Nobel scientific awards. We are still spending more on R&D than our four top economic competitors among the Free World nations combined.

And though there has been much hand-wringing over the slowdown in productivity gains in this country compared with certain others, U. S. productivity is still greater than that on any of its rival nations — half again that of Japan, for example.

But the key element in this has been the plowing back of money into research and development that has, as Sen. Glenn put it, transformed this nation "from a scientific backwater in the early part of this century into the dominant technological force in the world."

In recent times private industry in this country has been cutting back on R&D budgets. Now the federal government appears to be bent on doing so even more severely — while other developed nations are greatly increasing their efforts to catch up with and pass us.

To let this happen is myopic, failing to look ahead and see the decline it is sure to mean. The pounds to be gained — or lost if we fail to maintain the effort — too far outweigh the pennies that would be saved.

TULSA WORLD

Tulsa, Okla.,

October 11, 1982

THE announcement of the 1981 Nobel Prizes for medicine points out an important aspect of America's technical superiority. For years, this country has offered the world the most fertile ground for scientific and technological research. It is crucial that despite federal budget cuts and overall belt-tightening, that continue to be true.

The Nobel Prize winners announced Friday were Harvard University professors David H. Hubel and Torsten N. Wiesel and Dr. Roger W. Sperry of California Institute of Technology.

Hubel and Wiesel were recognized for their work relating sight stimulation in infancy to future vision. Sperry was honored for work demonstrating the division of labor in the brian.

Though all three men conduct their research at U.S. universities, only Sperry is a native American. Hubel was born in Windsor, Ontario, attended McGill University in Montreal and is now a naturalized U.S. citizen. Wiesel is a citizen of Sweden.

Since they were first given out in 1901, Americans have received about one-third of the awards. But of the non-Americans awarded Nobel Prizes, many conducted their research at U.S. colleges and universities.

In the past decade, however, America has cut back on the percentage of its Gross National Product devoted to research and development.

More money, especially from business and industry, needs to be channeled into technlogical study. Oklahoma saw a positive sign in this respect last week with a $30 million gift to the University of Oklahoma for an energy research center.

This kind of private giving may the only hope for preventing a reverse "brain drain" that would send American scientists scampering to other countries in search of support they cannot find at home.

The Hartford Courant

Hartford, Conn., February 22, 1982

There is nothing unique in this country about a private industry, like the Celanese Corp. of New York, awarding a university like Yale a $1.1 million grant.

It's done all the time, not only at private, but at public, institutions. The University of Connecticut, for example, has current research grants from such companies as Dupont, Monsanto, and Shell Development Corp., totaling more than $1 million.

One of the distinctive aspects of the Celanese grant, however, is that it is for basic research, which might or might not produce anything commercially useful for the industry. The research will be on enzymes, which are complex proteins produced by living cells and that produce biochemical reactions.

Most of the basic research at universities has traditionally been conducted with money from the federal government. That money has been slowly drying up in recent years, while an increasing proportion of government resources has been allocated to more pragmatic research.

There has been no shortage of money available for defense-related research, and the current arms buildup, including the renewed interest in chemical and biological weapons, could mean even more funds for scientists interested in that kind of thing.

The Celanese grant, although the company no doubt hopes for an eventual payoff, is a boon to basic biological research. University officials obviously want to see more industries provide such grants.

At the same time, such grants are a symptom of the growing obfuscation of the scientific mission at universities.

Still unsettled is the vehement controversy over the propriety of the increasing involvement of universities in the profit-making field of genetic engineering. And Harvard University, which last year rejected a proposal to start its own genetic research company to develop commercial products, nevertheless has accepted controversial, multi-million-dollar contracts from other private industries.

Faculties at other schools, such as the Massachusetts Institute of Technology and Stanford, are also wrestling with the difficult question of where to draw the line on corporate influence on the direction of research.

Meanwhile, the integrity of university science is being threatened by the federal government, which is not only a heavy investor in directed research but, in the name of national security, would like to stanch the free flow of information about research results that could be of any use to the Soviet Union.

In sum, the trends seem to be toward research that is either directed or influenced by pragmatic corporate and government goals, toward the increasing commercialization of science on university campuses and toward more secrecy to preserve some commercial, military or political advantage.

We don't know where all this is leading, but it doesn't seem to have much to do with the traditional role of the university as the only institution where curiosity and disinterested inquiry would find continuing support. It doesn't seem to much reflect the traditional definition of science offered by Dr. Linus Pauling, among others, as "the search for truth."

The Chattanooga Times

Chattanooga, Tenn., June 21, 1982

In a speech at UTC last week, Rep. Marilyn Bouquard performed a public service in calling attention to the fact that the Republican-backed budget resolution which passed the House recently contains "dramatic and disastrous" reductions in federal support for government research and development. At a time when the United States' ability to retain its impressive record of scientific and technological achievement is in doubt, it hardly makes sense to prune funding for research.

Mrs. Bouquard reiterated her belief that government has a "legitimate and vital role" in R&D, and pointed out that "the private sector is too ill-equipped or too undercapitalized to take the risks for many projects which have only a long-term payoff." The latter is especially true now, when high interest rates make the cost of long-term debt prohibitive. And if the trend of reduced private and federal spending for research as a percentage of the gross national product continues, the United States could find itself losing ground steadily to other nations, including the Soviet Union.

The proportion of the GNP spent on research and development in this country has dropped more than 20 percent since 1965, she said. Yet by contrast, R&D investment by the Soviet Union, Japan and West Germany has jumped by 21 percent, 27 percent and an astounding 41 percent, respectively. The figures for amounts invested in basic research by private industry and the federal government are equally depressing.

It is bad enough, as Mrs. Bouquard made clear, that reduced R&D funding could cost the U.S. its lead in developing essential technologies for the decade ahead. Even worse is the possibility that a policy of starving fundamental scientific research could imperil our national security.

To help head off both dangers, Mrs. Bouquard called attention to need for a renewed commitment to enhance the teaching of science and mathematics in the public schools, a campaign that will require stiffer educational standards — and more money to attract and retain qualified teachers. The success of the former hinges on the latter. But the latter is feasible if we get away from the insistence that raises for teachers be made across the board, regardless of ability, and concentrate more ensuring that valuable teachers now being lost to private industry are kept in the classroom, there to help train future scientists.

What Price Secrecy in Technology & Science?

Does secrecy hinder scientific research, or does it protect a country's valuable technological discoveries from "theft" by commercial or military rivals? In an open society like America's, keeping information secret is distasteful, tolerated only when there is a proven need. Secrecy hinders a scientist's work, for if a researcher cannot learn what his colleagues are doing, he runs the risk of needlessly duplicating experiments or wasting time on a fruitless inquiry. However, the world of technology is a world of competition, and industries and governments are not above trying to take "short cuts" by obtaining a rival's results. It is therefore natural for a company to want to keep secret the plans for a profitable invention, just as it is natural for a government to want to keep military secrets from an enemy. Individual scientists, too, keep secrets, lest another colleague beat them to an important discovery and carry off the honors.

In early 1982, Central Intelligence Agency Director Adm. Bobby Inman warned scientists that the U.S. faced a "hemorrhage of the country's technology," in particular, to the benefit of the U.S.S.R. Speaking to the American Association for the Advancement of Science, he asserted that in the military sphere, "the bulk of new technology which they have employed has been acquired from the United States." Inman warned that continued leaks would lead to government censorship unless scientists cooperated by setting voluntary limits on works published in scholarly journals.

Houston Chronicle

Houston, Texas, November 11, 1982

Science, in its purest form, is the systematic acquisition, analysis and dissemination of knowledge. It is best served by the free flow of information within the nation and the global community, and any attempt to censor or control this knowledge would inhibit the work of researchers and the advancement of science in the world.

The deputy director of the CIA, Adm. Bobby R. Inman, concerned about the flow of scientific and technical knowledge from the United States to the Soviet Union, recently warned U.S. scientists of a "tidal wave" of public outrage and legal restrictions if scientists do not agree to voluntary review of their work by U.S. intelligence agencies prior to the start of research and prior to publication of the findings.

Scientists working on secret defense projects are already subject to normal security and classified information procedures, but to place the flow of all scientific publication under the control of the CIA and other secret intelligence agencies, voluntarily or otherwise, would seriously damage the quest for knowledge and inhibit scientists from pursuing projects in fields that are vital for the nation's economy and security.

Scientists do not work in a vacuum. Their work depends on knowledge and techniques acquired from many sources, and their work is given validity under a system known as "peer review," in which the published findings of scientists are tested and analyzed by others.

It is true that a totalitarian police state such as the Soviet Union enjoys certain advantages: It can share the knowledge of the Free World without having to share advances made within its borders. But this advantage is achieved through methods that the United States and free societies cannot afford to emulate.

The deputy director of the CIA is professionally occupied with the security of the nation, but to place the work of researchers in the fields of "computer science, electronics, lasers, crop projection and manufacturing procedure" under an inhibiting system of censorship would be more damaging to the cause of national security and prosperity than the sharing of non-classified U.S. technical knowledge and pure research with the world.

Roanoke Times & World-News

Roanoke, Va., January 13, 1982

THE RUSSIANS are getting a lot of technical information from the United States and putting it to use in their military buildup, says Adm. Bobby R. Inman. The deputy director of the CIA thinks a key means of stopping this is for scientists to let U.S. intelligence agents examine their papers before they're published. They should do this voluntarily — or else.

That was the message Adm. Inman delivered recently to a panel session at the annual meeting of the American Association for the Advancement of Science. He said congressional investigations now in progress will demonstrate that as the Soviets have expanded their military, "the bulk of new technology which they have employed has been acquired from the United States."

Part of his remedy would be an intelligence review of scientists' work to see if any of it should be stamped secret. If scientists don't agree to this, he predicts a "tidal wave" of public outrage and of laws restricting their work.

Apart from his blatant attempt to throw fear into the scientific community, the admiral's approach is wrong on a couple of counts. For one thing, it implies that scientists are somehow responsible for what he calls a "hemorrhage of the country's technology." They're not.

The Soviets get technology from the West mainly by purchase of our goods and by reading our technical publications. In most instances there's no way to predict or control use. A computer and its programs can be employed in many ways, in both military and civilian sectors. Maybe the United States would want to choke off sales of such equipment — although that seems doubtful — but could a free country effectively police all of the hundreds of publications in which technical information is printed? Would it want to?

Another problem is that keeping scientific knowledge secret for very long is virtually impossible. No country has a monopoly on brains or resources. It frequently happens that scientists in different countries, who don't even know of each other's existence, arrive at similar findings near the same time.

It can make sense not to broadcast information on especially sensitive matters with a strictly military application, like the H-bomb formula; but even data in so narrow an area as this cannot be indefinitely bottled up. The kind of lid Adm. Inman wants to clamp on scientific information could never spread wide enough or hold tight enough to be effective. It is undignified and inappropriate for him to threaten scientists with a backlash in public opinion. The public understands this situation better than he thinks.

WORCESTER TELEGRAM.

Worcester, Mass., January 22, 1982

A Senate subcommittee has given its blessing to a piece of legislative mischief that would seriously endanger the free flow of information about government activities.

Worked up by the administration and Sen. Orrin Hatch, chairman of the Senate Subcommittee on the Constitution, the bill would eliminate much of the public oversight of federal agencies by limiting access to currently available information on a wide variety of activities. Among other things it would allow the government to suppress information in of its files on grounds of "personal privacy," including secret lists of government contractors and consultants.

One of the most objectionable features of the Hatch-Reagan bill is that it would establish, for the first time in American history, the concept of a government copyright on information developed at taxpayers' expense.

Under present law, government-generated research reports and technological and other information developed by the government at public expense are available simply for the cost of finding and copying them. That's as it should be; the public pays to have the information collected and the public ought to be the owner of it.

Since the government was founded, it has never been permitted to assert a proprietary interest in the information it has gathered. Congress has specifically barred the government from asserting copyrights for its information and from charging royalties for it.

The Senate bill would establish a dangerous principle in setting high royalty fees for citizen use of any "commercially valuable technological information" requested under the Freedom of Information Act.

If government technological information can be copyrighted, it is only another step to setting royalty fees for government research reports and studies and financial analyses.

There are other dangerous provisions of Hatch-Reagan. The government would be allowed to keep secret from the public the details of virtually all lawsuits it settles, including settlements accomplished because of improper political pressure or conflicts of interest. It would be allowed to keep secret from the public information about criminal violations committed by federal investigatory agencies. It would be allowed to suppress law enforcement records where there is no legitimate investigation, including illegal disruption of domestic political groups. It would be allowed to suppress information that "would disclose the identity of a confidential source."

The bill is headed for the full Senate Judiciary Committee where it should die a quick death.

Pittsburgh Post-Gazette

Pittsburgh, Pa., March 15, 1982

The Reagan administration has decided to try to throw a blanket of censorship over American high technology that might find its way to foreign countries, particularly the Soviet Union. But the administration's medicine could be worse than the disease.

Admiral Bobby Ray Inman, the deputy director of central intelligence, proposes that even some private scientific and technical research — *not* pursued with government financial support — should sometimes be covered by a state secrets law.

Admiral Inman wants to reduce the accessibility of substantial amounts of highly technical data through published sources that Eastern bloc scientists can now read in U.S. libraries. Reasoning that U.S. technological advantages are vital to national security, the admiral proposes a government screening of articles published about research related to computers, semi-conductors and other high-technology products and processes.

Many American scientific researchers fear that the censoring of basic research would be carried out by a monstrously inefficient bureaucracy that would stifle the very research on which American high technology depends, because superior technology arises from a free exchange of ideas. As Peter J. Denning, head of Purdue University's Department of Computer Science, noted recently: "If you want to win the Indy 500, you build the fastest car. You don't throw nails on the track."

The Inman approach is logical and springs directly from a view that the United States is giving the rest of the world too much to read about basic human knowledge. But history also makes it clear that scientific innovation inevitably becomes a commonly held patrimony. Equally important is the fact that the problem of high technology being passed to the Soviets has as much to do with the failure of the government to regulate the export of sensitive products as with the gleaning of information by Soviet scholars from U.S. publications.

The outstanding "loss" of American technology to the Soviets in the 1970s occurred when the Nixon administration approved the sale of machine tools to grind the precision ball bearings now being used in the inertial guidance systems of the most accurate Soviet intercontinental ballistic missiles.

For a vexing problem, the solution is not censorship of basic research but more vigilance by the government in its own sphere.

The Houston Post

Houston, Texas, January 27, 1982

One of the toughest security problems facing the United States and its allies is how to keep our high technology with potential military applications from reaching the Soviet bloc. Two recent initiatives by the Reagan administration indicate the seriousness of its commitment to tighten control over the transfer of our most advanced technology to the Soviet Union and its satellites.

At a meeting in Paris last week, the United States won an agreement by our Western European allies and Japan to redefine guidelines for technological exports, ranging from ball bearings to metalurgical processes, behind the Iron Curtain. Though the meeting was termed a success, many differences reportedly remain over how strictly the 30-year-old guidelines should have been redrawn.

Shortly before the Paris meeting, a top Central Intelligence Agency official proposed that U.S. scientists working in certain sensitive fields voluntarily submit their research for censorship by intelligence agencies. Adm. Bobby Inman, deputy CIA director, told a convention of the American Association for the Advancement of Science that there is a "hemorrhage" of this country's technology and that Soviet military advances of recent years have been based largely on the work of U.S. scientists. Inman suggested that scientists in certain fields submit their work both before research begins and before publication.

The reaction of the scientific community to the censorship idea ranged from skeptical to hostile. "What alarms scientists about the (Inman proposal)," said William Carey, executive officer of the AAAS, "is that once science accepts the government's right of prior restraint, the programs are carried out by individuals in the national security establishment. They resolve questions, where there is doubt, on the side of censorship rather than the freedom of scientists."

A White House spokesman said the administration is not considering a mandatory government review of scientific papers. But Inman wants scientists working in computers, electronics, lasers, crop projections and various manufacturing processes to submit their work to intelligence agencies.

The Soviets and their allies have engaged in a long, intensive campaign to obtain the cream of Western technology through outright purchases, theft or simply by reading scientific journals and government documents that are open to the public in free societies. But while the openness of our system makes accessibility to much scientific data easy for Soviet espionage agents, it also facilitates the exchange of information and ideas within the scientific community. It is that freedom of communication that has helped make our technology superior to the Soviet bloc's.

Classifying scientific material on the basis of its national security value would mean passing judgment on a huge volume of research. While there are certainly legitimate uses for the top secret classification for some sensitive material, an extensive review program by intelligence agencies could slow and, in some cases, stifle potentially valuable research. The Soviet Union has paid a far higher price for its pathological practice of secrecy than we have with our openness. If we adopted overzealous practices to keep our high tech research out of their hands, we could ultimately become the big losers.

The Des Moines Register

Des Moines, Iowa, January 25, 1982

During the Red Scare of the 1950s, many Americans were obsessed with the need to protect scientific secrets from the Russians. The fever of that obsession has abated today, but the malady lingers on.

In a recent address to the American Association for the Advancement of Science, Admiral Bobby Inman, deputy director of the Central Intelligence Agency, said Soviet military advances in recent years have been based in large measure on the published work of U.S. scientists.

To prevent this "hemorrhage of the country's technology," Inman urged scientists to voluntarily submit their research for censorship by intelligence agencies "prior to the start of research and prior to publication." Unless they agree to such "reviews" of their work, scientists will encounter a "tidal wave" of public outrage, resulting in tough, restrictive laws, Inman warned.

A similar tack was taken in a recent letter to the editor of Science magazine by Deputy Defense Secretary Frank Carlucci, who asserted that U.S.-Soviet information exchanges are often one-sided, with the Americans supplying more information than they receive.

Inman's proposal is more deserving of a tidal wave of protest than the problem he hopes to remedy. A spokesman for Stanford University makes the telling point that censoring publications to keep information from Moscow will keep it from Americans as well. This would punish them as much as those the censorship is meant to keep in the dark.

Inman, Carlucci and others should realize, too, that scientific progress requires, and a free society means, that the free exchange of ideas will not be impeded, except in extraordinary circumstances such as war. The United States will not protect its science or preserve its democratic traditions by adopting censorship and mimicking other tactics of totalitarian societies.

Keeping scientific secrets is a chancy business by its very nature. C.P. Snow, eminent British novelist, scientist and government official, put the matter bluntly: "There are no secrets in science, and very few, and those short-lived, in technology."

The attempt to preserve them is not only obsolete, un-American and anti-scientific, but futile. The United States can only damage itself by adopting such a policy. Let American scientists do what they do best: find and report the truth.

The Hartford Courant

Hartford, Conn., January 19, 1982

Adm. Bobby R. Inman seems to be the kind of man who would have kept the invention of the wheel a secret because an adversary might one day invent a cannon to mount on it.

He suggested virtually the same thing at the recent meeting of the American Association for the Advancement of Science in Washington.

Scientists, he said, ought to submit their work — "prior to the start of research and prior to publication" — to U.S. intelligence agencies for censorship of anything the agencies think could be used by the Soviets to the detriment of the United States.

That took a surplusage of temerity even in this administration, whose officials frequently exercise the art of extravagant expression.

With a few obvious exceptions, no scientist who cherishes his independence would voluntarily, as Adm. Inman suggested, submit his work for evaluation by spies. And any mandatory program of government review, besides being a nightmare to administer, would run into constitutional problems.

Adm. Inman would have a review system apply to everything from computer hardware and software, to electronic gear and techniques, lasers, crop projections and manufacturing procedures. He would, in short, turn the entire scientific establishment into an arm of the state, just as the Soviets have, as a means to prevent the Soviets from gaining any advantage from U.S. technology.

He also ignores the fact that the reason American technology and science are still without equal in the world is because they can still be pursued in the spirit of free inquiry. To staunch the publication of the results of research would limit the advances that are frequently based on prior work.

If Adm. Inman's proposals were ever implemented, they would turn science into a cloying and secretive enterprise. And then he wouldn't have to worry about leakage of American science and technology to the Soviets. There wouldn't be any worth leaking.

The Boston Globe

Boston, Mass., January 28, 1982

Citing cases in which the Soviet Union supposedly gained militarily by acquiring US high technology, either equipment or information, the Reagan Administration seems headed toward more comprehensive controls over the scientific community.

While the country obviously protect genuine military secrets, Congress and the public should be wary of secrecy policies that will hobble scientific research and undermine further technical advances that build a significant American leadership.

The alarm was sounded earlier this month by Adm. B. R. Inman, deputy director of the Central Intelligence Agency, in an address to the annual meeting of the American Association for the Advancement of Science; he cautioned scientists on the need for more stringent security reviews of their work to prevent exploitation by the Soviet Union. It was echoed two weeks later in an essay by Caspar W. Weinberger, Secretary of Defense, published in the Wall Street Journal.

In each case plausible arguments were offered for increased awareness of the issue by academic and corporate scientists and engineers, especially in the fields of weaponry and communications. In Inman's address, the field was broadened somewhat to include cases where "certain technical information could affect the national security in a harmful way. Examples include computer hardware and software, other electronic gear and techniques, lasers, crop projections, and manufacturing procedures."

Much of the information to which they allude appears in scientic journals or is built into equipment available on the open market. It is read and purchased not only by the Soviet Union, sometimes through straws in other countries, but is also read and purchased by Americans for their own use – and growth.

Weinberger in particular has been actively urging American allies to take seriously the dangers of allowing the Soviet Union access to such information and products. The idea is apparently to construct a technological membrane through which no sensitive material might pass.

Given the enormous numbers of channels through which such information and products pass all over the world, the task seems impossible without sharply curtailing both legitimate communication within the scientific community and interfering with normal commercial activities – to the detriment of both.

It is important to bear in mind that the Inman-Weinberger proposals are not directed primarily at information about such long-standing secrets as thermonuclear weaponry. They are directed at discussions at the fringe of computer development and use; at manufacturing techniques for miniaturization that has led to the explosion of computer-on-a-chip technology; at programming for a host of applications. All of them are widely used in commercial applications as mundane as elaborate computer war games.

Such developments flourish in an atmosphere that combines competition with free flow of ideas and information. The world proposed by Inman and Weinberger, although they promise no excesses, has a decidedly different cast – one of self-policing if possible and bureaucratic policing if necessary. If the latter develops, as will almost surely be the case, then penalties will attach to those deemed in violation. Scientists and engineers will undoubtedly spend (waste?) some of their time looking over their shoulders for the censors.

Creativity may not dry up in such a world, but it impossible to believe that it will not be diminished. "Secrets" will still not be kept much better than they are today, in all likelihood. In that event, we will have the worst of both cases, to the detriment of the most dynamic sector of our scientific and technological society.

Newsday

Long Island, N. Y., January 22, 1982

There may be the germ of a good idea in a rather vague proposal made recently by Bobby Inman, the CIA's deputy director. It's aimed at denying hostile foreign powers access to scholarly research that could help them advance their military technology. But it needs to be thought out much more carefully.

We can see the advantage of warning scientists that publication of some of their research might threaten the nation's security in ways they hadn't anticipated.

Yet we're also deeply troubled by any plan that would involve sending scientific papers to Washington for review before publication. Unless it were entirely voluntary, it would put at risk Americans' rights to speak and print whatever they wish. By hampering scholarly discussion, it might do far more harm than good to scientific research. It would require a new bureaucracy of science monitors. And even if it began as a voluntary system, it could easily develop a momentum that would nudge it along, a step at a time, toward compulsory censorship.

Several of the Reagan administration's actions also suggest that skepticism is warranted: Within the past few weeks, it has sought tight new controls on official contacts with the press, proposed curtailing access to government files and suggested barring foreign students from research projects that might result in technology leaks to their home countries. Taken together with Inman's proposal, all this suggests an administration with a budding fetish for security, and not much concern for the price.

Still, we can easily imagine an independent researcher unknowingly turning out a paper on some subject—lasers, say—that would help an enemy develop a new weapons system or counter an American one. Inman claims that upcoming congressional testimony will reveal shocking instances of just such inadvertent exposures.

If those examples do indeed demand some corrective, then it *might* be acceptable to set up a screening bureau and ask that research in specified areas be sent, voluntarily, for advisory review.

But the safeguards would have to be clear: The subjects covered would have to be inextricably linked to security concerns. Submission would be up to the researcher; review would be prompt, and—to avoid conflict with First Amendment guarantees—researchers would be free to reject any suggested changes.

And even then we'd be none too comfortable. Whether the discomfort would be worth living with depends on how convincing a case can be made for any kind of security review.

THE CHRISTIAN SCIENCE MONITOR

Boston, Mass., March 30, 1982

The free flow of information and ideas is vital to the advancement of science and technology. No nation has provided better proof than the United States. To maintain America's leading contribution to the world in these fields, the freedom of inquiry and research must be protected against new pressures for secrecy coming from both the private and governmental sectors. Fortunately, protective efforts are underway, with at least some cooperation from these sectors themselves. Such efforts must be vigorously pursued not only in the name of constitutional rights but in the name of the very commercial and national security needs supposed to be served by the new secrecy.

Note that officially classified "secret" information is not the issue here. The scientists and universities trying to protect their freedom are not trying to escape the responsibility for protecting classified secrets. They are trying to prevent restraints on the fruitful exchange of information under heavy financing by private companies seeking to keep results of research to themselves – or under initiatives by government agencies seeking to keep even unclassified information away from potential adversaries.

Last week a conference of university presidents and genetic engineering company executives took a step toward clarifying responsibilities in commercial arrangements between private companies and educational institutions. They came up with guidelines for fashioning agreements that, among other things, "do not promote a secrecy that will harm the progress of science." Here is something to build on – the more effectively so if the discussion expands to include independent voices concerned about the commercial impact on universities accepting more and more agreements under the economic gun.

This week governmental pressures for secrecy of unclassified information came to the fore again when Adm. Bobby Inman, deputy director of the CIA, appeared at a congressional hearing. He renewed his January call for the scientific community to place voluntary restrictions on itself or risk governmental restraints.

Already a number of universities have had to resist Washington efforts to change their practices on foreign scholars' access to unclassified information. Stanford, for example, was asked to restrict certain kinds of technical data from Chinese students and a Soviet robotics specialist – one whose work, incidentally, is said to have significantly aided American progress in the field. Stanford refused, noting that no secret or classified research is permitted on the campus.

Two panels of public and private officials have been set up to consider the matter of scientific freedom and national security needs. They properly want to avoid a collision course between government and the researchers it depends upon to ensure that the nation's defenses stay up to date.

Their work is important, because freedom of ideas and security are not opposed but inseparable. A group of university presidents put it well some time ago when protesting an effort to impose export restrictions on university research activities:

"Restricting the free flow of information among scientists and engineers would alter fundamentally the system that produced the scientific and technological lead that the government is now trying to protect and leave us with nothing to protect in the very near future. The way to protect that lead is to make sure the country's best talent is encouraged to work in the relevant areas, not to try to build a wall around past discoveries."

The Oregonian

Portland, Ore., January 22, 1982

Can the government inhibit the flow of technology valuable to the Soviet military without crushing legitimate scientific exchanges of mutual benefit? Whatever the answer to this question, the administration is determined to modify scientific exchange agreements made under the umbrella of Nixon-era detente.

Soviet science is strong in many basic fields, including metallurgy, theoretical physics, condensed matter physics, astrophysics, geophysics and cancer research. The conspicuous Russian weakness is an inability to put new technology into production due to an inefficient industrial and economic structure. But the Soviet educational system is placing far more emphasis on science and more demands are made of its students than in the United States, permitting Soviet research and development to grow rapidly.

With the improvement of the industrial system, the Soviet Union, quite apart from the military value of having Western technology, would be able to use technology in consumer competition that could prove ruinous to some Western economies.

The Defense Department has complained with some justification that student exchange programs, scientific conferences, technical information centers and travel permits are being exploited by the Soviets. In the case of students, the Russians send older persons with higher degrees to American universities, where they take courses in sensitive military fields, while our younger students are more apt to be studying literature in Moscow.

The United States will never entirely stop the theft of technical information or the transshipment of technology sold to other Western nations. But the Soviet job of obtaining sensitive information can be made more difficult.

We need to insist on equality of exchanges. We open our documents of the National Technical Information Services to the world, while a similar institute in Moscow, set up on a model of NTIS, limits publications to internal sources in the Soviet Union.

We cannot afford to lock up our scientists and prohibit all foreign exchange, else we cripple our own research and put a chill on the expansion of scientific thought. That the Soviets are behind us in so many fields is simply because they long had a policy of scientific isolation.

Let us not repeat the Soviet mistakes. But we cannot permit the easy obtainment of the intellectual and technical products of a free society when it is the clear Soviet intention to put such valuable knowledge to military use and to limit scientific exchanges with the West.

DAYTON DAILY NEWS

Dayton, Ohio, January 20, 1982

Pentagon censorship of American scientific papers is not an acceptable or wise way to keep Russians from stealing information. The government can and should to tackle the problem in other ways.

The problem itself is real. This nation's technological findings are being ripped off in a big way.

Weinberger

The Soviets save billions of dollars gathering or stealing information rather than developing it. What America's scientific press doesn't provide free, the KGB buys or steals outright — microcircuits and other high technology, shipped to the Soviet Union, analyzed and reproduced.

One Defense Department official suggested that scientists get Pentagon approval to publish new findings so sensitive data won't be cast into the public. "No way," roared U.S. scientists, who don't want somebody in Washington telling them what they can share with or find out from colleagues. Certain sensitive information *is* kept as secret by government and industry, but the proposed censorship would be far more extensive than current practice.

Defense Secretary Caspar Weinberger has proposed better answers: Involving NATO and other allies in a new-technology control system, to be considered at a conference in Paris next month. Establishing voluntary security committees in sensitive factories and encouraging industry associations to promote internal safeguards.

A reduction of Soviet diplomatic personnel is a possibility, if this country loses more to Soviet agents here than American agents get inside the Soviet Union. That, of course, is something only high-level insiders know.

America's open-forum liability also is its strength. The scientific information that Russians collect also is used by American and other free world scientists and industries. That's how the industrial democracies got ahead and how they stay ahead.

By contrast, the Russians are secretive about even benign information; often one scientist or engineer does not know what a colleague is doing. That's one reason the Russians have to steal America's discoveries.

This country has to learn to be discreet without stifling its own science, and as tough as possible in fighting theft.

Los Angeles Times

Los Angeles, Calif., February 7, 1982

The Reagan Administration thinks the Soviet Union is learning too much from unclassified scientific and technical information published in the United States. That information, it says, is often used to advance Soviet military power, to the detriment of American power. The Administration wants steps taken to reduce Russian access to this research. It is hard to see how that can be done, however, without unacceptably inhibiting the vital free flow of information within the American scientific community.

Adm. Bobby R. Inman, the deputy director of Central intelligence, has urged scientists to cooperate voluntarily with the government's approach. Failure to do so, Inman has warned, could lead to far more onerous restrictions on published scientific research than might otherwise be the case. Inman has suggested that scientists might have to submit certain papers to the government for pre-publication screening in an effort to satisfy national-security concerns. What he is talking about is giving the government the power of censorship over unclassified material. That censorship would be limited in application, perhaps, but would be censorship nonetheless, and would run head on into both guarantees of free speech and the necessity for scientists, in their own interests and in the national interest, to communicate freely with each other.

A precedent of sorts exists for official pre-publication review of non-classified scientific articles. This involves the relatively narrow field of cryptanalysis, the study of how to break codes. A year ago, scholars in the field began voluntarily submitting research papers to the National Security Agency before proposed publication. The agency had requested the right of advance review out of concern that information harmful to its activities of protecting secret U.S. communications and breaking codes of other countries might inadvertently get into the public domain. Scholarly cooperation with the agency may have been influenced by the threat that without it a law would be sought to prohibit open publication of cryptanalytic research.

The government now indicates that it wants to go well beyond this single and narrowly defined area. Inman says there is official worry about freely published information in a wide range of fields. These include computer science, electronic equipment and applications, lasers, crop projections and certain manufacturing techniques. Publication of unclassified material in these fields would presumably be made subject to prior official approval. Once again, the threat of a law to compel such screening has been raised as an alternative to voluntary cooperation by researchers.

Can a showing be made that the Soviet Union has made some gains in military technology because of unclassified information that it has gleaned from U.S. publications? Inman says evidence supporting that view exists. But he will not publicly disclose what that evidence is, although he does say that disclosure could be made in secret to appropriate congressional committees. Inman's argument is that, if the United States lets the Russians know that it knows what they know, the sources for the U.S. intelligence might be traced and exposed

Undoubtedly there is something to that contention. The problem is that it does nothing to help resolve the matter at issue. The government bases its case for pre-publication screening and possible censorship on what it sees as the detrimental results of Soviet access to freely published and unclassified U.S. scientific and technological research. But it will not publicly disclose what those results are. By imposing this lid of secrecy, the government effectively precludes essential and informed public discussion about the merits of its claim.

It stands to reason that the Soviet Union has learned useful things from overt information-gathering activities in the United States. But many scientists—including, interestingly, some dissident Russian scientists—strongly question the lasting value of these acquisitions. It is one thing, these scientists note, to get insight into technological advances; it is something else to exploit that information. The Soviet Union's industrial base is nowhere near as broad or as flexible as the United States'. Its ability to adapt copied technology is limited. Its own internal secrecy often works to retard technological development.

Science thrives when scientific communications are unimpaired. It would be self-defeating if the American government's concern about what the Russians might learn served to inhibit the free interchange of ideas in U.S. science and technology. It would be self-destructive if our open society became semi-closed in the name of protecting freedom.

The Providence Journal

Providence, R. I., June 6, 1978

The federal government, it seems, has arrogated the power to declare university-sponsored, publicly funded research findings outside the public domain because their disclosure "might be detrimental to the national security."

The University of Wisconsin, which sponsored research on computer security under a three-year $89,728 grant from the National Science Foundation, is fighting mad. It has announced that it will challenge a secrecy order issued by the patent and trademark office of the U.S. Department of Commerce. "We are obviously very concerned about the precedent established by the suppression of the study and the threat it poses to academic freedom," said the assistant chancellor for university relations.

The National Science Foundation is an independent federal agency, which allocates research funds appropriated by Congress. According to university officials, there was no warning connected with the grant that the Wisconsin computer project was sensitive; nor was there any provision barring disclosure of the findings. Only when the university sought to patent certain techniques derived from the study did the government order the information kept secret.

Genesis of the order was a "defense agency," a Patent Office official disclosed. Does this mean that any bona fide research project is subject to federal shutdown? Had the Milwaukee campus of the University of Wisconsin been asked to conduct classified research for the government, it would have refused under a policy of several years' standing.

If Washington should insist that it has carte blanche authority to suppress scientific findings — *after the fact* — whenever it believes the data "might be detrimental to the national security," it will collide with the cherished concept of free and independent inquiry, the outcome of which would be interesting, indeed.

The university has every reason to challenge the secrecy order and to expect the National Science Foundation to support its fight. Surrender to a federal policy of ex post facto restraint approaches the unthinkable.

The Washington Post

Washington, D. C., April 8, 1982

CIA DEPUTY DIRECTOR Bobby R. Inman stirred up quite a controversy a few months ago with a warning to scientists that they had better accept a voluntary system of pre-publication censorship. If they did not accept such restrictions, Adm. Inman predicted, scientists would be held responsible by public opinion for the "hemorrhage" of U.S. technology to the Soviet Union, and they would be "wiped away by a tidal wave of public anger."

In a recent congressional appearance, Adm. Inman regretted the tidal wave metaphor, but stood by his prescription. He is trying, he said, to "goad" scientists into taking action before the government is forced to. Restrictions would cover a sweeping range of research from crop projections to "manufacturing procedures"—this despite his acknowledgment that inadvertent disclosure of technological assets through communications, publications and conversations among scientists and engineers accounts for a "very small part of the problem." Seventy percent of scientific and technological losses occur through espionage, Adm. Inman estimated, while legal and illegal industrial transfers account for most of the rest.

Since this country relies heavily for its national security on its technological edge over the Soviets, even relatively small losses would be worth stemming if that could be done at an acceptable cost. The trouble is, it can't. Outside of the present administration, there are few who believe that a sweeping system of government pre-clearance of scientific research could even be imposed. If imposed, it would require legions of highly trained bureaucrats (that is, scientists and engineers who would be much more productively employed doing their own research than reviewing someone else's) to enforce. And if, somehow, such a system were created, the costs, in stifling and delaying U.S. technological advance, would be many times larger than the value of what would be denied to others.

The government already has more means for controlling export losses than it can effectively manage. There are several different export control lists covering weapons and sensitive technologies. These can be invoked against publication of listed technologies. There is a new 800-page "Military Critical Technologies" list under development. There is the Invention and Secrecy Act, which allows the government to impose secrecy on a patent application without justification and with limited opportunity for appeal. And there is the Export Administration Act, whose definition of "export" has been interpreted to cover "oral exchanges of information with foreign nationals in the United States."

The government should focus its attention on narrowing the critical technologies list to usable dimensions and on closing the loopholes and reducing the confusion, delay, overlap and error that surround the administration of the various export control lists. If these things can be done effectively, the case for imposing controls on research will disappear.

THE SACRAMENTO BEE

Sacramento, Calif., December 16, 1981

No one seems to be certain just what the government is trying to accomplish with its mounting effort to restrict the activities of foreign scholars at American universities. The effort includes requests for information from various universities about the work of visiting Chinese scientists; a demand that the University of Minnesota restrict the research activities of a visiting scholar; a notice sent to the University of Wisconsin that it must have an export license for a computerized weather analysis system the university is developing for a Chinese institute; declarations by the Pentagon that scientific exchanges are enhancing Soviet military power; and similar actions. In many instances, the government has intervened in situations which were not in any way connected with national security. All, of course, involve non-classified research.

The government's actions have caused growing concern on the campuses. Earlier this year, five major university presidents wrote the administration protesting restrictions on "the free flow of scientists and engineers." More recently, David A. Wilson, executive assistant to University of California President David Saxon, told a Defense Department panel that the rules are too vague and broad. "It becomes impossible," he said, "for people engaged in academic research and scientific activity to have any confidence in their ability to determine what is controlled and what isn't."

It is, in fact, a difficult problem. There is no question that American science is far less restricted than its Soviet counterpart, and that Russian, Chinese and other foreign scholars have access to training and laboratories of a sort that would never be accessible to Americans behind the Iron Curtain. But as many scientists are trying to remind the government, it's also for that reason that Western science is so much more sophisticated than its Soviet counterparts: Without the free flow, the ability to publish, to test and to exchange ideas, the whole purpose is paralyzed.

The government's efforts to restrict the activities of foreign scholars here are based on the presumption that their training amounts to an "export" of technical assets and is therefore subject to government licenses and controls. The attempt to do that, however, is so fraught with difficulties that at best it can have little beneficial effect. If the work of foreign scholars in non-classified research is restricted, what about the publication of results in scientific journals? And if the government restricts the work of Russian or Chinese scientists, what about the work of Frenchmen or Germans or, indeed, Americans, who might later sell their services to international corporations dealing with the Russians?

The government is now re-examining its security requirements in these areas; a Pentagon committee is expected to have new recommendations by mid-January. The committee will have to deal with real problems, particularly with respect to the export, legal and illegal, of electronic devices and other highly sophisticated equipment not available elsewhere. But it must also realize that for the most part broad restrictions on the flow of information — the circulation of ideas among scientists and scholars — can be imposed only at enormous expense and, in the long run, with only marginal effect.

The Birmingham News

Birmingham, Ala., March 8, 1982

It is fairly clear that the Soviet Union is tuned in to this country's great output of scientific, medical and technical data. Recently it was disclosed that Soviet intelligence had even penetrated the staff of the congressional watchdog agency, the General Accounting Office, and was obtaining data from the agency before it was even published in reports.

Few in government doubt that Soviet agents are gleaning valuable information on science and technology as they regularly examine government files such as those on patents, the Library of Congress and hundreds of scientific journals and reports available for pennies and at almost no effort.

But what to do about this kind of exploitation is easier asked than answered. Some senior defense and intelligence people believe that scientists working in certain fields should curtain off themselves and their work to avoid the prying eyes of Soviet agents.

If American science does not voluntarily censor itself, Adm. Bobby R. Inman, deputy CIA director, said recently, its traditional resistance to "regulation of any kind" will be crushed by a wave of concern by the public and Congress. The admiral is probably overestimating both public and congressional concern.

University-based computer scientists have already agreed to voluntary pre-publication review by government agencies of research papers of possible value to the Soviet Union on making and breaking codes.

But the country should go slow on restricting — even voluntarily — the free flow of scientific information. While it may be difficult to prove, it is generally believed that the essential difference between the Soviets and the United States regarding progress in science and technology is the free flow of information about research. In other words, it is the free exchange of scientific information that makes the United States superior, and without that free flow, the situations in both countries would be relatively equal.

We do have advanced technologies in the form of both equipment and information that we want to keep secret as long as possible. But security techniques already exist for accomplishing that end without closing off our most productive scientists from new information.

So, greatest care should be exercised before we voluntarily surrender the essential practice that gives the United States its great advantage.

Universities & Companies Rule on Research for Profit

New technology, especially genetic engineering, was developed to a great degree by university research. The enormous profit potential of these discoveries quickly attracted corporate interest. Universities, chronically short of cash, were at first happy to take part in marketing their departments' discoveries. However, it was inevitable that the demands of the marketplace would clash with the freedom of academic inquiry. A corporation that gave a university millions of dollars in research grants for a particular discovery might, for example, demand restrictions on university policy in return. Five universities and 10 corporations met March 26-27, 1982 in California to draw up a set of rules governing corporate-sponsored university research. They agreed that faculty members who worked for corporations should not be restricted in their teaching duties or their commitment to scientific progress; that the nature of the research should be determined by scientific, not commercial, criteria; that corporations should not receive exclusive licenses to the research results; that universities should not own or control companies that employed faculty members; that universities should establish clear guidelines to prevent conflicts of interest, and that free exchange of information should not be "unduly" restricted by a company's desire to keep trade secrets from competitors. The universities agreed that the publication of research findings could be delayed to give a company time to obtain a patent.

Participants in the conference included: Stanford University, Harvard University, the Massachusetts Institute of Technology, the University of California, the California Institute of Technology, E. I. du Pont De Nemours & Co., Eli Lilly & Co., the Gillette Co., Genentech Inc. and Cetus Inc.

THE DAILY HERALD

Biloxi, Miss., March 29, 1978

When your tax dollars support scientific research, does your government have a claim to the fruits of that research's discoveries?

It would seem that the government, since it shares in the expense of the research, should also share in the financial benefits that might flow from it. Unfortunately, the issue isn't that simple. The federal government stumbled, accidentally, into some of the complications earlier this month.

General Services Administration issued a regulation that would allow universities and nonprofit organizations to profit from discoveries made with federal research money. The regulation drew immediate fire from two quarters.

Sen. Gaylord Nelson, D-Wis., complained that the GSA policy conflicted with a congressional study; Ralph Nader's Health Research Group attacked the policy as a giveaway of government-funded research.

Next year, the government will spend more than $3 billion on research programs, so that the resolution of the conflict is significant.

The policy has been, generally, that universities conducting the research have been allowed to license their inventions to private firms for further development, testing and marketing. Money from the licenses has been plowed back into research and development.

Health Research Group admits that government has a proper function in encouraging research. But it feels some arrangement should be worked out so that the government recoups at least some of its investment.

The Society of University Patent Administrators, on the other hand, feels that much technology developed under government grants would never reach the public without the licensing system. Private firms are needed to undertake costly clinical testing and market development.

There is some merit in both viewpoints.

GSA has now indicated a willingness to suspend enforcement of its new policy for 120 days. Sen. Nelson's Subcommittee of the Select Committee on Small Business is planning hearings next month to learn more about the issue.

Government sponsored research should be motivated by a desire to be of benefit to all Americans and not for profit. But should the fruits of that research include items with profit potentiality, the government should not be excluded from sharing in that profit.

THE PLAIN DEALER

Cleveland, Ohio, August 28, 1978

The drafters of Internal Revenue Service rulings have long been noted for their ability to split the hair of a frog four different ways. But they have gone too far in their illogical attempt to strip tax-exempt status from the American Institute of Physics and the American Chemical Society.

IRS, in its misplaced zeal to attack the two institutions, even came up with different "reasons" for attempting to revoke their exempt status.

The American Institute of Physics is a cooperative service for nine societies and publishes more than 40 journals for physicists in specialized fields. Apparently IRS believes that because the institute serves as a clearinghouse for research physicists rather than initiates research, it cannot have the value of a research organization in the strictest sense.

As for the American Chemical Society, which publishes 18 journals of specialized interest, IRS said it charged more for subscriptions for its journals to nonmembers than to members. There would appear to be nothing wrong with that. But IRS believes it is grounds to attack the society.

Both organizations serve vital roles in America's research capability, which should not be destroyed capriciously by the IRS. Neither organization has the same tax problem as the landmark "National Geographic magazine" case of untaxable advertising revenue.

In 1968 the tax law was changed to require the National Geographic Society to declare as taxable its advertising revenues, as "unrelated business income." But it retained its tax-free status in other areas, partly because it makes research grants and provides public services.

The American Institute of Physics could make an annual research grant and maintain a minuscule public museum, thus retaining its nonprofit status. And the chemical society could charge everyone the same amount for its journals even though it might mean that the research scientists who need the esoteric publications the most may be the most unable to afford them.

But why should they? These requirements are not central to their vital public function of helping to keep America technologically superior. The IRS rulings should be reversed, even if it takes an act of Congress.

Houston Chronicle

Houston, Texas, October 29, 1980

Some Wall Street analysts' eyes grew as large as saucers recently when the stock of a little-known California company, Genentech, Inc., rose from $35 to $89 on the first day it was publicly traded, and subsequently settled to about $54. Genentech's appeal? Rights to several discoveries in the field of genetic engineering, this year's "hot" new technology.

Now comes word that Harvard University is considering entering the field commercially, through minority ownership of an engineering company. The Harvard administration reportedly is debating whether to create a commercial outlet for marketing biological patents held by the university. In effect, Harvard is considering whether to compete with the Genentechs and other commercial genetic engineering concerns for profit.

The temptations for a university, especially a private one, to rush into "genetics-for-profit" type research are strong. Even Harvard, with the nation's largest university endowment, is feeling the pinch of inflation and poor stock market performance.

At the same time, commercialization of research on university campuses holds the promise of creating a "fast track" between the university lab and the marketplace, a pleasing prospect not only for industry, but also for the consumer.

But there are also several cautions. Already Harvard faculty members have raised the question of a possible confict of interest between the traditional free flow of knowledge in the academic world and possible obligations to keep secrets the way industry scientists often are required. Commercialization also has an undetermined potential to force subtle, or perhaps not so subtle, changes in the way research money is distributed, to the possible detriment of scientific progress.

To those could be added a reservation about what linking prestigious universities, such as Harvard, to such a controversial field as genetic engineering would do to quash legitimate debate. It is conceivable, for example, that proper moral and ethical concerns over the direction of genetic engineering could be brushed aside, not because they lack merit, but because they are misconstrued as a criticism of a presitigious institution.

Harvard's is a decision that is being watched carefully by other major universities with similar patent rights, who could be expected to follow the Harvard lead. As such, it is one to be taken with the utmost care, in the full knowledge that there are pitfalls as well as potential rewards.

Herald News

Fall River, Mass., June 17, 1980

Subcommittees of the House Science and Technology Committee have begun hearings on the links between universities and the genetics industry. The subcommittees will be venturing on new ground in terms of ethics simply because the implications of genetics are by no means clear as yet, to say nothing of what the commercial potential of the genetics industry will be.

One thing is clear, however. That industry depends for its development on the research conducted by several of this country's centers of learning. The temptation for those centers to engage in outright commercial contracts with one company or another is very strong in times like these when every university and hospital is feeling the financial pinch.

Harvard, after much soul-searching, decided not to enter any such contract because of the possible compromise it might entail in terms of pure research. This has not been the case everywhere, however.

Just last month Massachusetts General Hospital accepted a grant of $50 million from a West German chemical company, which in return for what might better be termed an investment than a grant will receive exclusive rights to commercial products the Mass. General's laboratory develops.

The questions that are being raised as a result of the connections between research centers and the genetics industry will not be easy for the House subcommittees to answer.

Massachusetts has already experienced the industrial benefits derived from the research centers for high technology that are located here. It is reasonable to expect that, sooner or later, similar economic benefits will be derived from the genetics industry now coming into being.

Detroit Free Press

Detroit, Mich., November 30, 1980

EDUCATION and commerce have never been totally separate. Major universities often invest in land, license patents and condone consulting by members of their staffs. But those activities have never stirred up the fuss of a proposal put forth by a biologist at Harvard that the university join him in starting a gene splicing company.

Harvard finally rejected the offer, but not before stirring up a tempest within the academic community, inside and outside Harvard, for many weeks. The offer raised several important questions, having to do with everything from the company's effect on the university's reputation for impartiality to whether profits would affect decisions about who is awarded tenure. They are questions that other universities will also have to face — and soon.

Unlike any invention since the small computer, recombinant DNA (gene splicing) has sent Wall Street into convulsions. Genentech, a pioneering company in the field, saw its shares go from $35 to $89 in a matter of minutes, before they finally settled down to nearly twice their initial price. Monsanto recently purchased a $20 million share in Biogen, another small gene splicing company.

A number of pioneering biologists, and their supporters, stand to become extremely wealthy. And Mark Ptashne, the Harvard biologist, was offering his university a chance to profit along with him.

Harvard was right, however, to say "no" — not because universities shouldn't make money, but because making money should not be one of a university's primary missions. And the field of genetic engineering is commercially so hot that at least one department within the university would risk being overwhelmed by the corporation it created.

It's only right, if profits are to be made, that a university be allowed to license its patents to companies that can use them. That is a substantially different proposition, however, than having the university itself set up a corporation — especially in an area as controversial as genetic engineering.

As one member of the Harvard faculty pointed out, if the university is commercially promoting the technology, to whom will the public turn for impartial advice? Or if a bad professor made a fortune for the university, would the university dare to deny him tenure?

Universities would do us all a favor by avoiding circumstances in which such a question would be raised. Harvard's decision sets the right standard.

Los Angeles Times

Los Angeles, Calif., April 9, 1982

The need to keep universities' research interests free of conflict with their financial interests has until recently been a truly academic question. But university laboratory discoveries—like gene splicing—that have far-reaching industrial uses, coupled with an increasing financial squeeze, have thrust the issue into universities' day-to-day governance.

The problem is of legitimate public concern as well. Financial support, not the pursuit of knowledge, could conceivably determine research subjects and even color the outcome of research that affects everything from the quality of tomatoes to the ability to create life.

Last month, the heads of five major research universities—Stanford, Harvard, Caltech, Massachusetts Institute of Technology and the University of California—and officials of ten corporations met at Pajaro Dunes near Watsonville to talk about the issue. The conference was billed as a beginning, and it was only that.

The statement issued by the conferees stressed the importance of keeping academic pursuits ahead of commercial concerns so as not to "diminish the role of the university as a credible and impartial resource." But the conferees gave no firm guidance to universities trying to determine whether they should patent professors' inventions, or to faculties grappling with conflict-of-interest issues involving professors who have financial ties to companies paying for their research. The meetings also did not include representatives of groups directly affected: the faculties and the public.

Obviously, no guidelines that the conferees could have written would have been binding on a diverse range of universities and faculties. Each university must establish its own principles for its own situation. But, since some universities have been slow to act themselves, some leadership is in order.

The University of California, one of the conferees, represents a case in point. In 1977, the university petitioned for exemption from a requirement of California's Political Reform Act that public officials file annual conflict-of-interest statements. The university said such reporting would inhibit professors' freedom to choose research subjects. The Fair Political Practices Commission, which oversees the reform act, granted the exemption with the understanding that UC would develop its own systemwide policy and enforce it.

UC academic vice president William R. Frazer acknowledges that the adopted policy fell short of what many people wanted. The policy required that each year UC faculty members report to the administration the nature of any outside consulting activities. The researchers did not have to name the firms involved nor the amounts of money earned.

A new policy to deal with the new interest in the issues was "evolving" at UC, but evidently not quickly enough for the commission. So last month the commission acted, making California the first state to require that professors at its public universities disclose personal financial interests in businesses sponsoring their research projects. That is perfectly legitimate information for disclosure, but in the interests of academic freedom it should have been the university—not the commission—that established the requirement.

The Pajaro Dunes meeting was an important first step. As Harvard president Derek Bok said, "We are at a very early stage in trying to come to grips with these issues. Over time, one would hope that some consensus as to what is the proper solution would emerge."

That consensus must include the ideas of the public and the professors themselves. As the conferees said in addressing another point, "Universities have a responsibility not only to maintain these values" of independent teaching and research "but also to satisfy faculty, students and the general public that they are being maintained." They can be satisfied only if they are fully informed.

Reasonable people may differ on how best to ensure academic independence, the conferees said, adding that it is essential "that each university establish some effective method." Otherwise, as the University of California has learned, it risks having someone else do it.

The Washington Star
and Daily News

Washington, D. C., November 11, 1980

On October 21, the president of Harvard, Mr. Derek Bok, proposed to the faculty a money-raising scheme to which some of them have reacted as if he proposed to change the university's motto from *veritas* (truth) to *cupiditas* (greed).

It may be more than coincidence that one week earlier to the day, a small corporation called Gerentech, capitalized four years ago at $1,000, was selling its first public stock offering on Wall Street for well over $500 million — and incidentally making the two university scientists who founded it multimillionaires overnight. At least on paper.

Gerentech is one of several companies poised to make big profits out of the new biochemistry, specifically the new techniques of "gene-splicing." Insulin and interferon (the anti-viral agent naturally produced by the human body and tentatively identified as a powerful cancer-fighting agent) are among the marketable products.

Of course, those who believe Wall Street to be incurably romantic at heart might have noticed that Gerentech has yet to earn its first penny — is, in fact, about $700,000 in debt — and that its prospects are clouded by a California court case that could alter accepted notions of the legal duties commercial firms owe to scientists whose discoveries they borrow and use for profit.

President Bok is proposing that Harvard set up its own company (Harvard, Inc.?) to use the university's own scientific research (chiefly, we assume, in the flourishing field of biochemistry) for profit. Or profit in the limited sense that its earnings would support the fabulous overhead of scientific research. There is a lot to be said for such an enterprise — especially, we imagine, by university administrators and the managers of their investment portfolios. But the reaction from Harvard's scientific community is reported to be unenthusiastic, or worse.

The objectors aren't, we assume, so unworldly as to assume that the discoveries made in university scientific research will never be exploited by somebody, somewhere, sometimes for great profit. In an ideal world those with important life-saving capacities would be made and marketed at cost plus a small overage to finance further research. But this world is not ideal. It isn't uncommon, in fact, for university scientists themselves to do a bit of lucrative research or consulting on the side.

The issue is what may happen when a university itself goes directly into science-for-profit. It is an issue on which the negative arguments are impressive. No doubt the purity of "pure" science can be exaggerated, but what distinguishes it from the applied kind is the allotment of time, energy and imagination to investigation for its own sake.

It is not hard to imagine Mr. Bok's enterprise skewing research priorities in subtle ways. Even professors like to play Monopoly. It is quite easy, moreover, to imagine universities that are in the business of commercially-exploited biochemistry (and if biochemistry, why not other kinds of science?) concealing what goes on in their labs, not sharing it broadly through publication and conversation in the age-old custom of academic science. One can foresee questions of priority (who found what first) translating themselves from issues of scientific honor and etiquette into furious litigation.

In short, as Nicholas Wade has observed in a recent issue of *Science*, "the powerful forces of the profit motive clearly have the capacity to strain and rupture the informal traditions of scientific exchange." The universities would be well advised to use a long spoon when they sup with this particular tempter.

President Bok, who like all chronically hat-passing university presidents is to be sympathized with, might solve his money problems by putting Harvard into science-for-profit. But he might also find he had begun a fateful trend in which academic science, by long custom an open process, became as secretive and as partitioned as the Manhattan project, whose considerable violations of scientific custom had war and national survival to justify them.

The Oregonian

Portland, Ore., June 11, 1982

There was a day long ago when industries looked upon campuses as repositories of fuzzy-thinking professors. But these professors built the atomic bomb, revolutionized agriculture, cured diseases, helped man reach the moon and, not to be trivial, developed fluoride toothpaste, Pacman and Gatorade.

The universities long ago ceased to be virgins in the marketplace. They have known commercial sin. The question now is what kind of new deals should they cut as government reduces its research support and as corporate sirens sing of million-dollar grants. Another question in Oregon is whether skimpy state financing policies are making the ethical issue moot. If the leading professors pack up and go elsewhere, Oregon will not have any sinful corporate offers to consider.

While there are ample precedents for campus-industry deals, the magnitude of recent offers, linked to the explosion in biotechnology and microelectronics, raises heavy ethical problems, many of which challenge traditional academic values. Some arrangements may threaten even the dispassionate reputation of science itself.

It is also a multinational problem. Foreign companies have moved to purchase U.S. brains, taking a shortcut in the hope of becoming world leaders in their fields. There is the recent example of Hoechst AG, a German pharmaceutical company, which has given a $70 million guarantee over the next 10 years to found a new department of molecular biology at Massachusetts General Hospital. That is big money.

The list of fat grants grows, not from corporate altruism, but from the company need to compete in world markets. Questions of patent rights, secrecy, peer review, conflicts of interest, competing companies in the same laboratory, freedom to publish results, equities of faculty in research firms and constraints on academic and scientific freedom are among the delicate issues being discussed under the money trees.

There are even grants to study the issues exposed by corporate grantsmanship. Universities are appointing committees, holding seminars and otherwise wrestling with their consciences. But unless ethical practices are solidly established by devices such as faculty review groups, scandals seem certain to erupt with all that money floating about. Confidence in science and the universities could suffer. Worse, scientific advances could be blunted.

An ominous pattern is emerging. The campuses with talented, prestigious professors are getting the fat research grants. Oregon's universities, whether at Eugene, Corvallis or Portland, cannot continue to get badly needed corporate funding if they suffer faculty losses because of higher education budget cuts.

This loss of talent cannot be tolerated. This seems obvious, but past legislatures have not understood it well. Oregon cannot attract a more diversified, technological economy if it has universities capable only of second-rate research work. Unless the trend is reversed, Oregon will lose the leadership it has pioneered so well with home-grown technology and industry.

The Birmingham News

Birmingham, Ala., March 28, 1982

The wave of the *present* is research-spawned high technology. And that wave will grow both in size and in force in the near future. Communities which are not aware or sensitive to this development in the economy will be left behind in terms of both growth and wealth. Cities and states which are aware of what is happening in industry and move to embrace research and technology will flourish.

These views are not revolutionary or even new. Analysts in almost every field over the past decade have proclaimed the birth of this new economic era. Of course, steel mills will continue to produce. And certainly coal will continue to be mined. Heavy industry will always be needed as long as we need heavy materials and equipment for the multitude of products we use in our homes and work.

But the character of the machines which do the work in heavy industry will change even more than they have changed in the past. And the job market will change with them. As a matter of history, low skill jobs are already disappearing and will continue to disappear at an ever-increasing rate in this decade as sophisticated machines go into operation in heavy industry.

However, sophisticated research and high technology industry are not endeavors a community goes into overnight. Some years of preparation are necessary.

Required is a concentration of multi-discipline expertise, such as is found only in great universities and research institutions. Among other requirements are far-sighted university administations and political leadership at all levels which understand at some depth that high technology industries are based in research capability.

Fortunately, Alabama, and especially the Birmingham area, has the necessary requirements to become a center for research and development of high technology industry in a number of fields. In some areas, such as medicine, micro-biology and health care, Birmingham may be uniquely qualified.

That was the consensus which was expressed at the recent conference at the University of Alabama in Birmingham.

We have in our midst in Birmingham a university that ranks in the top 20 research universities in the nation. We have a highly respected research institution in Southern Research Institute. We have three great institutions at Auburn, Tuscaloosa and Huntsville with proven track records in research and research capability.

According to one conference speaker, Birmingham alone has 10,000 engineers in various fields, a strong knowledge base for new technology industries.

We have land needed for a research-industrial park of sizable proportions in abundance. We have $10 million in tax funds set aside by the Legislature to begin such an endeavor.

So why don't we get cracking?

Well, some problems are yet to be solved. We have yet to create a research-high technology mentality in the state. More officials, more businessmen and more rank-and-file citizens need to develop expectations of a high technology future. This is a knowledge problem, an educational problem, that must be addressed immediately by the state's leaders in higher education, business and government.

Land must be acquired. The city of Birmingham has 500 acres available, enough for a start. But according to experts who attended the UAB conference, acreage in the magnitude of 5,000 to 7,000 acres should be considered for future growth. Huntsville has already moved ahead on a research complex. It has recently expanded a 700 acre plot to 3,000 acres.

Decisions on how the $10 million appropriated by the Legislature will be spent must be made. So far, no consensus has emerged on where and how this seed money will be targeted. While the Legislature will have the final say, experts among the universities should come together to present a coherent, suitable plan or the development the money makes possible may be dissipated by piecemeal allocation.

What we need now may be a self-starter group, appointed perhaps by Gov. James, to forge a plan and to call in those who can expedite the plan in terms of land acquisition, promotion and the search for industries which need the expertise already available in Birmingham and Alabama.

And while in the process, we need a legislative group which constantly studies the future needs of an economy based in research and technology in terms of the demands it will make on higher education.

It is high time that legislators recognize that higher education is the only resource we have for providing the expertise that is being demanded and always will be demanded to build the kind of economy that will provide the quality of life that is within reach of all Alabamians.

The time to move is at hand. The future belongs to those who embrace it. If we fail to grasp the opportunities which so obviously lie before us, the future will belong to more imaginative, more enterprising and more energetic cities and states, and we will be left with only the work others are too busy to handle.

America Leads in DNA Research, but Future Outlook Is Cloudy

Research in recombinant DNA (deoxyribonucleic acid), also called gene-splicing, produced remarkable advances in the early 1980s. The production of interferon, a hormone with potential use against cancer, and the development of a vaccine against hoof-and-mouth disease quieted some of the public's fears about the implications of the new technology. Further reassurance was provided by the National Institutes of Health in 1979 with their guidelines for government-funded DNA research. The NIH code, which was observed voluntarily in most privately-funded projects, sought to keep new life forms from "escaping" the laboratory and to give the public a role in monitoring recombinant DNA research. The guidelines were relaxed in 1980 and 1981 to the point where fewer than 20% of DNA experiments required the highest degree of security and fewer than 10% required prior approval by screening committees. A proposal in April 1981 to make the screening process voluntary for government-funded experiments was rejected in February 1982.

(Controversy arose over the screening process in the case of Dr. Martin Cline of the University of California at Los Angeles in 1980. His work involved the use of recombinant DNA as a possible treatment for thalassemia, a fatal genetic blood disease. For more than a year, two separate screening committees at the university debated his proposal to test his treatment on humans. Cline finally sought and received permission to conduct tests on two terminally ill patients in Israel and Italy, with the full cooperation of the patients, doctors and hospitals involved. Unfortunately, his attempt at a cure was unsuccessful. Although the patients were not harmed by his treatment, the NIH withdrew all government support from his research.)

Despite U.S. successes in the field, recombinant DNA research was not immune to foreign competition, according to government analysts. In 1981, the President's Office of Science and Technology Policy warned that, "At the moment, the United States leads in the fundamental science of genetic manipulation, but that lead could be short-lived. International competition is heavy."

The Boston Globe

Boston, Mass., March 22, 1980

The importance of federal funding for basic research and the ability of American scientists to break new ground was again proven with the news this week that researchers at MIT have found a way to produce abundant amounts of the antiviral substance interferon at one-twentieth of the present cost.

Now, in addition to a form of interferon produced in very limited supply by Finland, scientists will have a new and plentiful source of interferon obtained by growing cells from connective tissue on minute beads of starch.

Because there are at least three variants of the substance, each biologically different, and because interferon's functions are still so little understood, much work remains to be done. But the speed with which developments have taken place is encouraging.

Interferon was first discovered 23 years ago when two scientists in London noticed that cells infected by one virus produced a substance that protected other cells from infection. Since then its study has depended almost exclusively on interferon obtained from white blood cells and produced in Finland.

In August 1978, the American Cancer Society spent $2 million to buy enough Finnish interferon to test it on 150 patients in the United States. This year the funding has been increased to $5.8 million, and $7 million in congressionally approved money may soon be available for the research.

Meanwhile, recombinant DNA technology has been used in Japan and Switzerland to isolate two different interferon genes. The Swiss researchers have claimed a low amount of activity in their recombined interferon but have not yet achieved a productive process. Japanese and American scientists are actively working on this technique at Harvard and elsewhere.

Now scientists at MIT, using a totally unrelated technique, say they can produce interferon in usable amounts, beginning as early as next summer.

Because that means that testing of the substance can be stepped up, it may be very good news for victims of particular forms of bone, skin, lymph and breast cancer as well as for sufferers of other viral afflictions such as chronic hepatitis and shingles.

Minneapolis Star and Tribune

Minneapolis, Minn., July 23, 1978

Most of the steam is gone from efforts in Congress to legislate safeguards for "gene-splicing" experiments. That is because science lobbyists have made a persuasive case that the hazards of such research are greatly exaggerated. It is also because the scientists have painted government regulation of their work as a threat to free inquiry.

With the first point we have little quarrel. It does indeed seem that microbiologists' own apprehensions led to needless scare-stories about doctored bacteria escaping from the lab. Now the matter is in calmer perspective. The scientists' super-strict self-imposed restrictions — in rules established for federally funded work — are being relaxed. Important questions remain about insuring safety in private research and whether to allow states and cities to set their own standards. But the urgency for legislation is much diminished. In fact, the news from the labs is benign, not alarming. Just last month a research team in Massachusetts transplanted genes from rats to make ordinary bacteria produce insulin.

But that does not settle the broader question about government regulation of scientific work. Essentially the question is whether non-scientists have a legitimate role in making value judgments and policy decisions about a realm of inquiry where they have no expertise. Two principles are potentially in conflict: One is that research must never be subject to the chilling threat of imposed preconceptions or official orthodoxies. The other is that in today's society science policy is inherently a public affair and public-policy makers should take knowledgable part in science decisions. Both principles are valid. But neither should be pushed to its purist extreme. To do so would either make scientists a sort of sacred priesthood, beyond the touch of lay criticism, or set up government as the arbiter of what questions can be asked.

How to avoid those extremes is the underlying challenge posed by current concern over gene-splicing research. Given the rate of scientific advance, it is bound to arise repeatedly in the future. That means hard work for both legislators and scientists. The legislators must learn to grasp the scope of highly technical investigations. The scientists must learn to interpret better for the laity the frontiers they are exploring. Good science policy will depend on both groups doing those unaccustomed jobs well.

The Houston Post

Houston, Texas, February 14, 1980

Now that the National Institutes of Health has lifted most of its restrictions on recombinant DNA, or gene-splicing, research, the door is open for industry to go on a veritable bacteria binge that could make the 1980s the bug decade. The bugs, in this case, are microorganisms that could revolutionize many manufacturing processes and give us an incalculable number of new products as well as existing ones made in a different way.

Microbiologists, teaming up with industrialists, may be able to engineer bacteria that can produce a wide array of items, ranging from medicine to fuel. Only recently Biogen, S.A., a firm headquartered in Switzerland, announced that it had used bacteria to manufacture interferon, an anti-viral substance so rare and expensive that it is available only for extremely limited experimentation. The Biogen development could eventually make interferon more widely available and less costly.

Though some questions remain, such as whether a living organism can be patented, the National Institutes of Health seems satisfied that any danger from recombinant DNA research to the public is remote enough to greatly relax its stiff curbs. The NIH instituted the restrictions a few years ago, when fears were expressed that the public might be exposed to deadly strains of bacteria created by such experiments. The restrictions were binding only for federally financed research, but they were submitted to voluntarily by private industry.

One benefit of the impending biochemical revolution might be the production of liquid fuels and other substitutes for petroleum in many products, such as plastics. Who knows? Our independence from OPEC may rest with a microscopic bug.

ST. LOUIS POST-DISPATCH

St. Louis, Mo., February 11, 1980

The relaxation of federal safety rules for recombinant DNA research should result in a dramatic increase in the investigations by nonprofit and commercial laboratories. The possibilities for medical advances and hence commercial applications from the research are rich. In that respect, the decision to lift the barriers to research may quickly usher in new generations of chemicals, hormones, insulin and disease-fighting substances. Still, there is a question as to whether the government has moved too far, too quickly.

In recombinant DNA research, bits of genes are spliced together and inserted into bacteria, which "manufacture" new genetic material. The first federal guidelines, set forth in 1976, created four categories of DNA research by risk factor and established various safety regulations for laboratory work. These applied to laboratories doing government-sponsored work, but private facilities have abided by them. The prevailing scientific view is that earlier fears of creating virulent new disease strains or other laboratory horrors have proven unfounded and that impediments to most research can be removed. Restrictions remain in effect on research into poisons or powerful disease organisms and special precautions still must be taken for work with organisms other than the common intestinal bacterium, *E. coli*.

Despite the existence of the guidelines, commercial genetic research has been impressive. A multimillion industry has arisen, led by small companies and encouraged now by the pharmaceutical, oil and chemical giants, which are aware that payoffs are likely to come in the fields of agriculture, energy, food processing and chemicals, as well as medicine. One does not need to oppose the lifting of the guidelines to argue that they have had something to do with the safety and responsibility of the research. Congress has never been successful in enacting legislation to cover genetic research. With the guidelines being lifted. is it not time to try again?

The Washington Post

Times Herald

Washington, D. C., February 4, 1980

NATIONAL INSTITUTES of Health Director Donald S. Fredrickson's decision to lift most of the guidelines under which scientists have had to perform recombinant DNA research is a major milestone in a precedent-setting attempt at self-regulation.

Seven years ago, when it first became possible to separate out genes from bacteria, viruses or higher organisms and insert them into other bacteria where their function could be closely studied, scientists immediately recognized the potential dangers. The new techniques, even in their most rudimentary form, obviously opened dramatic new vistas in molecular biology and medicine and were certain to be widely employed. So a group of the most prominent researchers in the field joined in a letter to the National Academy of Sciences expressing their concern that these new recombinant DNA molecules "may prove hazardous to laboratory workers and to the public" and might require formal regulation.

From that first letter, and a three-year long series of conferences and studies that followed, emerged the NIH guidelines, which established minimum safety conditions for different types of recombinant DNA experiments. The conditions ranged from those normally found in any carefully run medical laboratory to the totally closed and sterile conditions that could be found only at the Army's old germ warfare facility at Fort Detrick. Some experiments were banned altogether

The guidelines have been a source of controversy and have been studied and revised almost from the moment of publication. As scientists gained familiarity with the new techniques, some felt that the dangers had been overdrawn. Others believed exactly the opposite, always postulating new dangers that had not yet been studied. While heated disagreements persist, a new consensus has developed that many types of these experiments are safer than had been thought—hence Dr. Fredrickson's decision to, in effect, remove the regulations from them.

A few scientists among those who first voiced warnings believe they made a mistake. They have been buried for years under mountains of paper work, experiments have been delayed until the necessary clearances came through and many experiments have not been done at all because clearances were not received—and all because of what now appear to have been unfounded fears.

We hope that will not be the prevailing view. Despite their flaws, the recombinant DNA guidelines have been the model of a responsible approach to a dangerous technology, and of cooperative action between government and the private sector. Had nuclear engineers, pesticide chemists and numerous others acted with similar caution and sense of public responsibility, everyone would have been much better off.

Roanoke Times & World-News

Roanoke, Va., June 1, 1981

When you tamper with DNA (deoxyribonucleic acid) you're tampering with the basic stuff of life. When you tamper with human DNA, you're tampering with human nature.

Dr. Martin J. Cline, a UCLA scientist, tampered with human nature. He experimented with recombinant DNA on two human subjects suffering from a fatal blood disease. He knowingly violated strict rules laid down by the National Institutes of Health, which has funded his research. He performed one of the experiments in Israel after assuring the Israeli government that he did not plan to use recombinant DNA.

Neither he nor anyone else can say what risks he ran for us all. But he deserves more than routine reprimand.

Dr. Cline introduced human DNA into a form of bacteria, thus "cloning" human genes. These healthy genes were injected into the patients — a 21-year-old woman in Israel and a 16-year-old girl in Italy. The hope was that the healthy genes would replace the defective genes that prevent their bone marrow from manufacturing red blood cells.

The experiment was not successful. The patients are still kept alive only by blood transfusions. Thus far no harmful effects have been identified.

But the use of recombinant DNA on human subjects is still uncharted ground. The technique has been compared with the splitting of the atom in its potential for good and evil. No one knows whether the potential for evil has been exaggerated.

But Dr. Robert L. Sinsheimer, chairman of the biology division of the California Institute of Technology, warns that natural barriers exist to genetic interchange between human cells and bacterial cells. To break down these barriers is to risk unpredictable damage to the evolutionary process. The NIH holds that more time and research are needed before the giant step into human experimentation. Dr. Cline jumped the gun by some three years.

His colleagues have called his experiments "totally irresponsible," "wrong," and "one of the most flagrant abuses in our memory." The NIH is considering terminating $600,000 in grants to Dr. Cline. This would be the heaviest penalty it has ever meted out.

The punishment would be appropriate. The evidence indicates that Dr. Cline can't be trusted to use prudently the awesome knowledge to which he has gained access.

Roanoke, Va., October 4, 1981

Can a steer be made to produce prime beef on buffalo grass? Can a human brain be implanted into the offspring of an ape? Can diabetes, sickle cell anemia and other hereditary diseases be wiped out through the insertion of a gene into an egg?

The good and the bad questions arising from genetic engineering are pressing upon us.

The genes of rabbits were recently transferred to the offspring of mice. The feat was accomplished by a team headed by an Ohio University biologist. The mice passed the rabbit characteristic to their offspring, and they passed it on to theirs.

The possibilities for animal breeding are immense. The world would benefit, for instance, if high-quality beef cattle could acquire the buffalo's ability to grow fat on grass or hay instead of expensive grains. It would allow many Third World countries whose soil won't grow quality grain to become beef producers.

In humans, the technology could lead to the elimination of hereditary diseases, which afflict about 4 percent of the babies born in this country.

But Gordon Rattray Taylor saw sinister possibilities too, which he described 13 years ago in his book, "The Biological Time Bomb."

Take the case of race horses. "Genetic surgery could push up their speeds until they no longer looked like horses." And in the human field, "Athletics could become a battle between geneticists, each seeking to endow his DNA with outstanding athletic properties."

Should we contemplate the breeding of a race of slaves through the genetic crossing of humans with animals such as apes?

Suppose an authoritarian nation were to try to use genetics to breed certain characteristics into its people, thus creating a "master race." Gene warfare — the introduction of genetic weaknesses into an enemy nation — presents frightening possibilities.

The prospect of three-parent offspring raises theological questions similar to, but more complex than, the questions surrounding artificial insemination. Suppose a straight-haired woman married to a straight-haired man wanted a child with wavy hair. Would she be committing adultery if she allowed another man's gene for wavy hair to be transferred to her ovum?

Theologians are beginning to wrestle with the problems.

Some worry that by artificially altering the human race, man is blasphemously "playing God."

But Dr. Lane P. Lester of Liberty Baptist College in Lynchburg told The New York Times that genetic engineering might be one way God is restoring perfection to a flawed human race.

Genetic engineering is a technology that is upon us, and mankind has found no way to stay the advance of such a technology, even if it wanted to.

Sir Macfarlane Burnet once remarked that "there are dangers in knowing what should not be known." There is also difficulty in determining which knowledge should remain unknown. Until someone engineers out of man the urge to discover the undiscovered, our task is to learn to live with the consequences as well as the benefits of our technology.

The Virginian-Pilot

Norfolk, Va., October 20, 1980

Laymen may not understand, and may even fear, genetic engineering, the new science of splicing genes from one organism to create something new. But Wall Street investors think they see a winner in firms that venture into this scientific wilderness.

When stock in Genentech Inc. was offered to the public the other day at $35 per share, demand was so heavy that the price shot up to $87 before leveling off at about twice the offering price. Not bad for a company that has lost $1 million.

Genentech, however, is one of a handful of companies laying the ground for a new dimension in drug production, possibly even production of interferon, the newly-discovered substance from human cells which medical science thinks may combat cancer.

Genetic engineering conjures frightening dimensions, too. It involves transplanting a gene of one organism into a bacterium and producing what some fear would be forms of laboratory life.

Investors are gambling, nonetheless, that this process can produce valuable substances such as insulin or interferon. Even the American Cancer Society has invested heavily on the latter's potential. Only a cheap method for producing interferon is essential for it to be practical.

The Street is betting that genetic engineering will be profitable. Mankind must hope the profits will be counted in progress against disease, not test-tube creatures of uncertain parentage.

The News and Courier

Charleston, S. C., November 1, 1980

Oh brave new world that has such people in it! — Shakespeare, The Tempest

The English writer Aldous Huxley, in his futuristic novel, *Brave New World*, tells of a time when babies are decanted in factories rather than delivered in hospitals. Human beings, by and large, are made to order. Huxley's "Alphas," "Betas," and "Gammas" are perfectly, or almost perfectly, adapted to the work planned for them by a bureaucratic elite.

Brave New World, together with George Orwell's *1984*, painted a bleak and terrifying picture of a world that would be shaped by a generation that was then enthralled by scientific and social experiment.

Much has happened since *Brave New World* and *1984* were on the required reading list at most colleges and universities. The secrets of the atom have been disclosed. Man reached for and touched the moon. A passion for equality led to enormous transfers of wealth, and steady erosion of liberty.

The old standards of excellence in education and the arts fell by the wayside. Social promotion became an acceptable substitute for scholarship. Drama became pornography. Music, an obscenely amplified trip. Television, which held such early promise, emerged as the Great Leveler Down.

Today, we stand on the threshold of Huxley's Brave New World. Genetic engineers are busy creating new forms of life designed for specific purposes. Developments in the field are explosive. Experiments on human genes already have been carried out in Israel and in Italy by an American doctor grown impatient with "safeguards" established in this country. Somewhere, even today, a new race of Alphas, Betas and Gammas is stirring in the creative mind of some man or woman in a white coat.

Mankind is poised for a great leap into — what? A paradise of our own making? An abyss?

No. Only into the future. Only into the Brave New World our children shall — must — inherit.

Newsday

Long Island, N. Y., September 8, 1980

They are fearsome possiblities: scientists mistakenly unleashing dangerous man-made bacteria on the world, or altering the very genetic structure of animals or even people.

And yet, as the National Council of Churches has warned, there is "no government agency or committee currently exercising adequate oversight or control over, nor addressing the fundamental ethical questions" raised by the rapidly evolving research in recombinant DNA.

The plea for more such oversight, which was seconded by spokesmen for the Catholic and Jewish communities, went largely ignored. In fact, the federal government had decided early this year to further relax, not tighten, its controls over laboratory experiments in genetic engineering. And Congress has sidelined pending legislation to extend controls by law.

The Council of Churches appeal was issued in July. In August, a University of California scientist seeking to clone material for disease control research accidentally created a smallpox-like virus. And last week Yale researchers announced they had transplanted alien genes in a laboratory mouse.

Nightmares come true?

As it turns out, no. The California accident, although a violation of U.S. guidelines for federally funded research, supported the official view that laboratory experiments are safe: A university investigation showed the dangerous virus never posed any hazard, even to the experimenters.

And, although Yale's mutant mouse carried a gene transplant, the gene appeared to have no effect on the mouse's functions. No mouse monsters there.

Were the religious groups worrying needlessly?

We think not: The problem is not what happened in those two college labs, but what might happen elsewhere.

In fact, genetic research holds great promise of practical—and profitable—application. It's moving rapidly out of university laboratories and into the commercial world, encouraged by the U.S. Supreme Court's unfortunate ruling that new life forms can be patented.

The government's finding that university DNA experiments are safe doesn't necessarily prove that commercial applications, involving far larger quantities of materials, are safe too. And businesses, wishing to protect their control over new processes, are more inclined to be secretive about their research than academics.

At this point, only researchers backed by U.S. funds are required to follow federal guidelines for DNA research, though many commercial labs have done so voluntarily. Given the uncertainties that still remain, we don't think that's enough. Congress should pass a Senate bill, now stalled in committee, that would extend federal oversight to *all* DNA experiments.

WORCESTER TELEGRAM.

Worcester, Mass., May 26, 1981

A West German chemical company has offered, and Massachusetts General Hospital in Boston has accepted, a $50 million research grant in molecular biology.

On the surface, the announcement looks like a remarkable deal with benefits to all mankind. Hoechst A. G., an international chemcial company with headquarters in Frankfurt, will fund basic research at one of the world's great hospitals, a teaching and research facility of Harvard Medical School. Hoechst will put up considerable money, guarantee MGH academic freedom and in return will have an opportunity to have an advance look at and use of the research findings.

Molecular biology research will involve study of recombinant DNA and such byproducts as interferon, insulin, hormones and antibodies. These in turn will enable scientists to develop products to benefit medicine, agriculture, the chemical industry and who knows what else.

But the implementation of the research is still a long way off. Not all scientists and doctors are convinced that medical research should be so closely allied with commercial enterprise. And some local residents may be aroused, as they were in Cambridge, where Harvard was forced to cancel plans to organize a company to produce DNA and derivatives.

On the other hand, the United States government is funding less and less basic research. If there are to be further discoveries to be made in molecular biology and the chain of life, there may be no other answer than to accept financial help from industry.

Rocky Mountain News

Denver, Colo., May 24, 1981

THE Japanese are beginning to enter the new field of recombinant DNA research — popularly known as gene-splicing.

Although a professor at Tokyo University's Institute of Medical Science estimates that Japan is at least three years behind the United States in gene-splicing, the country's Ministry of International Trade and Industry plans to spend $150 million over the next decade to support corporate research in biotechnology.

If itty-bitty, efficient Japanese-made genes ever start invading this country and biting into the market for made-in-U.S.A. brands, don't say you weren't warned.

The Dallas Morning News

Dallas, Texas, July 10, 1981

Good news is coming regularly from private research laboratories on medical miracles that are being developed from new gene-splicing techniques. The latest is from Genetech and the U.S. Agriculture Department on a new vaccine for controlling hoof-and-mouth disease in livestock.

Strict import restrictions have controlled the disease in the United States since 1929, but it still ravages cattle, sheep and pigs in much of the rest of the world. Blisters develop in the animals' mouths and feet, weakening and reducing their agricultural value.

Officials say that the new vaccine, which does not need refrigeration, will substantially advance efforts to control the disease in underdeveloped countries. That means more food for people.

Not many years ago, some scientists opposed research into gene-splicing, conjuring up pictures of creation of Frankensteinian monsters. Fortunately, only voluntary controls were levied, and today even these are considered unnecessary.

The major result of the research is breakthroughs that can provide benefits for many of the world's hungry. As this latest example shows, the risk was worth it.

DESERET NEWS

Salt Lake City, Utah, June 10, 1981

Should there be a national policy on genetic research? If so, what should that policy be and who should set it?

These questions have gone largely unanswered for nearly a decade. That's how long it has been since the controversial gene-splicing technique known as recombinant DNA emerged from the laboratory into the public limelight.

This week a congressional committee started down what could easily be a long road toward some tentative answers. The House Science and Technology Subcommittee heard testimony from some of the leading figures in the academic world, including the presidents of Stanford University and Massachusetts Institute of Technology. The consensus so far, if there really is one, inclines toward cautioning Congress against hasty attempts at regulation.

That stance reflects more than just the usual resistance to federal intrusions against academic freedom. It also reflects the fact that regulation of genetic research and its commercial applications involves a foray into what is largely strange and unfamiliar territory.

Just a few years ago, some scientists were warning that genetic research might create uncontrollable new diseases. While DNA still has its critics, these early fears have largely died down.

One reason they died down is that the early fears were an over-reaction, and there is now general agreement that the potential risks were grossly exaggerated. Another reason is that the National Institutes of Health adopted a complex set of rules in 1976 governing genetic research and banning certain types of experiments.

But some scientists complained that the guidelines did not cover private industries or laboratories with independent funding. Others said the guidelines were overly restrictive.

The upshot is that there are now various legislative proposals to relax the NIH guidelines, review patent laws dealing with new forms of life, and set up a presidential commission to take up the ethical questions raised by genetic research.

The ethical questions involved in tinkering with genes are immense. Scientists are said to be on the threshold of using genetic engineering to produce superior breeds of cattle. Some understandably find it disturbing that genetic engineering in cattle is technically no different than in humans. The fundamental question is where do we draw the line as we manipulate biological processes, and who makes that decision.

The eagerness that some corporations have shown in exploiting gene-splicing also raises fears that ethics might be sacrificed for profit.

Still another complex set of considerations involves the patent laws. Last year the U.S. Supreme Court ruled that new organisms produced by genetic engineering may be patented. Some scientists believe the ability to patent genetic research will reduce the need to maintain secrecy about experiments and encourage a more free exchange of information, enabling the public to learn more about what is going on.

But others insist that genetic patenting might even enhance secrecy. Previously there was a free exchange of information and organisms among scientists. "But now if you can patent a strain and make a bundle," one scientist contends, "the organisms won't be publicly available until a patent is granted."

Still other observers worry that patenting of new plants produced by genetic engineering could give a few big corporations monopolistic control over seeds and reduce genetic diversity in plants.

At this point, there are far more questions than answers about regulation of genetic engineering. As Congress explores this field, it should keep in mind that federal control may not just prevent abuses. Rather, such control could also slow research and prevent new advances in science unless we're extremely careful.

The Boston Globe

Boston, Mass., May 14, 1981

Observers of the Boston City Council on even its best days would not blame scientists if they failed to show much pleasure at the prospect of having the Council regulate their research projects. Although that is exactly what the Council will be doing if it passes a proposed ordinance to regulate recombinant DNA research, even the scientists can see benefits in a measure which provides valuable protection for the city.

The safety issue has been raised in Boston because Genetics Institute, a firm headed by Harvard biologist Mark Ptashne, plans to lease laboratory space at Boston Lying-In Hospital to conduct research aimed at producing interferon, the promising anticancer drug, and insulin. Genetics Institute has voluntarily agreed to comply with safety guidelines issued by the National Institutes of Health, and has even incorporated the proposed ordinance – modeled on ones recently adopted in Cambridge and Waltham – into its lease agreement.

The ordinance, proposed by Councilor Raymond Flynn, will give the guidelines the force of law. Among the safeguards which will be most wecomed by people living near Mission Hill in the Harvard hospital area, are those which regulate the disposal of waste materials. Although scientists believe that the alien organisms they create will be too specialized to survive in the outside environment, no one is likely to want to test that theory and safe disposal rules will be in everyone's interest, particularly as a warning to any commercial laboratory tempted to take a few shortcuts to save a dollar or two.

Responsible scientists and others involved with the exciting – and potentially profitable – field of genetics research and production also recognize that it is better for them to locate in a community with reasonable standards already in place, rather than take their chances in an unregulated community where a crisis could trigger passage of overly restrictive regulations.

Fortunately, most of the DNA scare stories have remained only stories; they will die more easily if people, particularly those living in areas like Mission Hill, where DNA firms will locate, are confident that reasonable standards of safety are being followed – even if adoption of the standards provides a hospitable climate for research.

The News and Courier

Charleston, S. C., September 17, 1981

News that Ohio University microbiologists have succeeded in inserting a rabbit gene into mice signals more apparent progress in the field of genetic transfer. The announcement was accompanied by the researchers' assurance that their experimentation is aimed at speeding selective breeding of farm animals; that inducing the physical traits of one animal into another is not the goal; that it is "most unlikely" the transfer process could be used on humans.

Controlled hybridization of animals has, of course, proved valuable. Crossing Short horn cattle with Brahmans produced, for example, a hybrid that thrives in hot and humid climes. Hybridization produces pigs with more lean and less fat, and chickens that lay more eggs but eat less food.

While benefits to be derived from speeding selective breeding in animals are appealing, there remains room for reservations about tampering too much with nature's schemes. To be avoided are such hybrids as rabbits that kick like mules and mice that roar.

AKRON BEACON JOURNAL

Akron, Ohio, September 14, 1981

DESPITE all the "gene tinkering" scientific work that has been done through recent years and all the exciting news accounts it has generated — the most recent about the introduction of a rabbit gene into mice by a team of biologists working at Ohio University and at a laboratory in Maine — science is still groping in almost total darkness in the field of genetic tailoring.

So far, at least, we are still a long, long way from the power to do the sort of "genetic engineering" in human beings that causes an almost superstitious uneasiness in most of us. In fact, none of the hundreds working in this field has yet achieved any demonstrable success in decreasing proneness to even one hereditary abnormality or weakness in humans, animals or plants — the prime purpose of genetic engineering.

The biologist who heads the rabbit-mouse experiment, Dr. Thomas Wagner of Ohio U, foresees a time when we will be able, through gene insertion, to give beef and dairy cattle a bison's ability to convert into meat and milk coarser forage than any on which cattle can now thrive.

And Dr. Joseph Jollick of Ohio U, a co-worker on the project, says researchers can probably achieve this kind of gene transplant among farm animals within the next five years.

This would allow substantial improvement of a breed within a single generation rather than through the many-generation process of selective breeding mankind has used for millennia to achieve desired characteristics in domestic animals.

But though the outcome of this experiment — transferring rabbit hemoglobin genes into mouse sperm cells just after their penetration of mouse egg cells — may foreshadow much wider-ranging powers later, it did nothing to improve the breed. Two of the 46 "doctored" mice born merely transmitted to their young and their young's young a slight anemia traceable to the rabbit gene.

The problem is that so far scientists know almost nothing of the linkage between specific genes and specific traits. These researchers, for example, had no particular interest in developing a strain of anemic mice. They chose this particular gene merely because it is one of very few known to be linked to an identifiable trait.

Because of this lack of knowledge, Mr. Jollick commented, it is most unlikely that the process developed with the laboratory mice could be used to alter human traits. This would require multiple genes, he said, and "You'd have to identify the genes involved and we're not even at that stage yet."

Scientists questioned by New York Times reporter Harold Schmeck for a recent story on genetic engineering, in fact, saw "little or no prospect" that anyone will *ever* transplant genes into fertilized human egg cells.

They did, however, express optimism that we will eventually be able to use related techniques against certain kinds of hereditary ailments; diabetes and sickle-cell anemia have been mentioned as examples.

The significance of the rabbit-mouse experiment is not easy for non-scientists to appraise. With varying techniques, others have earlier "crossed" cells from differing mammals by laboratory introduction of genes — a human interferon gene into a mouse ovum, for example.

The rabbit-mouse experiment apparently did at least two things that are new and different:

- It focused on a fleeting moment in the reproductive process when cells are the least likely to attack and destroy anything foreign coming at them: the moment when both egg and sperm cells *must* be ready to tolerate something alien to do their job.
- And it succeeded, for the first time, in transferring a traceable trait that would "breed true" for three or more generations.

But we remain a long, long way from a time when gene scientists can manufacture human geniuses or monsters in their laboratories, or wipe out any of the thousands of human ailments and abnormalities known to be linked to heredity.

What we have is a very preliminary basis for hope, not fear.

The Dispatch

Columbus, Ohio, September 19, 1981

OHIO UNIVERSITY researchers have successfully mixed rabbit and mouse genes in a series of experiments that will help alter the future of scientific investigation into the very essence of life.

We hail their triumph and stand in awe at their scientific agility. But we are concerned by society's inability to keep pace with and evaluate progress in genetic research. Genetic research could dramatically change life on earth, yet the layperson knows pitifully little about it, and understands even less.

Genes, and their particular arrangement within a species, determine animal characteristics and abilities. They make a cow a cow, and make it different from a dog, or a human.

For years, scientists have worked to isolate genes that are beneficial in one animal and to implant them in another animal in order to derive new, more beneficial traits in the host animal.

OU researchers accomplished this feat for the first time in history. They produced mice with the protein-generating capability of rabbits. "Application of this technique could dramatically shorten the time necessary to selectively breed species of animals with food-producing characteristics," said Thomas Wagner, the molecular geneticist who led the OU team.

Continuation of this research could lead to the creation of animals with vastly higher food-production capacities. This could have a significant effect on world hunger, Wagner points out, since there are "vast regions of the world where the only thing that grows well is grass and the only way to convert grass into a usable food source for humans is to have animals eat the grass."

No one objects to the beneficial possibilities of genetic research. What raises concern is that the techniques developed could be used in ways generally regarded as not beneficial. Would some distant society breed out "undesirables" through genetic engineering? What would Hitler have done with this knowledge? Could an accident produce a bacteria for which their is no known control? Could researchers, not nearly as careful as the OU team, produce a creature that would serve to limit food production?

These are weighty questions prompted by OU's research breakthrough. The consideration of the ramifications of the research must go hand-in-hand with the progress of the research. It is not meant to diminish the researchers' work. They do their job well and and take their work seriously.

It's time society as a whole started taking their work seriously, too.

The Providence Journal

Providence, R. I., June 2, 1981

When tinkering with genes, scientists have to be careful. The controversial new field of genetic engineering, dealing with those tiny components of cells that determine heredity of all living things, carries a potential for danger. Seen in this light, recent disciplinary action against a California scientist who overstepped federal regulations is both understandable and warranted.

The scientist is Dr. Martin J. Cline of the University of California at Los Angeles, who sought to treat with genetic experimentation two patients suffering from grave hereditary blood defects. It might be argued that he did no harm. Indeed, the National Institutes of Health (NIH) in its report on the incident did not suggest or imply that Dr. Cline had endangered the patients or that his research was intrinsically improper.

Why, then, the fuss?

The larger issue, which transcends Dr. Cline's activities, is whether genetic research with an enormous potential for good or ill should be subject to the government's close supervision. The consensus in both the scientific and lay communities is that careful regulation is very much in order. This type of research carries so many overtones of risk and imaginative enterprises in the realm of human cloning that only strict adherence to the rules can avoid appearances as well as specific instances of impropriety.

Guidelines for genetic research and federal regulations to protect human subjects of research have been established. Without strict enforcement these rules would become a charade. The scientist whose style they crimped could wink at the provisions and go his own way.

Making an example of Dr. Cline, therefore, is a preventive measure to be observed by every gene-splicing laboratory in the country that receives federal funds. A committee of the NIH has found that Dr. Cline violated the guidelines and deserved disciplinary action. Said Dr. Donald S. Fredrickson, NIH director, "My examination of the report of the committee and of the larger record upon which its decision was based leads me inexorably to agreement with the conclusion that Dr. Cline has violated both the letter and the spirit of proper safeguards to biomedical research."

The recommended penalties would include a requirement that Dr. Cline clear certain aspects of his work with NIH in the future, increased difficulty in obtaining future research grants and the possibility of losing a large portion of $600,000 in grants he has been awarded.

It is not pleasant to see adverse attention brought to the work of a reputable scientist like Dr. Cline; but neither would it be in society's interest to gloss over serious infractions of the rules so that others would be encouraged to depart from accepted norms. Strict enforcement now can prevent untoward consequences in the future.

The Seattle Times

Seattle, Wash., June 8, 1981

THE filing of charges by the National Institutes of Health against a California physician has refocused attention on an enduring question concerning the future of gene splicing and other applications of so-called genetic engineering:

Are existing government controls sufficient to protect the public against genetic experiments that might go awry?

Federal guidelines have been in place for several years, but they apply only to federally funded research.

Large federal grants held by Dr. Martin J. Cline of California could be withdrawn if the government affirms its case that he broke the research rules. Last year, Dr. Cline withdrew red blood cells from two young women suffering a fatal genetic disease of the bone marrow and inserted "cloned" genes in hopes they'd multiply and fight the disease.

One reason the procedure violated the guidelines is that it never had been proved effective in laboratory animals, let alone humans.

The Christian Science Monitor, meantime, reports that a sharp increase in genetic engineering projects in the private sector is behind a growing demand for the application of federal guidelines to private as well as public research.

Gone by now are the science-fiction visions of destruction by new life strains escaping from a genetics laboratory. But risks are present in genetics research — the release of potentially harmful organisms, for example — and Congress would do well to re-examine the adequacy of existing controls.

The Ann Arbor News

Ann Arbor, Mich., June 1, 1981

RECOMBINANT DNA research, which involves re-arranging genes into new forms of life, was the leading subject of campus debate at the University of Michigan half-a-dozen years ago.

The subject has scarcely made its way into the news, outside of science journals, since it was realized that keeping the subjects of DNA experiments alive in laboratories is more of a challenge than preventing them from escaping.

NOW, the long-feared possibility of a serious mistake involving DNA research has become general public knowledge. No researchers at the U-M or in this part of the country were involved, and it appears no one was injured. This involved some luck, because the case involves research on humans – an Italian and an Israeli – who were used as test subjects by a University of California at Los Angeles researcher, Dr. Martin J. Cline.

His error was not in the mechanics he used in developing genes he injected into the subjects, in hopes of developing a treatment for a rare blood disorder. His error was to reject a decision by UCLA colleagues that more research on animals should come before experiments with humans, and then following a procedure different from one approved by Israelis who were aware of Dr. Cline's work.

Scientists and federal administrators responsible for general supervision of recombinant DNA work in this country are not letting this incident slip by. Dr. Cline faces a strong possibility of losing $600,000 in research grants, and being required to seek approval for any future DNA research and grants applications directly from the National Institutes of Health.

For the general public, this means that a field of research full of promise, in areas ranging from cancer research to food production, is being pursued with no break from the standards of control scientists set when they started.

THE ATLANTA CONSTITUTION

Atlanta, Ga., June 1, 1981

Genetic engineering research — commonly called gene-splicing or recombinant DNA research — already has created a new era in biological research which raises both legal and moral questions in the minds of most Americans. Rigid guidelines outlined by the National Institutes of Health are to be followed by all conducting research in this area. When violations of these regulations occur, severe disciplinary action should be taken.

A noted California blood specialist has been accused of the most serious violation of federal guidlines for genetic engineering on record. Dr. Martin J. Cline violated the guidelines when he inserted man-made genes into two young foreign women with a fatal blood disorder. The purpose of his experiment was to try to overcome grave hereditary blood defects by transplanting copies of a normal gene into each patient to serve the function that their own defective genetic endowment was lacking.

The NIH has outlined broad disciplinary action against Cline, and it should be carried out. It is difficult to make the decision to restrain science or hamper its research by restricting federal funds for work that may well prove not only feasible, but beneficial to human health and life in future years. Still recombinant DNA research is a most sensitive area. The government guidelines were designed to ensure that the research work would be done in a manner that is above reproach.

The controversy of recent years concerning gene-splicing originally centered on fears that the work might beget dangerous abnormal microbes, upset the balance of nature or even impell some unscrupulous persons into attempts to redesign human heredity in ways that defied predictions and might prove disastrous. That fear still exists. Research in this area can either be a blessing or a horror story.

The NIH became aware of Dr. Cline's research with human patients last year, after the attempted treatment provoked public controversy. Cline has since resigned the chairmanship of an important division of the UCLA Center for the Health Sciences, but has remained on the faculty as an endowed professor.

The special institute committee that investigated Cline recommended in part that he be required to clear certain aspects of his future research proposals with the NIH. Other parts of the disciplinary actions may make it difficult for Cline to get new research grants, and might eliminate a large portion of his current $600,000 in research money.

While this is a stiff penalty imposed on one of science's leading minds, it is necessary. Guidelines for the fast-advancing field of recombinant DNA research probably will not keep step with the research itself. Public fears and dangers surrounding the research will not vanish if the rules continue to be violated. There must be no exceptions to the guidelines.

Atlanta, Ga., June 22, 1981

When gene-splicing research is conducted according to rigid guidelines outlined by the National Institute of Health, some wonderful scientific discoveries can be made. The recent development of an effective vaccine against foot-and-mouth disease is a perfect example of how the technique can boost the new era in biological research. The disease is one of the world's most economically serious infections of livestock.

Still, researchers should continue to be warned by authorities that there are dangers surrounding the controversial research. Fears still exist among the public that gene-splicing techniques could produce abnormal microbes, upset the balance of nature or even impel some unscrupulous persons into disastrous attempts to redesign human heredity.

The development of the vaccine was a collaboration between scientists of the U.S. Department of Agriculture and Genentech Inc., a California firm that is among the leaders in gene-splicing research and development. It is believed to be the first production — through gene-splicing — of an effective vaccine against any disease in animals or humans.

Foot-and-mouth disease, formerly called hoof-and-mouth disease, is a virus that seldom kills, but produces sores in the mouths and feet of cows, sheep, and other cloven-hoofed species. The virus weakens the animal and reduces its agricultural value. Agricultural officials predicted that the use of the new vaccine could mean an annual saving of billions of dollars, and an increase in the world's meat supply.

Although the United States and some 30 other countries are free of the disease, it exists in some parts of Europe and is widespread in regions of Africa and Latin America. Since World War II there also have been serious outbreaks of the disease in Mexico and Canada. The vaccines now being used are not effective in these regions, and in many cases have been implicated in outbreaks of the disease. These countries will benefit greatly from the new vaccine discovery.

This is indeed a blessing in the area of gene-splicing research, but the potential for abuse remains. Federal officials already have recommended disciplinary action against one researcher who violated the guidelines in conducting his work. Officials should continue to watch researchers and slap stiff penalties on violators. One unscrupulous act, and the next story heard from gene-splicing could be one of horror.

Atlanta, Ga., September 14, 1981

It has been only eight years since gene-splicing technology was developed — a process that mixes the genetic material of two or more species. Since that time, some scientists have warned that such experiments might create new organisms that could unleash uncontrollable diseases. If this is so, then how can a National Institutes of Health advisory committee dare discuss abolishing federal regulations on the research?

The advisory committee voted 16-3 last week to end all the regulations on experiments that mix the genes of different creatures. Instead, the committee said it would rather see research steered by voluntary guidelines like those governing other possibly hazardous experiments in biology.

Although the gene-splicing technique still has its critics, much of the public controversy seems to have died down. The main apparent reason for public silence on the subject was the institution of rigid federal guidelines that the committee now wants to abolish. The fact that researchers were bound by the guidelines seemed to calm public fear. Without those regulations, protests from the public almost surely will return. The committee's proposal will be published and sent out for public comments before a final vote is taken.

The religious community continues to raise ethical questions regarding genetic research. Religious leaders say "new life forms may have dramatic potential for improving human life. . . . They may also . . . have unforeseen ramifications. Control of such life forms by any individual or group poses a potential threat to all of humanity. Those who would play God will be tempted as never before."

Scientists fighting for relaxation of the NIH guidelines argue that they were initially adopted in response to ignorance on the subject. They now say there is a wealth of experience in the field and that the current and more realistic appraisals of the research have led scientists to believe that it is less dangerous than originally thought. They claim that this new evidence has not been applied in deciding whether federal guidelines should be abolished.

Few will argue the benefits of gene-splicing. Dozen of corporations have poured millions of dollars into genetic research, claiming the products of this research could revolutionize agriculture, greatly simplify control of pollution and lead to cures for diseases like cancer.

However, the fact still remains: it only takes one unscrupulous act by a researcher to caused a catastrophe. Why take that chance?

The Dispatch

Columbus, Ohio, February 14, 1982

A GOVERNMENT agency has wisely decided to alter a previous decision and keep mandatory certain restrictions on genetic research experiments involving desoxyribose nucleic acid (DNA).

The Recombinant DNA Advisory Committee, which works with the director of the National Insiututes of Health, says it decided to change its earlier decision to ease restrictions on experiments after receiving comments from scientists and the public.

Genetic research deals with the very mysteries of life: how living things are made, grow, reproduce and interact with one another. Experiments in this field seek to alter natural systems and find ways to improve those systems for various reasons. At Ohio University recently, for instance, scientists were able to produce mice with the protein-generating capacity of rabbits — a technique that "could dramatically shorten the time necessary to selectively breed species of animals with food-producing characteristics," according to one scientist involved in the work. Such breakthroughs could have a significant effect on the world's hunger problems.

But DNA research is a relatively new field and the government and scientific organizations have imposed limits on what can be done in the laboratory until more is known about both the experiments and their possible effects. No one wants DNA experiments to get out of hand and produce dangers to human, animal and plant life.

The advisory committee had felt that enough experience had been gained to ease some of these restrictions, and it proposed making them voluntary. But concerns voiced by many over the possible dangers involved has caused the committee to reconsider. The committee is now expected to ease some restrictions but to keep them mandatory.

It's good to be cautious. There is still much that is unknown about the way genes work, and the best approach to this subject is the safe, slow approach.

THE DAILY HERALD

Biloxi, Miss., February 10, 1982

A government advisory group has recommended that the National Institute of Health not abandon its restrictions on genetic research. It is sound advice

Scientists are becoming more confident of their abilities to control recombinant DNA experiments, those concerned with gene-splitting. Experience indicates there is evidence to support that confidence. But there is also at least one instance which indicates the value of restrictive guidelines.

Three years ago, the consensus of knowledgeable scientists was that the then-current federal restrictions on gene-splitting experiments were adequate. Despite that consensus, there was a considerable concern among the public that restrictions had to be stringent; the concern was greatest, naturally, in those communities where the pioneer gene-splitting work was being done.

Scientists have produced some remarkable results in their dabblings into the new science, taking genetic material from one organism and adding it to another to give it characteristics it would not normally have. Working with bacteria, scientists have developed the ability to produce large quantities of rare compounds such as drugs, hormones and vaccines.

The progress has been made under tight safety guidelines

which the National Institute of Health established in the early 1970s. The rules were mandatory for all government-sponsored research, but were followed voluntarily by private industry. Several times since the guidelines were adopted, they have been relaxed.

There still exists the potential for serious biohazards to be created in the gene-splitting experiments.

Last March, remember, a former biology professor at the University of California at San Diego conducted cloning experiments with a rare African virus. The professor, Ian Kennedy, was found to be in violation of several federal guidelines for genetic research. He was charged with improper laboratory practices and subsequently resigned his post to labor in the private sector. A university biosafety committee reported that Kennedy may have falsified data and had failed to keep an accurate record of his work.

While this is the only publicized incident of serious violation of the guidelines, it is sufficient to make the point that the potential for biohazards will continue to exist as long as scientists conduct pioneering research in this field. Safety concerns should not be allowed to subside and the guidelines should not be abandoned.

Japan Shows Interest in U.S. Fusion Projects

The development of energy from atomic fusion, the combining of atoms instead of splitting them apart (fission), holds great promise for the future of nuclear energy. Fusion, unlike fission, generates little radioactive waste, uses plentiful hydrogen rather than scarce uranium and produces more energy at a lower cost. A major step forward was achieved in August 1978 at Princeton University, where a test reactor managed to produce a temperature of 60 million degrees Celsius, well on the way to the 100 million degrees Celsius required for a fusion reaction. In September, Japan concluded an agreement with the U.S. to cooperate on research in fusion and other energy technologies. Japan's commitment to nuclear energy was well established beforehand. According to a National Science Board report, *Science Indicators 1980,* Japan had allocated 66% of its energy research funds to nuclear energy in 1978, down from 74% in 1977, but well ahead of the U.S. proportion of 44% for 1978. Japan also embarked on a $1-billion project to build a "tokamak" fusion generator, scheduled for completion in 1985.

THE DAILY OKLAHOMAN

Oklahoma City, Okla., June 26, 1978

JAPANESE Prime Minister Fukuda has made the United States an offer which we shouldn't refuse because it could lead to fantastic benefits for both countries.

Briefly, Japan is willing to contribute up to $1.6 billion for a joint research and development project to speed the advent of fusion power. Our Department of Energy is reportedly receptive, not only to the Japanese offer but to the concept of expanded international cooperative energy research.

Experimental work already is in progress here and in Europe and Japan on so-called tokamak reactors. The goal of scientists since the dawn of the atomic age has been to find a way to harness the unlimited energy potential of the fusion process which in its military application, produces the dreaded Hydrogen Bomb.

The problem is to find ways to induce and contain the fusion of hydrogen or tritium atoms, capturing the tremendous release of energy and transforming it into electricity. Temperatures approaching those of the sun's surface are needed to start the process.

Pooling of the best scientific and engineering minds in the advanced nations of the world in a cooperative program of fusion energy development could hasten the day when virtually unlimited and non-polluting energy would be available for all mankind on the planet. The potential is staggering to contemplate.

It is clearly in our national interest to conclude a cooperative research and development agreement with Japan, and the Carter administration should waste no time in accepting the offer.

Lincoln Journal

Lincoln, Neb., August 15, 1981

On this page last week was a fascinating article originally published by and in a major Japanese newspaper, detailing the Japanese advance in the field of cost-cutting industrial robots.

Everyone agrees that Japan now leads the world in development of such assembly-line machines. When you add to that the knowledge the Japanese are relentlessly dedicated to coming up with improved robots and wider robot applications, Japan's lead is not likely to be threatened.

Industries the world over, including those in the United States, increasingly will be purchasers of Japanese robots and/or Japanese technology.

It is a technology which, incidentally, originated in the United States but failed to take deep root here.

In another critical area, we may be witnessing pretty much the same pattern taking formative shape.

Marsha Freeman, research director of the Fusion Energy Foundation, says that although Japan's technological leadership became convinced rather earlier than its American counterpart that fusion could be a primary energy source in the oil-depleted 21st century, U.S. scientists pioneered in fusion experimentation. They are still ahead of anybody, with the possible exception of the Russians.

But Ms. Freeman suggests Japan is committed to moving ahead.

In another decade, she says, Japan could outpace the U.S. in fusion development work. And as a new century rounds into view, the U.S. could be importing Japanese technology and equipment for fusion power plants in the U.S.

If this all has a kind of familiar ring — and it does — you get to wondering what kind of vibrations said ring stimulates among America's competitive captains of industry.

How willing are they, or the nation, to make long-range giant research and development investments which may or may not pay off when there always is either a quarterly corporate statement which must be better than the one before it, or another election coming in which candidates must pander to calls for reduced government spending?

In today's climate of opinion, those may be regarded as unfriendly sorts of questions. They probably aren't, however, in Japan, looking strictly at the record.

BUFFALO EVENING NEWS

Buffalo, N. Y., August 16, 1978

Dr. Glenn Seaborg, for many years head of the former Atomic Energy Commission, once called controlled fusion the "most difficult scientific-technological project ever undertaken by mankind."

Scientists have known since the 1930s that if the enormous technical challenges to controlling nuclear fusion could be mastered, the power it would generate could ensure an inexhaustible energy source for a world whose fossil fuel reserves are declining.

Theoretically, the deuterium atoms in only one gallon of seawater could yield energy equivalent to 300 gallons of gasoline — provided science laboratories could find a way to safely harness the sun-like temperatures necessary for the fusion of atoms. Until recently, most scientists have assumed that the solution of the heat problems in any practical power-producing fusion reactor is doubtful any time before the 21st Century.

It still may be, given the magnitude of the challenges in confining temperatures that would melt any known metal. Yet experimental research has been moving ahead in various laboratories here and abroad, and Princeton University scientists have now achieved what energy officials hail as the most significant step since the beginning of fusion research.

Using a magnetic field to confine the blazing gases in a prototype fusion reactor, the Princeton researchers are said to have established the scientific basis for eventually using the pressures to run steam turbines producing electricity.

Too many developments and discoveries remain to be made before nuclear fusion can be regarded as a potentially feasible supplement for fossil fuels, solar power and the splitting of uranium atoms.

Even so, the potential advantages of fusion, in contrast with the problems handicapping reliance on fission in the present nuclear power technology, are so great as to well merit all the federal research dollars that scientists can effectively use.

For compared with both fossil fuels and nuclear fission, the fusion process could take over the job of generating power at only a fraction of their cost. It promises an inexhaustible source of energy without the hazards of nuclear waste or proliferation of plutonium, thus easing both environmental and national security concerns. It would free depleted petroleum reserves for use in the manufacture of chemicals, plastics and fertilizers, as well as for transportation.

One breakthrough does not mean that fusion can be hailed as the energy wave of the future, but it does suggest that the rewards of an unlimited and renewable genie for doing the world's work may be closer than scientists have assumed.

The Providence Journal

Providence, R. I., August 18, 1978

The promise of almost unlimited energy through thermonuclear fusion — the same process the sun uses — is so attractive that it must be brought to fruition, but the technological difficulties may be the most challenging that man has ever faced.

Just think: The heavy-hydrogen isotope for the fuel in a fusion reactor can be extracted from sea water.

In this kind of reaction, positively-charged nuclei or ions are compressed or hurled at each other with tremendous force. The squashing makes them give off heat. It is the opposite of nuclear fission, in which atoms are split, and yields 10 times as much energy. We achieved fusion in the awesome hydrogen bomb experiments of the 1950s. Since then, scientists have been trying to harness the reaction for peaceful purposes.

To gauge the point that the scientists have now reached in this quest, imagine that you are Orville or Wilbur Wright, just enjoying the triumph of the first sustained air flight at Kitty Hawk in 1903. A spectator walks up and describes his concept of a huge, supersonic aircraft that can whisk hundreds of passengers from New York to London in a few hours. You understand what he proposes, but when he asks if you can start to build such a machine, you gulp and hesitate.

A pause for sober reflection, as well as pride in American know-how, is merited by the recent announcement that Princeton University's laboratory model fusion reactor has achieved a temperature of 108 million degrees Fahrenheit. This is well within the range of heat that has to be attained *before* a fusion reaction can be initiated in the magnetic "doughnut" that Princeton is using. Some other laboratories are experimenting with magnetic mirrors.

But magnetism as a means of confining the fuel in just the right way may not be the route we eventually go. In various other experiments, lasers or converging beams of ions are being used to compress fuel pellets and make their nuclei fuse.

To call the Princeton achievement a breakthrough, as a Department of Energy spokesman initially said, is oversell. It may be a breakthrough when scientists get more energy out of a fusion reactor than they put in. This may occur in the coming decade, but we will still be looking only at a scientific demonstration, not a commercially useful producer of electricity.

Any high technology is likely to have environmental impacts. Fusion may have several. For example, what would be the effect of super-high magnetism on living creatures if we follow the doughnut model?

Despite reservations that any reasonable person can think of, the expenditure of $500 million a year on fusion research, a figure advocated by the Carter administration, is warranted if it will lead to a true breakthrough in this utterly fascinating exploration.

The Oregonian

Portland, Ore., August 16, 1978

A small milestone has been passed on the long and difficult road ahead for the development of a practical fusion reactor, capable of producing electric power from hydrogen isotopes abundant in the oceans.

While no major "breakthrough" has occurred, a significant step has been taken. Dr. Melvin B. Gottlieb, director of the Princeton University laboratory where a fusion temperature experiment took place several weeks ago, said that some information released to the press over the weekend was "not entirely accurate." He said science still has a long way to go to prove the feasibility of fusion.

Unlike the current nuclear power reactors that use fission to produce heat and energy, the fusion reactor would use the same processes that take place in the sun, or in the hydrogen bomb. The work at Princeton has proceeded on technology borrowed from the Russians in the late 1960s.

At a press conference Monday, called by the Department of Energy officials, Gottlieb explained that a powerful pulse of energy was aimed at a gas inside a small, doughnut-shaped device called a Torus Tokamak. A temperature of 60 million degrees centigrade, or about four times hotter than the sun, was achieved. But the temperature lasted only a tiny fraction of a second and the experiment absorbed 100 times the energy created by the reaction, he said. This temperature is well above the theoretical threshold of 44 million degrees needed to achieve a fusion reaction. It represents an important milestone.

The reaction failed by a wide margin to reach the energy break-even point. Gottlieb suggested the break-even point might be reached in the 1980s, but he said it would take another 20 or more years to produce a commercial fusion reactor.

Thus, energy salvation-by-fusion is still not around the corner — not even the turn of the century corner. Other forms of electrical energy, particularly nuclear power reactors and their fuel-producing breeders, will be needed for a long time.

We can only hope that fusion can be brought on the line in time to save the world from an acute energy shortage that will develop as oil and gas supplies are pumped down early in the next century.

Scientific milestones are always helpful, particularly when an agency needs to defend its budget, which is the case with fusion where congressional critics have been successful in getting Department of Energy research funds for fusion reduced.

The Honolulu Advertiser

Honolulu, Ha., November 3, 1978

Experts tend to think that when thermonuclear fusion is perfected for peaceful uses, our worries about energy shortages will be as extinct as the dinosaurs.

Fusion is the process of hydrogen bombs. More fundamentally, it is the process by which energy is produced in the sun and other stars. It involves a nuclear reaction in which nuclei combine to form more massive nuclei with the simultaneous release of energy.

In the race to harness that process for peaceful uses, the United States now appears to have zipped far ahead of the Soviet Union.

RESEARCHERS around the world currently are pursuing fusion programs along two major lines. One involves the use of lasers to trigger small thermonuclear explosions with bursts of light. The other involves the use of powerful magnetic fields to bottle up the superheated hydrogen plasma in which fusion occurs.

Los Angeles Times correspondent Robert Gillette, who has been writing extensively on Soviet fusion, reports enormous problems for the Soviets in both areas.

The major Soviet fusion laser program is centered on the Delphin machine at Moscow's Lebedev laboratory. Western experts who have recently been there say the project appears to have made little progress in about a year.

In contrast, the largest American laser fusion machine, the $25 million Shiva at Lawrence Livermore Laboratory near San Francisco, is working better than expected.

As for magnetic fusion, the U.S. is constructing a $230 million Tokamak reactor at Princeton University. It is expected to start up in 1982.

In contrast, Soviet plans to build a giant Tokamak in the early 1980s have slipped to the end of the decade. A smaller version is under construction but isn't expected by Soviet or American scientists to equal the Princeton device.

All of this suggests that American physicists will be the first to achieve "energy breakeven," a term referring to a fusion reaction that produces as much energy as is used directly to trigger it.

Crossing the breakeven point would widely be regarded as the clinching test necessary to show that controlled fusion is scientifically feasible. But about 20 years of work would probably be necessary before electricity could be generated economically from the fusion process, though this work would primarily center on engineering rather than theoretical physics.

SOME AMERICAN scientists fear that because the Soviets are lagging, the U.S. Congress and administration may become less willing to fund fusion research at the present rate.

That would be short-sighted, because fusion would have enormous intrinsic value to us as an energy source regardless of the rate at which Soviet research proceeds.

Chicago Tribune

Chicago, Ill., February 10, 1978

The Carter administration has proved itself a pretty poor judge of bargains by denying Chicago's Fermi National Accelerator Laboratory the funds it needs to maintain its pre-eminence in nuclear research. For the $35 million the laboratory had sought this year to build a Tevatron, it would have obtained a machine worth perhaps $1 billion with immeasurable potential in the field of energy. The administration has allowed the laboratory only about $4 million in new money for 1979 barely enough to pay for inflation.

Without adequate financing, the laboratory stands to lose both the Tevatron and the services of its director, Dr. Robert R. Wilson, who is given credit for much of the laboratory's past success and has submitted his resignation as a result of the administration's shortsighted decision.

The Tevatron [from TEV for trillion electron volts] is an advanced accelerator capable of probing far deeper into the secrets of nature than we have been able to probe before. Developing this capability will involve as a byproduct the development of new technologies in superconductivity which could eliminate present losses in the transmission of electrical power. Applied to electric power companies, the saving in energy would be about 10 per cent. The machine would be almost certain to pay for itself many times over in savings of energy.

A Tevatron would cost about $1 billion to build from scratch, but the Fermi laboratory can build one for $35 million by "piggy-backing" it on the existing accelerator, a four-mile circular "race-track" for nuclear particles 35 miles west of Chicago.

A number of other countries are working on similar advanced atom-smashers, and without the Tevatron it is more than likely that the United States would lose its lead in nuclear research, and Chicago would lose a distinction it has enjoyed in this field ever since the first chain reaction was produced under the bleachers of Stagg Field back during World War II.

The Carter administration has already handicapped American progress in nuclear power by denying funds for the breeder reactor. To lose both the Tevatron and the services of Dr. Wilson would surely deprive Chicago and the Fermi laboratory of the pre-eminent positions they have enjoyed and would make it harder to attract the caliber of physicists needed to take advantage of the laboratory's existing equipment, which has already cost about a half-billion dollars—and this was below budget.

We sympathize entirely with the administration's search for ways to cut spending. But surely there are wiser economies to be made than this.

THE CHRISTIAN SCIENCE MONITOR

Boston, Mass., April 14, 1978

For more than two decades, physicists who dreamt of taming hydrogen fusion – the power source of the stars – pursued an ever-receding goal. Now they have the goal in sight and may soon have fusion running in the laboratory. It seems ironic that skepticism should threaten their efforts at this point.

Some theoretical studies suggest that the main line of development now being followed will only lead to industrial juggernauts that no utility could afford. Those who accept this reasoning, including some high energy officials in the Carter administration, suggest cutting down on fusion development and sending it back to the basic research stage. They want to learn more about the fundamental physics involved and to explore a variety of technical options.

Admittedly, getting fusion going in the lab would be only a first step toward the ultimate goal of a full-scale practical power plant. Nevertheless, it would show that physicists have at last mastered the difficult art of containing a gas of electrically charged particles at the fantastically high temperature of 50 million degrees C or more and of getting it to produce energy. It would give them the opportunity to begin practical tests of the factors that must be taken into account in designing actual power plants. This would give the theorists a better data base on which to rest their projections.

Thus it seems premature and self-defeating to be faint-hearted about fusion at this stage. Throttling back funding and reorienting research could waste much time and money that have gone into the most successful – and most seriously questioned – line of attack.

This involves fusion devices that control the hot gas with magnetic fields in doughnut-shaped machines called tokamaks. Many hundreds of millions of dollars have been invested in tokamaks in the United States, as well as in a number of other countries. Some of these machines now built or on order, such as that recently dedicated at the Massachusetts Institute of Technology, promise to ignite a self-sustained fusion reaction within the next few years.

This is the work that would be most seriously curtailed if the skeptics have their way. Dr. Edwin E. Kintner, acting director of the Office of Fusion Research at the U.S. Department of Energy, has warned researchers in this field that the threat of cutback is serious. He says he and other supporters of fusion are having trouble persuading responsible officials that it is wise to continue to pursue the successful magnetic fusion research line vigorously.

To us, this wisdom seems obvious. We trust President Carter and Energy Secretary James Schlesinger will listen more to the arguments of successful experimenters than to the doubts of skeptics who prematurely prophesy economic disaster.

Reagan Administration Alters Science Policy

President Ronald Reagan came to office in 1981 determined to reduce federal spending. In the first annual report of the Office of Science and Technology Policy under his administration, issued April 1982, Reagan acknowledged "the important role of the federal government in supporting our scientific enterprise." However, he maintained that "some things can best be done by the private sector." Accordingly, reductions were made in the federal science budget, and the remaining funds were earmarked for projects that "justify the cost to the federal taxpayers." The report emphasized that an "important criterion for the public support of applied research . . . must be pertinence—pertinence to the nation's economic and social goals and needs." Reagan's preference was for practical research directed toward practical ends, rather than the quest for knowledge for its own sake. It was "applied" technology versus "basic" research. Applied technology is something the public easily understands. It has immediate impact: labor-saving devices, improvements in health care, transportation or communication. Basic research is difficult, if not impossible, for the non-scientist to understand. It is such areas as physics, animal behavior, anthropology—disciplines that have no immediate application to daily life. However, basic research is the foundation on which applied technology rests. We could not have television or space flight without physics, or be able to preserve endangered species without knowledge of animal behavior. Basic research depends on trust, the belief that in the long run, all knowledge is valuable.

THE ARIZONA REPUBLIC

Phoenix, Ariz., November 25, 1980

THE National Science Foundation suddenly finds itself interested in graphoanalysis.

It sees the handwriting on the wall.

Congress has lost patience with some of NSF's grants for scientific esoterica.

With the arrival of cost-conscious new senators and representatives, the foundation can expect tougher questions about the way it parcels out $1 billion a year.

NSF-sponsored research has accomplished a certain amount of good.

When money's tight, however, taxpayers are angered by NSF spending:

✓ $350,000 for a survey showing 48 percent of Americans believing in the devil.

✓ $36,500 to investigate the "Evolution of Song Learning of Parasitic Finches."

✓ $84,000 to study "passionate love."

✓ $81,000 to look at the social behavior of the Alaskan brown bear.

The new emphasis at NSF, says its director, will be on engineering, which has had only sporadic support within the foundation during more than a decade of congressional criticism.

The mood now favors industrial innovation, which made the nation a world power and global provider of better and longer lives.

European and Asian counterparts of NSF have been advancing technology.

To reverse this nation's downward industrial trend will take more than applied research, and less dabbling in scientific never-never lands.

THE SUN

Baltimore, Md., January 22, 1981

So far as we know, John Kenneth Galbraith, the economist, coined the word "technocrat," which he defined as a member of a technologically oriented elite in business and government. Mr. Galbraith distrusts technocrats, and so, it seems, do many other Americans; surveys show a growing uneasiness about science and technology, probably aimed not so much at these endeavors as at the way they are used and managed.

The growing distrust is not a groundswell yet. Especially among "attentive" (better-informed) Americans in a recent National Science Board survey there was still a strong belief that science and technology are "making our lives healthier, easier and more comfortable." At the same time, even many of the "attentive" respondents worried that science and technology were changing American life too quickly and in ways that could break down value systems that should be preserved. Less well-informed Americans were even more concerned with these two factors; and between 1957 and 1979, the percentage of "nonattentives" with clearly favorable views towards science declined from 87 to 66 percent.

Respondents tended in general to rank humane goals for science and technology highest. Health research and development of energy resources were at the top of a list of goals worthy of tax dollars. Educational research was third. While space research was considered a largely innocuous activity, it was not seen as particularly useful; it was near the bottom of the list of tax-worthy activities. So, too, was basic research, described in the survey as "discovering new knowledge about man and nature." Though there seems now to be a growing constituency for military applications of science, the poll showed a public bias for scientific and technological activities considered both practical and humane.

One aspect of the survey results that concerns us is the lack of strong support for pure research. This kind of scientific inquiry may not yield immediate results, but it will provide longterm answers to the most important humanitarian questions, such as how to cure cancer. The public preference for the humanitarian in science and technology is commendable; but its preference for the practical betrays a failure to understand that the groundwork for meeting humane goals is usually laid by "pure" scientists.

Detroit Free Press

Detroit, Mich., December 1, 1980

THE DECISION by the National Science Foundation to focus more strongly on applied (as opposed to basic) research was probably inevitable. The nation seems a lot more interested these days in discovering ways to "reindustrialize" America, for instance, than in remaining on the frontiers of research into high-energy physics.

Even so, the NSF decision is disturbing, especially if it means a sharp cutback on the funds going into basic research.

NSF is the only large agency in the federal governement with a primary mission to support scientific research with no clear or near-term payback. The plans of the new director, John B. Slaughter, to alter that mandate may leave many researchers short of important support funds. Though Dr. Slaughter indicated that new funding for engineering and applied sciences could be found without cutting the basic research budget, that does not seem very likely with an administration pledged to cut federal spending.

Moreover, a substantial share of the federal budget, especially the Defense Department's share, already goes to engineering and applied research — not to mention the tremendous sums poured into those areas by private industry.

Basic research does not enjoy that support. And the danger in the new orientation is that we may find ourselves sacrificing the long-term gains from basic research to respond to nothing more than the latest political trend.

THE SACRAMENTO BEE

Sacramento, Calif., December 7, 1980

John B. Slaughter, director of the National Science Foundation, bowing to pressures from Congress, industry, the engineering profession — and now the incoming Reagan administration — has announced a shift in NSF's original purpose. Instead of concentrating on basic research, which so often has proved the wellspring of important technological breakthroughs, the foundation will divert more of its subsidies to engineering and applied science aimed at spurring industrial growth.

The pressures on Slaughter are understandable. In large measure, the failure of American industry to keep apace with the West Germans and Japanese in research and development has dulled this nation's competitive edge in world markets. Further, the current plight of the U.S. economy has put new focus on the need for technical innovation to boost productivity. While most engineering and technological development has been done historically by private industry, a sort of loose partnership has existed between the private and public sectors in broad fields of applied research, and there's no reason why that should not be encouraged. But it should remain in the domain of private industry.

There is a distinct danger in the shift of emphasis for the work of the National Science Foundation. Basic research, despite the many fortuitous and often serendipitous dividends its findings have offered to applied science and engineering, has usually gone begging for funds. That's why an enlightened Truman administration in 1950 created the National Science Foundation — to assure continuing American progress in pure scientific exploration. It was basic research, after all, that led to splitting the atom; that made possible the transistor and its semiconductor revolution in information storage-retrieval and communications; that cracked the genetic code and thus opened the way for genetic engineers to synthesize insulin and the cancer-fighting substance, Interferon.

To divert NSF's emphasis from that focus may well produce short-term gains in applied technology which will help industrial productivity. But if, in response to immediate national priorities, the foundation's work is deflected from its most important purpose, the nation may pay a price later on — as, indeed, already is happening in Japan, where the lack of original, basic research has begun to narrow technological innovation. Japan's remarkable technological gains, incidentally, derive almost entirely from the breakthroughs made possible by U.S. scientific groundwork.

Slaughter's announced new course comes partly in response to congressional legislation to create a National Technology Foundation which no doubt would offer NSF strong competiton for government funds. Slaughter may feel if you can't lick 'em, join 'em, and at least preserve under one roof both basic and applied research. Yet where those things have had to compete for money in the academic world, pure research labors under the handicap of offering no immediate practical benefit. The NSF was created as a shield for this important work. Compromising its charter would risk the loss of invaluable, longer-range dividends for all of humanity.

Part III: Education

Few people would deny that there are serious problems with American education. In the past 15 years, there has been a substantial decline in average student achievement levels in all subject areas. Plenty of reports have been written, blaming schools for giving students too much choice of subjects (although 15 years ago, reports criticized schools for forcing unwanted subjects on students); blaming television for destroying children's attention spans; blaming parents for failing to supervise their children, and blaming the federal government for refusing to increase funds for state and local school districts.

Whatever the causes of declining student achievement, the implications for American science and technology are already clear. Scientists say the public's mistrust of their work stems directly from the education system's failure to give students a proper science background. "Creation science," the presentation of the Bible as scientific theory, is a prime example of how fundamental confusion over the nature of science undermines its position in the public schools. Hostility to genetic research is largely a result of misunderstanding or ignorance of biology.

At a National Academy of Sciences conference on education in May 1982, Paul DeHart Hurd, professor emeritus at Stanford University, pointed out that "only 34% [of high-school graduates] have completed three years of math and only 8% complete a course in calculus. Most seniors have had a biology course; a little over a third have had chemistry, but less than a fifth have had three years of science." Higher education was no better; a March 1982 report by the National Research Council criticized colleges and universities for failing to give non-science majors an adequate background in science. America's future leaders, politicians and voters leave school at all levels without any preparation for a society increasingly involved with technology. As NAS President Frank Press observed, "How can we expect our children to have successful lives in an intensely competitive and highly technological world if they don't have the education to understand it?"

General Education Decline Endangers U.S. World Position

By the beginning of the 1980s, there were few Americans who doubted the existence of serious problems in the public school system. Education Secretary Terrel Bell named a panel in August to study the problems of U.S. schools and recommend ways of raising standards. The 18 members of the National Commission on Excellence in Education were instructed to examine secondary schools and colleges and suggest ways of improving student achievement. Headed by President David Gardner of the University of Utah, the commission included representatives of business, politics, education and civic affairs. Bell said the nation's schools had been too preoccupied with "bringing the bottom up," which he called a praiseworthy aim. However, he continued, it was time to focus more on "challenging the outer limits of abilities and talents." Bell called for "encouraging adoption of policies and standards that will hold in high esteem the attainment of the highest order of literacy and academic competence." Bell said the commission's recommendations would not be submitted to him but to "the American academic community, to governing boards, to state legislative bodies and to others responsible for general control, support and supervision of schools and colleges." He said the commission would pursue its 18-month study whether or not the Department of Education remained in existence.

THE SUN

Baltimore, Md., January 4, 1981

As measured by educational testing standards, today's high school graduates aren't as smart as yesterday's—not by a long shot. A study by the State Board for Higher Education of Scholastic Aptitude Tests shows a continuing decline in verbal and mathematical skills. Since 1972, the average state SAT scores for college-bound seniors has dropped 35 points in the verbal test and 22 points in mathematics. Two years ago, the Maryland mean score on the SAT-verbal test slipped below the national average. The state has been below the national mean in mathematics testing since 1975. Last year, there was no sign of improvement in either category.

This is an alarming trend that should alert local school systems to the importance of intensified efforts in basic-skill courses. In just six years, the number of pupils whose low verbal scores indicated a need for remedial education increased by nearly 50 percent.

Among graduates intent on studying mathematics, scores on both the verbal *and* math tests were substantially lower than two years ago. Even more alarming, the percentage of college-bound students who realized that they needed special assistance in reading, writing or mathematics declined sharply.

Thus, it is incumbent upon Maryland's colleges to fortify and broaden their introductory English and math courses and to see that students who need help can get it. Many with low SAT scores come from inferior school systems or from families where education is not a priority. Their presence in college, though, attests to their eagerness to learn and to overcome these barriers. A solid foundation in the basics is absolutely essential. That is the best way of assuring a higher retention rate for the colleges, both at the undergraduate and the graduate level, and a better chance in life for the students.

The Morning News

Wilmington, Del., January 14, 1981

The Department of Education may not make it much beyond its first birthday this spring, since the incoming Reagan administration appears bent on abolishing this new department. Such short shrift will make it impossible to assess whether the cause of Education with a capital E benefits from having a cabinet secretary all its own or whether it does just as well when a somewhat lower-ranking education commissioner pleads education's cause.

One thing is certain though. Nationwide, there is need for a strong, forceful advocate for education, regardless of that person's administrative rank. For the signs abound that Education U.S.A. is in poor repair.

Every year, high school students come in with lower and lower scores on standardized tests for mathematics and English. A recent series of articles in the News-Journal papers revealed that college students wishing to pursue a career in education are often among the academically less qualified. The discipline problems in schools are legion. Inflation has had a negative effect on school programs.

Some improvements may be on the way. In Delaware, as in several other states, students must now pass competence tests before they can obtain high school diplomas. A few states have instituted teacher competence tests. Stress on the fundamentals of reading, writing and arithmetic has improved instruction in some elementary schools.

But much more remains to be done. Secretary of Education Shirley M. Hufstedler charged in the New York Times magazine that "some of our more severe educational deficiencies threaten the nation's security." She said that every year more than three million Russians are graduated from high school with two full years of calculus. American schools turn out fewer than 125,000 high school graduates a year with one year of calculus instruction. Repeatedly, it has been said that Americans are falling behind in science — and again school curriculums as well as college entrance requirements have a lot to do with that.

When it comes to foreign language instruction, American schools have always lagged behind their counterparts in other parts of the world. Again comparing us to Russia, Mrs. Hufstedler points out that in the Soviet Union there are almost 10 million students of English, while in the United States there are only 28,000 students of Russian. Similar figures could be cited for Chinese, Japanese, French, German and even Spanish. The last is particularly alarming considering the prevalence of Spanish in the Western Hemisphere.

The point isn't really whether we have a secretary of education or a commissioner. What does matter is that intellectual discipline must be returned to the American educational system. Without that, America will not be able to compete in the world.

The News and Courier

Charleston, S. C., May 4, 1981

The fact that 30 percent of the education majors at 14 colleges and universities flunked an entrance examination that will become mandatory in 1982 might go a long way toward explaining why Johnny Pupil can't read and why Scholastic Aptitude Test scores continue to decline. If the teachers themselves can't read — and the entrance exam revealed that 40 percent of these would-be teachers can't — small wonder their pupils can't.

It is possible, however, to read too much into this first sampling. The S.C. Educator Improvement Task Force which conducted it didn't announce just where the exam was given and how many flunked at each school. That kind of information might have put an entirely different light on the whole thing. In past years, for example, compilations of National Teacher Examination scores have showed graduates of some private colleges and universities did extremely well while those from the public institutions performed miserably. It would be helpful now if the Task Force disclosed where the failures are getting their alleged educations.

That the Task Force will indulge in any such candor is unlikely. It is equally unlikely, we assume, that this examination will ever again be administered to any aspiring teacher. The trend in such things today is to water down examinations until almost anyone who wanders in from the street can make a "passing" grade. Otherwise, the cry of racial or cultural bias will be raised and the heat will force the examining body to wilt.

The Kansas City Times

Kansas City, Mo., January 9, 1981

Strictly speaking, to push a child along through the educational halls without teaching the child to read should be rated an immoral act. Illiteracy handicaps a person in every facet of living, from social contacts to work, from mobility in the community to conducting mundane business keeping body and soul together. It is, perhaps, only when the eyes begin to blur that those who read everything automatically begin to realize what a precious gift we were forced to accept decades ago.

But illiteracy is no new problem, caused by the new social consciousness of public schools or drives to desegregate. Judging from the census estimates that 100,000 area adults cannot read above a fifth-grade level, the problem has dogged citizens and experts for decades. We just didn't notice. It's not the kind of handicap people volunteer to talk about. Victims suffer in silence. They hustle in guilt, trying to compensate for the lack of the taken-for-granted skill.

A recent story in *The Kansas City Star* quoted experts pointing the finger at a variety of culprits: men dropping out of school to join the military, prejudice against educating girls, television and even part-time jobs.

Too often a search for reasons ends up sounding like lame excuses. Could we dare hope an era has started when citizens will no longer be content to let youth drift along, blinking an eye year after year while they fall farther behind their peers in reading ability? When we are willing to look below the surface for reasons for poor grades, quarrelsomeness, delinquency, fumbling in any areas where reading is the key to smooth performance?

The shameful acknowledgement that millions of persons are illiterate in this country which prides itself on an egalitarian public education system is not an easy one. The penance is not 10 lashes with a grammar book. It must be a resolve to reform, to turn around the weak wills and rehabilitate the practices that allowed the situation to develop. Not only a lifelong pleasure and an inexpensive way to travel the world, reading is a useful tool. Americans should demand their right to its possession.

FORT WORTH STAR-TELEGRAM

Fort Worth, Texas, January 29, 1981

During his "state of the state" message to the Legislature last week, Gov. Bill Clements drew spirited applause when he called for a return to the basics in Texas' public schools.

"Nothing is more on the minds of the people of Texas than public education," Clements said, "and there is nothing needed more in this area than a return to the basics of reading, writing and arithmetic and stronger discipline in the classrooms."

Among Clements' goals along these lines are competency tests for prospective teachers, elimination of "social" promotions, establishment of guidance centers to handle disciplinary problems and requirement of summer remedial programs for students who do not meet minimum academic standards.

We urge the Legislature to do all in its power to see that the governor's aims become reality. Test scores indicate that far too many students are graduating from our high schools without a proper education, and one thing the state, and the nation, can certainly do without is an illiterate with a high school diploma.

A recent story in *The Wall Street Journal* emphasizes how desperate the situation is. The story revealed that it is becoming fairly common practice for larger firms across the country to conduct classes in remedial English or mathematics for employees or promising job applicants.

Because so many students are being graduated from high school deficient in the basics, America's businesses are having to try to accomplish what the schools could or would not do.

And that's not right.

It's also expensive. Consumers eventually pick up the tab for such educational programs in higher prices, so it is a case of double economic jeopardy. We pay school taxes to educate the students, and when the job isn't done properly, we pay through higher prices the fee for remedial education of our schools' graduates.

The *Wall Street Journal* story read like a horror tale, detailing accounts of job applicants or workers recently out of high school who could not spell the simplest words or do the simplest arithmetic problems. Many are regarded as being potentially hazardous to their fellow employees because they either can't read or can't understand basic safety instructions on the job.

So, hang in there, Gov. Clements. Keep the pressure on. Texas schools need to emphasize the basics and graduate only those students whose educational progress merits it.

Herald News

Fall River, Mass., May 4, 1981

The National Association of Education Progress has found that, on a country-wide basis, the reasoning ability of youngsters between the ages of 13 and 17 has declined.

The decline in reasoning ability was directly related to a decline in the ability to read.

There is no real news in the announcement of a decline in reading ability, but the National Association of Education Progress is giving us all important information by correlating the loss of ability to read with a decline in reasoning power.

It stands to reason that the one capacity affects the other, but in the recent past, there has been a misguided tendency among educators to dissociate them. The association's report makes the connection clear.

It also stresses the value of remedial reading courses among younger children, and the importance of continuing and expanding those courses.

The cutbacks in federal funding for education are scheduled to make fewer of these remedial reading courses available, especially for youngsters in economically blighted areas.

Should these cutbacks occur as scheduled, we may anticipate that far more young people than need be will be handicapped in terms of both reading and reasoning capacity.

The failures of education in recent years have angered the public, but the justified anger should not carry over into making matters worse instead of better.

A democracy in the age of technology requires an educated electorate more than ever before. Every effort should be made by the government to see to it that a sound basic education, stressing both reading and reasoning power, is available to all.

THE BLADE

Toledo, Ohio, May 28, 1981

ALTHOUGH there are many reasons for the recent decline in the performance of high school students on scholastic aptitude tests, a sizable share of the blame must be laid at the door of permissive academic requirements.

The Toledo board of education has taken a commendable if overdue step toward strengthening the high school curriculum by adopting a proposal to require two additional academic units of study for graduation and upgrading offerings in basic subjects such as mathematics, science, English, and history.

Under the new curriculum students entering Toledo high schools next September will have to take an additional year of math, a full-year course in world history in addition to present offerings, and an additional unit of science. The number of electives will be cut from 9 to 7 even though the number of required units would increase.

Superintendent of schools Hugh Caumartin called the proposal "a giant step toward improving academic achievement in the district." Actually, the changes are not that revolutionary, but they do put more rigor into the academic curriculum, and there is some hope that the new curriculum also will encourage more students to enroll in additional foreign language and fine arts courses.

The changes are made possible by the return last September to a six-hour day for high school students, beginning with entering freshmen. This year both the freshman and sophomore classes will be on the six-hour day.

These changes, which were recommended by a 31-member study committee, do not automatically bring about a higher-quality education. Teacher qualifications and preparations, improved and updated textbooks and other teaching materials, and discipline in the classroom are equally important variables, and most of them cannot be changed overnight. But the curriculum revisions are decidedly a move in the right direction.

The Union Leader

Manchester, N. H., January 31, 1981

The different me and my parents thay are more relaxs Thin I am. So I like to Be more Taets thin thay are Thay Drink more Thin me But I smok more Thin Them. I'm different Thin my parents Because I like To in Jouy Life more Thin Thay Do . . .

Part of a freshman English essay, University of Nevada-Reno, 1980.

The above is not by a grade school student or a high school student. It is part of a freshman essay, written by a University of Nevada-Reno student in 1980! Probably the same inability to spell and write a logical sentence could be found on almost any college campus in the United States.

This, fundamentally, is because the public education system, which is costing us more and more, is doing an ever more inadequate job. This is because of the lack of discipline in the schools and a refusal to insist that certain basic standards be met before a student goes on to the next grade. We have been prisoner for many years of the psychological nonsense that if you don't promote little Johnny or little Jane, even though they don't deserve to be promoted, it will hurt their egos and make them psychological cripples.

It is so bad that large U.S. corporations are spending hundreds of millions of dollars to educate their employes so they can function properly in a good-paying job.

Think of the complete waste which in the end the consumer pays for. First of all, the taxpayer spends billions on education which is suppposed to do the job. Then it has to be done all over again by businesses which, of course, charge for the cost of that second education in the prices they charge for their goods. They have to.

In a recent Wall Street Journal article, a number of industries were cited. For instance, at JLG Industries, Inc., a manufacturer of cranes and aerial lifts in Mc Connellsburg, Pennsylvania, "poorly educated workers are our No. 1 problem, the main factor slowing our growth," says William Barnes, vice president of finance. One employe who didn't know how to read a ruler mismeasured yards and yards of steel sheet. "He wasted nearly $700 worth of material in one morning," Mr. Barnes said.

Also, JLG purchased electronic equipment to help regulate inventories and manufacturing schedules. But employes who fed the equipment incorrect five-digit numbers were "wreaking havoc," Mr. Barnes said. As a result, wrong inventory numbers sent the wrong spare parts from the company's warehouse to manufacturing shops, disrupting both inventory control and manufacturing schedules.

So far JLG has spent nearly $1,000,000 just to correct those incorrect entries, wiping out the savings that the equipment was supposed to provide!

Also, it is not a matter of just blue collar workers. At Mutual of New York, an estimated **70 percent** of the insurance firm's correspondence must be corrected and retyped at least once because typists working from dictation recorders don't know how to punctuate sentences and often misspell words. Typists who don't understand English and whose work has to be corrected over and over again are defeating that purpose.

It is also noted that clerical workers who speak ungrammatically and have difficulty communicating even simple information to company clients may be costing Mutual some lucrative business.

It seems also that many workers don't know how to use the telephone. They let it ring and ring and when they finally answer they just say, "My boss ain't here." They can't answer the simplest questions.

Undereducated workers can be dangerous, both to themselves and to others. One illiterate worker, employed at an unspecified company cited in a study, was killed because he couldn't read a warning sign. And an assembly line worker at a large industrial equipment manufacturer almost killed several co-workers when a heavy piece of metal that he hadn't attached properly to a machine — because he couldn't read the assembly instructions — flew off.

A National Conference Board survey indicates that 35 percent of 800 companies said they provided some instruction in subjects that are "really the responsibility of the schools."

There's the picture. Quite evidently, the public school system has failed. We cannot expect to produce a successful nation if we don't lay its foundation with our young people in the school system.

An ignorant, illiterate electorate is what they have in Iran and other backward countries. Right now the United States, with its emphasis on equality and degrees for everyone, without regard to quality, seems to be headed in that same direction.

The Houston Post

Houston, Texas, June 12, 1981

When a big city system like the Houston Independent School District produces children in first grade through fifth who score above the national average on the Iowa Tests of Basic Skills, it means a large number of them are scoring very highly indeed — in the 99 percentile of the nation. They boost the average that is lowered by children with visual perceptual problems, children entering English from any of 95 different foreign languages, or children who are naturally late bloomers.

The fact that all six elementary grades scored at or above the national average while all the grades from seven through 12 scored below stands for a tremendous turnaround in one of the biggest school systems in the United States. Present-day junior high and high school students had the misfortune to pass through the early grades at a time when the schools were being used to right old wrongs.

In the late 1950s the nation began to admit that our public schools were not truly public. They were giving excellent education to some children in some cities and some neighborhoods, while neglecting the handicapped and offering schooling that was anything from mediocre to poor to non-whites in many parts of the country. The courts and the governments pressed for long-needed reforms. Reform brings movement that is both stimulating and unsettling. The schools have come through a rough decade but will be the better for it.

In the same period, educational theorists thought it important for the elementary school child to be well adjusted and accepted by his peers, even if it meant promoting him before he had learned to read. High schools and universities felt the pressure to be relevant and to relinquish much administrative authority to the students. Standards sagged, requirements were eased or dropped. Film-making and communications tended to edge out foreign languages and logic.

A generation of young people emerged with not enough education to be employable. Spelling and writing the English language and simple arithmetic are essential in most fields of endeavor. Now educators across the country are being allowed to turn the focus once again to the basics and to quality education. It is no longer thought elitist to expect the best and to require proof of competence.

Unfortunately, the change for the better is coming when the Reagan adminisration is pushing for a 25 percent cut in federal spending for public schools. Education Secretary Terrel H. Bell estimates that this will mean a 2 percent budget cut for most public school systems. Simultaneously, the administration tends to favor a tax credit for parents who put their children in private school. Should this pass Congress, public education will have received a double blow. Instead, citizens in every community should be working *for* the public schools as our grandparents did. We must insist that public education be the best available. Otherwise, we will weaken a foundation pillar of self-government.

Buffalo Evening News

Buffalo, N. Y., May 13, 1981

A recent survey of reading skills around the country revealed considerable progress in elementary grades, especially among black pupils. The survey confirms that disadvantaged pupils do indeed benefit from remedial-instruction programs.

Despite these hopeful aspects of the report, however, surprise and concern were aroused by the findings that overall reading ability in the higher grades, including junior and senior high school, has actually declined over the past decade. Of course, in the higher grades, measurement of reading ability goes far beyond the mechanics of sounding out words to such matters as understanding and drawing conclusions from what is read. This is where students are falling down.

Over the 1970-1980 period, nine-year-olds in the national survey increased their reading scores by an average of 3.9 percent. In the same period, the scores of 17-year-olds fell by nearly 1 percent. The surveys were made in schools throughout the country by the National Assessment of Education Progress, a federally sponsored monitoring project.

There is no agreement by educators on the cause of these mixed results. Some cite the same factors that are often mentioned for the decline in educational performance generally — television, less emphasis on examinations, watered-down curriculums, social disruption and so on. Some think the emphasis on basics in recent years may have diverted schools from the teaching of more advanced skills.

Dr. Dorothy Strickland, an education professor at Columbia University, said: "We need to ask ourselves what can be done at the early stages of reading development to help produce more thoughtful, critical readers who are better prepared to shift to the more complex reading tasks of the middle grades."

A similar problem has been detected in mathematics. Many students master the fundamentals and then have trouble with the more complex applications. Perhaps the same kind of concentrated effort that has been successful in increasing basic skills in reading and math should now be applied to the higher grades, where these skills must be harnessed in more advanced studies involving comprehension, reasoning and creative thought.

The Chattanooga Times

Chattanooga, Tenn., January 24, 1981

According to *The Wall Street Journal*, hundreds of businesses around the country are now offering basic courses in English and math for employees who didn't learn those essentials while in school but who graduated anyway. The institution of such courses strikes us as a serious indictment of education. And while business leaders may have their own ideas in the ongoing debate over what factors have led to a decline in test scores for high school graduates, their decision to set up the courses is based on something more tangible: the direct cost to business of education's failure.

While the employees themselves will benefit from the courses, so will business. That's because when an employee's lack of education prevents him from doing his job correctly, the costs incurred in lost production and waste skyrocket. A Pennsylvania manufacturer, JLG Industries, reported, for instance, that one worker ruined more than $700 of steel sheeting in a few hours merely because he couldn't read a ruler correctly. And a New York insurance firm said that many of its typists, working from dictation tapes, regularly misspell words and commit punctuation errors; as a result, much of the company's correspondence must be corrected and retyped.

One problem is that business offices are turning more and more to new technology to eliminate much of the routine tasks previously done by low-skilled workers. But for that equipment to be cost-effective, businesses need workers adept at the sophisticated skills necessary to operate it. JLG Industries had to spend more than $1 million merely to correct mistakes when it found that workers were unable to punch in the five-digit numbers which its new electronic equipment used to improve inventory control and manufacturing schedules.

The Conference Board, a business research association, surveyed nearly 800 businesses in 1976 and discovered that 35 percent had met the problem by setting up remedial training courses in basics like English and math. It's not just business' problem; the armed services — and even colleges — have been forced into remedial education. Whatever the explanation for the failure of young people, it should not be up to business to train new workers in fundamentals they should have learned in school.

We're hearing a lot these days about special tax incentives for business, reindustrialization — and considerable talk about cutting governments' budgets. Budget cutters need to remember, however, that education is vital to our young people and to the country. Improving it is, well, just good business.

National Study Finds Science Achievement Decline

A study prepared by the National Assessment of Educational Progress during the 1976-77 school year found that high-school seniors continued to decline in mathematics and science achievement. The report concluded that although "the decline in science performance is slowing among students at ages 9 and 13," the performance for 17-year-olds was worse than the levels recorded in the first two NEAP studies, done in 1969-70 and 1972-73. In 1969, the NEAP noted, there was a "high level of public consciousness of science during the early days of space exploration," but more recently, "an overemphasis on the 'basics' in the elementary classroom has diminished the amount of time available for the study of science. . . ." The report complained that "science courses are often elective after the 10th grade. The consequence is that many 17-year-olds do not continue the pursuit of science education after their sophomore years." Recommendations for improving the status of science education included better preparation of science teachers and a broader view of science, which would include social and economic implications of science and the connections between science and technology.

DAYTON DAILY NEWS

Dayton, Ohio, July 13, 1978

The latest thing the younger generation is stupid about, in case you are keeping score, is science. A federal outfit that keeps track of high school students' knowledge says 17-year-olds know less about science than when similar tests were given in the late '60s and early '70s.

(Which presumably means we not only are getting stupider but are getting stupider fast. Do you ever suspect that if the young were really as ignorant and unable as is being widely insisted, about half the folks under, oh, 25 would be sitting in corners drooling?)

Whether these test comparisons are correct or not, the report brings up a thought. Most high school science courses — mathematics courses, too — don't meet the need of most students to understand science and math but rather are geared to the minority that intends to *do* science and math.

Anyone of average intelligence can understand and appreciate science and math, but it takes special talent to do either of those well. As a result, fewer than 10 per cent of high school students take physics and chemistry courses.

Yet no one can claim to be an educated, or even an aware, member of this society without understanding science and math — their great figures, the mental disciplines required, the contributions they have made to society, how they have developed, the issues they create for political decision-making.

Oddly, there are no math appreciation and science appreciation courses. An education system that reasonably believes students can learn to appreciate good music and have sensible opinions about it without playing the piano insists that they must do algebra or calculus and solve physics problems and calculate chemical formulas in order to appreciate math and science.

This isn't to say the skills courses should be scrapped. But might not more high school students be served better if, instead of having to choose either to learn science and math or not learn them, they also had the choice of learning about them?

PORTLAND EVENING EXPRESS

Portland, Ma., August 12, 1978

"Mathematics, rightly viewed," Bertrand Russell once wrote, "possesses not only truth, but supreme beauty—a beauty cold and austere, like sculpture."

For most students getting out of schools these days, Russell's metaphor is distressingly prophetic—they can understand modern math about as well as most people understand modern sculpture.

In fact, a new name has been grafted to categorize the emerging generation of "new math" illiterates spared the chore of learning by rote and now baffled by checkbooks and timetables—they are said to be suffering from "innumeracy."

For whatever meager consolation it's worth, innumeracy is not a uniquely American malaise. Like a spreading swarm of killer bees, it's been upsetting the natives in several countries.

The London Times editorialized a few months ago about its victims, berating the British educational system for "turning out too many children grossly ill-equipped for the everyday needs of adult life."

In its zeal to have students understand the larger operative principles of mathematical function, the new math has failed to teach them how to apply those functions.

It's a disaster softened by only a single—possibly coincidental—circumstance best expressed as an inverse ratio. As the number of innumerates increases, the price of pocket calculators decreases.

It's the 1978 equation for higher education—brain power battery power.

The Des Moines Register

Des Moines, Iowa, August 29, 1978

Both houses of Congress have agreed to spend federal funds to improve mathematics skills. Congress in 1974 voted money to improve the reading skills of elementary school pupils. The new measure extends the remedial program to math and covers secondary schools and adults who are out of school as well as elementary pupils.

The bill authorizes little more for the expanded program in both reading and mathematics than was authorized for the reading program alone this year — $147 million vs. $144 million. But only a fraction of that is being spent.

The program is for testing and providing special instruction to overcome deficiencies in reading and math.

The program is administered by the U.S. Office of Education through contracts with state and local governments. The new act provides for allocation of funds to state education offices and for state administration of local programs. It puts emphasis on programs to encourage and train parents to become better helpers for their children in the home.

That Congress should have to provide money for such a basic program is one of the ironies of the "Proposition 13" mentality. Denied money locally for programs parents say they want, educators naturally turn to Washington.

Conservatives in the Senate criticized the 1978 amendments for extending federal control over local school districts. But opposition to adequate state and local support for schools is making a federal role ever more important.

THE TENNESSEAN

Nashville, Tenn., July 8, 1978

WHEN the "basics" of education are mentioned, most people think about reading, writing and arithmetic and deplore their supposed decline in the public schools.

But there is another basic in modern education — science — which seems to be declining as fast as the three R's.

The National Assessment of Educational Progress reports that knowledge of science among the nation's 17-year-olds has fallen steadily in the decade since the U.S. landed men on the moon.

The same survey found that the physical science scores of 9- and 13-year-old students have also declined over the last several years.

Mr. Arthur Livermore of the American Association for the Advancement of Science Education says decreasing enrollment in high school physics and chemistry courses may be to blame for the 17-year-olds' poor scores.

"Less than half of the high schools in the country even have physics courses," he said.

The rising costs of education and the resistance of taxpayers seem to account for the decline of science education.

Teaching science, with all the laboratory equipment and other materials required, is extremely expensive. When hard-pressed school systems are faced with the necessity of cutting back to meet pinched budgets, the more costly science courses often get the ax first.

This is an unfortunate trend. Important as the basics are, science education cannot be neglected by a nation with pretensions of remaining a world leader in the scientific age.

It is doubtful if any other major industrial nation in the world is skimping on science education. Americans were shocked in the 1950s when the Russians became the first to put a satellite into orbit.

That inspired a new wave of interest in science which culminated with the U.S. putting a man on the moon in 1969.

Now the trend seems to be in the other direction, with science once again being downgraded because it is too expensive to teach.

Although some may still deplore the findings of science and would like to substitute various doctrines in its place, there is nothing more basic to national survival in the modern world than thorough grounding in the physical sciences.

The Washington Post

Times Herald

Washington, D. C., September 16, 1979

SURELY EVERYONE agrees that a 13-year-old child—a seventh grader—ought to be able to tell you what two-thirds of 9 equals. But, in fact, it appears that fewer than half of all 13-year-olds can answer that one. This fragment comes from the National Assessment of Educational Progress, a federally supported research organization that gives tests and reports the results—generally depressing—to parents and taxpayers. Over the past five years, the National Assessment has found, there has been a perceptible drop in the ability of American schoolchildren to do simple arithmetic.

That conclusion is, unfortunately, altogether consistent with the long decline in the scores that high school students have been getting on the College Board's Scholastic Aptitude Test. After much analysis, a College Board committee concluded two years ago that the decline had a lot to do with the peculiar cultural currents of the late 1960s and early 1970s. Since the end of the Vietnam War, the atmosphere in the schools and colleges has become a great deal more businesslike—but the SAT scores have dropped again this year. It was a very small drop, but it was visible. Could the turbulence of the 1960s have any effect on the progress in arithmetic of 13-year-olds today? None directly—but, of course, some of the college students of that period are now junior high school teachers.

The poor results of the math test are also consistent with the grades in the basic competence tests now being introduced as a requirement for high school graduation in many states, including Virginia and Maryland. Competence testing makes it clear that these children's difficulties are not limited to formal textbook drills, but extend to the everyday exercises of the paycheck and the light bill.

Some specialists speculate that the deterioration is related to the back-to-basics movement in school curricula. That suggestion seems utterly implausible. The return to basics has been recent, and not sufficiently widespread, to account for a decline as general as this one.

It's an oddity: American prosperity increasingly depends on advanced technology. Business increasingly depends on highly sophisticated accounting systems. Everybody's personal taxes and household finances are getting more complex—not to mention the indescribable Metro fare formulas. And through it all, children's ability to deal with numbers is declining.

The only bright sign is the country's rising interest in this kind of testing, as a measure of what goes on in the classroom. The very existence of the National Assessment is a welcome indication that people are now prepared to hold their schools to a tighter measure of performance. These tests constitute a legitimate report card for the country's system of education. In math this year, the grade is somewhere between a C-minus and a D-plus.

The Cincinnati Post

TIMES STAR

Cincinnati, Ohio, September 26, 1979

For the umpteenth time since the sputnik-induced anxiety of the late 1950s, education experts have tested American schoolchildren and found them wanting.

A report by the government-sponsored National Assessment of Educational Progress says that the computational skills of elementary and secondary school students aged 9, 13 and 17 have grown "perceptively worse" since a previous study in 1973.

They're learning the rules of math pretty well, says the report, but they aren't learning how to apply this formal knowledge to practical problem-solving.

Most students, it was found, could accurately add a string of numbers. But they faltered on problems involving several steps or requiring more reasoning.

For example, when asked how many kilometers a car would travel in one hour if it went at a speed of eight kilometers in five minutes, only 23 percent of 13-year-olds and 56 percent of 17-year-olds came up with the correct answer of 96 kilometers.

A panel of educators who interpreted the survey data put a lot of the blame for the students' poor showing on the "back-to-basics" movement.

We find this surprising. Back-to-basics was a reaction to such innovations as the "new math," which was spawned in the post-sputnik scramble for educational panaceas and was later criticized as being too abstract and theoretical.

Now, if the experts are right — and we are far from being convinced that they are — we have come full circle back to 1957.

Kids who allegedly were being shortchanged in high school then have become the parents of some of the kids who are allegedly doing poorly now. More effective ways to teach mathematics — not to mention reading and other skills — still must be sought.

The Houston Post

Houston, Texas, October 18, 1979

Whatever happened to science as a popular field of study in high schools? The National Research Council has found that nationally science and mathematics teaching has declined to a deplorable level in our secondary schools. As a result of this lack of emphasis on basic science, too few students are being inspired to major in science as they enter colleges and universities. We cannot simply wait for outside stimuli such as occurred shortly after the Russians launched their first sputnik and spurred the U.S. into the space science race.

That international competition among the early earth-orbiting satellites stirred imaginations of young and old. The national Physical Sciences Study Committee at that time became very active in providing scientific films, textbooks and revitalized courses in high schools. Many of today's scientists and engineers were steered in the direction of their present careers by those educational projects that took them beyond basic studies at an early age. But all that dates back to a decade following the late 1950s. The manned space flights and moon exploration provided the last major glamorization of science. But the last moon walk is now nearly a decade in the past. The special science programs, started in the late 1950s and 1960s, have been largely phased out. They have not been replaced on any broad-scale basis. Even to revive the old programs would probably be less than adequate. Education, especially in modern science, must constantly change with the times.

The present need is for educationally-concerned scientists to design and promote new programs for teaching science at the high school level and not wait for government help to do it. Business and industry have a large stake in the number of scientists that will be turned out in the future. The National Research Council proposes establishment of regional science-mathematics teaching resource centers to help teachers and schools improve courses. It also recommends that the National Science Foundation sponsor programs to develop new courses. During the current dearth of glamor in scientific undertakings, we need to resort to the hard sell of science to help students with potential identify their talents at an early age.

The Chattanooga Times

Chattanooga, Tenn., October 1, 1979

Quickly now, what's the sum of 54 and 21? And here's another problem: What is ⅔ of 9?

Few adults would have any difficulty with those simple mathematical problems and probably assume that most junior high school students could answer them easily. The assumption would be wrong.

According to results published by the National Assessment of Educational Progress, more than half of that nation's 13-year-olds — seventh graders — are unable to figure out that 54 plus 21 equals 75. That news is depressing in itself. What's worse is the NAEP's report that for the past five years, American schoolchildren have become demonstrably less proficient in simple mathematical ability.

Fewer than half the 13-year-olds and only 13 percent of the 9-year-olds (mostly fourth graders) knew that ⅔ of 9 is 6. Two-thirds of the 13-year-olds and less than half the 17-year-olds could express 9/100 as a percentage (9%).

In one sense, the NAEP's findings should not be all that surprising. The decline in mathematical ability among elementary and junior high school students is, unfortunately, consistent with the overall poor showing by high school students as measured by the College Board's Scholastic Aptitude Test. The steady drop in SAT scores has slowed somewhat but not enough to overcome concern among parents and educators. Whether the requirement that high schoolers pass competency tests before graduation will have any bearing on correcting this problem is unclear but even that is an inadequate answer. It is not much help to the school or the student for the latter to learn late in his senior year that he lacks certain fundamental skills, among them a knowledge of basic math to help him, say, balance a checkbook.

Like reading, mathematics is a skill that must be learned "from the ground up," so to speak. Parents have the right to expect, for instance, that by the end of the fourth grade, their sons and daughters should have a firm grounding in basic mathematical functions (addition, subtraction, etc.) and are being introduced to more difficult concepts, like fractions and percentages. The NAEP's report shows that, nationally, the record in these areas is both discouraging and ominous.

It is ominous, of course, because the United States is a highly technological society in which business, industry and even individuals must rely on a knowledge of mathematics to function successfully. The decline in children's mathematical abilities suggests that unless corrective measures are undertaken, a vital aspect of America's future could be compromised.

Business and industry leaders have been calling for such measures for years now for the simple reason that it costs money to train new employees in skills they ought to have learned in school. Parents must be equally insistent, for the obvious reason that, having entrusted their children to educators for training in fundamentals, the youngsters don't show up in junior high school unsure of the sum of 54 plus 21.

White House Report Warns of "Scientific Illiteracy"

A White House report released Oct. 22 said the U.S. was in danger of "virtual scientific and technological illiteracy." Prepared by the National Science Foundation and the Department of Education, the 230-page report declared that "there has been, over the past fifteen years or so, a shrinking of our national commitment to excellence and international primacy in science, mathematics and technology. . . ." It cited "a nationwide trend toward reduction of high-school graduation requirements," a reduction of college and university math and science requirements and the "current focus on 'basic skills' " as factors in the science and math decline. The report added that "the Soviet Union, Germany and Japan . . . are educating a substantial majority of their secondary-school population to a point of considerable scientific and mathematical literacy, in part because they apparently believe that such literacy is important to their relative international positions." The report warned that the U.S. faced "a loss of our competitive edge," because of a "serious shortage" of high-school mathematics and science teachers and college-level computer-science and engineering teachers. "Those who are the best seem to be learning about as much as they ever did," the report warned, "while the majority of students learn less and less."

Houston Chronicle

Houston, Texas, October 27, 1980

There is a natural, human tendency to react strongly when respected groups, such as government-sponsored study commissions, reach drastic conclusions. Surely there is cause for general concern over the results of a recently released study on the state of American science, but our response must be more than handwringing.

First, the bad news. The White House-commissioned report confirms that American science and mathematics curricula at the elementary and middle school levels have fallen behind those of some of our major competitors, including the Soviet Union, West Germany and Japan. It notes, too, that the country faces a serious shortage of qualified high school math and science teachers.

The report adds, however, that the general dip has had little or no effect on the performances of math and science majors. And it notes, reassuringly, that American eminence in basic research is "secure," with American scientists continuing to dominate in the fields of published research and achievements, such as the Nobel prize.

There is a "science gap," then, but it is not a possible shrinkage in the number of American expert scientific minds compared to the world. It is, rather, a growing gap between American scientific experts and the American public. And, on consideration, there is good reason for concern.

That concern, broadly, is that the next generation of American policymakers may be ignorant of science, with obvious and potentially damaging consequences. A body of legislators without even a rudimentary knowledge of biology, to take just one example, is not likely to understand either the values or the pitfalls of genetic research. So, too, with space and nuclear power, and a host of other fields.

But there is no cause for panic. As was demonstrated after the last crisis of this sort, when the Russians launched Sputnik and beat the United States into space, the country can be mobilized to meet such a challenge. We were first on the moon, and in a number of other areas in space technology.

The logical response is a recognition of the problem and calm consideration of possible solutions. We can't afford to be a majority of scientifically ignorant people even if our elite continue to be dominant.

The Evening Gazette

Worcester, Mass., June 20, 1980

Connecticut ninth-graders read relatively well, but they have trouble adding and subtracting.

Commissioner of Education Mark R. Shedd has announced that the first state-wide proficiency testing of ninth-grade students shows that 93.1 percent achieved minimum standards in reading, 92 percent in language arts (punctuation, grammar, using reference sources) and 88.6 percent in writing. But only 78.2 percent achieved the minimum standards in mathematics. Nearly 10,000 Connecticut ninth-graders are unable to handle numbers properly.

Shedd said the problem is "very serious and very critical," and noted that it is city youngsters in particular who lack the necessary skills. The "minimum standards" are not a passing grade but an even lower level below which a student cannot function effectively.

There are 169 school districts in Connecticut; in eight municipalities every student scored above the minimums on reading, writing, and language arts, but none recorded a perfect score for mathematics. The town of Avon came close with 99.7 percent of its ninth-graders meeting the minimum standards.

New Haven, home of Yale University, had the worst score, with only 28.8 percent of its students meeting minimum levels. Hispanic students who are in special language courses were not tested.

Ninth-graders will be tested each year from now on for their proficiencies in the four areas. The General Assembly has already mandated the tests and requires school districts to submit plans for remedial work where needed.

In addition, the Department of Education has named a special task force to study the problems of math deficiencies and make recommendations for improvement. The study will include teacher certification, math curricula and the possibility of in-staff training for math teachers. Elementary school teachers need have only three college credits in mathematics for certification. Connecticut public schools have remedial reading teachers but do not have remedial math teachers.

Connecticut is aware of its problem and is taking steps to correct the deficiencies. Shedd said, however, that Connecticut is not far out of line with the rest of the nation. Probably there are about the same percentages everywhere of students who cannot add, subtract and multiply or understand basic mathematics.

It's a sad situation and one that every American should be concerned about.

THE ATLANTA CONSTITUTION

Atlanta, Ga., October 28, 1980

Now comes a report that says we Americans are headed toward scientific and technological illiteracy. This is based on a study of elementary and high school programs and the amount of science therein. It says we lag behind Japan, Germany and the Soviet Union.

The report said that in the Soviet Union youngsters take courses in algebra in the 6th and 7th grades, advanced algebra and trigonometry in grades 8 to 10.

We can't say we think we're becoming stupid in the sciences as a nation but we would like to see more emphasis placed on it than is now the case.

The Cleveland Press

Cleveland, Ohio, May 30, 1980

A University of Chicago professor reported the other day the startling and worrisome information that Soviet youths are 10 times better prepared in mathematics and science than their American counterparts.

Professor Izaak Wirszup, an internationally known mathematics expert, called the findings of studies he directed a "grave threat" to U.S. security, a challenge to this country more formidable than the Soviet Union's launching of Sputnik in 1957.

Under the Soviet Union's educational mobilization, youths are required to attend school for at least 10 years. Before they finish secondary school, said Wirszup, they've had three years of arithmetic, two of arithmetic combined with algebra, five of algebra, 10 of plane and solid geometry, two of calculus, five of physics, four of chemistry, one of astronomy, five of biology, five of geography, three of mechanical drawing and 10 of workshop training.

Another mathematics educator, Professor Edwin Hewitt of Seattle, commented on the findings: "When will the U.S.A. wake up to the fact that the Soviets, despite their frightful political system, are overtaking us in education and technology? And what will the nation do about the problem when it is finally recognized for what it is?

Good questions.

Minneapolis Star and Tribune

Minneapolis, Minn., July 14, 1980

You may have heard the phrase and even suffered the affliction known to pop psychology as "math anxiety." Many people have the notion that they just aren't much good at working with numbers, and that math beyond arithmetic — say, algebra, calculus or trigonometry — is far too difficult even to consider trying. When we were in school, oral tradition transmitted that view strongly. In matters mathematical it was almost a badge of honor to plead stupidity. Boys believed that numbers must be numbing, and girls accepted their math ignorance as invincible. By senior high, in fact, many girls had learned well that math could not be learned. That's why "math anxiety" has entered the language largely by way of feminist concerns for curing and preventing it.

Of course it's all absurd — both the general fancy that math must be mysterious except for the super bright, and the special silliness that adds sex as a barrier on top of brains. But at least the notion that girls can't do math is easily exploded. In Minneapolis schools, teachers have moved straightforwardly to overcome the traditional prejudice. Last year, 46 percent of high-school calculus students were young women; three years earlier the percentage was 25. So far, so good. But in 1979 how many students got to calculus at all? Exactly 109, out of 3,000 graduates. That is 3.6 percent, in a city where employment hinges heavily on high technology and where customers come from far to shop for computers. Absurd or not, math anxiety is still a force to reckon with.

That thought came forcefully to mind as we recently read about teaching math and science in the Soviet Union. In many fields Soviet science — sluggish, timid and bureaucratic — shows little to emulate. But a decade of reforms in Soviet schools has brought an emphasis on math and science that Americans should not ignore. Besides favoring less rote and regimentation, the Soviet reforms have impressively expanded math and science requirements. Policy calls for most Soviet youngsters to take several years each of several different natural sciences. And calculus is taught as standard fare, equally expected of students headed for higher education and students headed for jobs in factories.

By comparison, American ambitions look modest indeed. This spring, for example, Minneapolis high schools doubled their requirement for math and science — from one year of either to one year of each. And even that small step was resisted by some as being elitist or risking an increase in the number of dropouts.

We make no case for copying or competing with the Soviet curriculum. A democracy's schools have values to serve that a dictatorship's do not. It is troubling, nevertheless, that Soviet educators show so much more confidence than most American school boards in their ordinary students' math and science abilities. We doubt that the reason is superior Soviet intelligence or a greater national aptitude for analysis and abstraction. More likely, Americans are just continuing the convenient cultural habit of indulging in math anxiety instead of tackling math. In the age of the computer, with new challenges to technology on every side, that is a tradition the country can do without.

DAYTON DAILY NEWS

Dayton, Ohio, October 27, 1980

The Soviet Union is in hot pursuit of the United States, and gaining. The effort by the Soviet Union and by other nations such as Japan and West Germany to develop mathematicians and scientists ought to stir this nation to improve educational programs and objectives. The way America's educational system works makes that difficult — but the warning lights are on.

The Carter administration, worried about the possibility of the United States losing its scientific prowess, commissioned a 230-page report by the Education Department and the National Science Foundation.

The report is out, and it shows that the United States' best math and science students continue to rank high in national aptitude tests. But the majority of students are doing worse than their predecessors did in math and science.

An elite can carry a nation far; one Albert Einstein can have more more impact than a thousand more ordinary physics students. But a shortage of work-a-day scientists, mathematicians and engineers can undermine a nation's ability to apply and build on the knowledge.

Already top computer science jobs go begging, the military is having a hard time getting enough personnel to operate its increasingly technical equipment, and business is being challenged on all sides in technological innovation.

Though the Soviets have many problems with their educational system — difficulties with the new math curriculum, job discrimination against ethnic, non-Russian graduates and some poorly run schools in the rural areas — their crusade is methodical and massive.

The Soviet Union's 10-year compulsory math and science program covers more than the U.S. system does in 13 years. The Russians take 10 years of geometry, for example, compared to one year for most American students. The Russians are graduating 5 million secondary school students with two years of calculus behind them, compared to 105,000 Americans who have taken one year.

The United States has several reasons for its relative decline. Americans pay teachers poorly; a good science or math teacher can double the salary in private industry. The U.S. educational system is not centralized; taxpayers shout for local control but don't provide support for educational excellence.

It is not necessary to make every American kid suffer calculus. But the educational system must be bolstered overall, and especially in the sciences that acquaint citizens with the ideas and disciplines reshaping the world and that prepare those with scientific talent to contribute to these important fields. This nation, faddish and free, must realize that, like it or not, it is in a long-distance race and the starting gun went off years ago.

The State

Columbia, S. C., November 25, 1980

A REPRESENTATIVE of the National Organization of Women was recently in town to persuade folks that math isn't a hard subject. Anyone can do it.

The trouble a lot of people have with math, she contended, is that they have been conditioned to accept wrong answers they may come up with as evidence of some grave personal failing or intellectual inadequacy. There is an analogy which is made to the role which women have been assigned in a male-run society.

That is an interesting hypothesis. However, we are not persuaded that there isn't a natural order of students who have differing aptitudes for math and for English, for science or whatever. Most will excel at one or the other, and only some will excel in all. That is our opinion, and not the result of verifiable research.

Be that as it may, there were two news stories recently concerning young students and math. In Peking, China, there's an 11-year-old farm boy who can solve complicated math problems in his head faster than accountants can using calculators.

And at Johns Hopkins University in Baltimore, a search was launched for 350 youngsters with extraordinary mathematical talent. These pupils, below age 13, are to be taken into a program to develop their skills to the highest degree possible. The purpose is to provide the United States with a cadre of analytical and scientific minds for the future.

No matter how long we have to stand at the blackboard, we'll always believe that math aptitude is a gift.

Post-Tribune

Guarding Your Interests Daily

Gary, Ind., April 21, 1980

Corporal punishment has been banned in many — possibly most — American schools for years, but the schools themselves seem on the way to becoming "whipping boys" again.

That's not necessarily a bad idea — if handled judiciously.

We don't like picking on the great American public school system just for the sake of having something to carp at. We do favor those with training in the educational process keeping a weather eye on that system and speaking out when criticism seems in order.

We have noted two important examples of that in the past few days. One was the piece by Thomas A. Shannon, executive director of the National School Boards Association on Page 1 of The Post-Tribune Viewpoint section Sunday. The other was an advance report by the University of Chicago's distinguished mathematician, Izaak Wirszup, to the National Science Foundation.

Shannon charged an inadequacy of instruction on world affairs and documented the growing need for such instruction.

Wirszup warned of an increasing emphasis on science education among Soviet youth and a corresponding lack in this country. While admitting "Soviet educational mobilization" was "not as spectacular as the . . . first Sputnik," he called it "a formidable challenge to (U.S.) national security . . . far more threatening than any in the past and . . . much more difficult to meet."

We would hardly welcome another wrenching of the U. S. school system of the sort that set liberal arts back after Sputnik. We think, however, both cited challenges are of the sort which require study.

Richmond Times-Dispatch

Richmond, Va., July 11, 1980

While lax educational standards in the humanities can cripple the cultural life of a nation, a nation's inattention to training in mathematics and science can invite a technologically more advanced enemy to deal it a coup de grace. This latter, fatal threat now looms large against the United States, according to Izaak Wirszup, professor of mathematics at the University of Chicago.

Dr. Wirszup's National Science Foundation report on mathematical learning in Russia and Eastern Europe, cited by *Times-Dispatch* science writer Beverly Orndorff in a June 22 article, declares grimly that Soviet students are ten times better trained in mathematics and science than American students. Because of high standards and requirements, students in Soviet high schools now are gaining more learning in science and math than most American students learn in high school *and* college.

"The disparity," writes Dr. Wirszup, "between the level of training in science and mathematics of an average Soviet skilled worker or military recruit and that of a non-college-bound American high school graduate, an average worker in one of our major industries, or an average member of our all-volunteer Army is so great that comparisons are meaningless."

Inferior mathematical and scientific training is combining with inadequate pay to bring about a grave shortage in skilled personnel to run the Navy's ships and to operate the increasingly sophisticated weapons needed by all the armed services. American vulnerability stemming from this educational lapse is compounded by the single-mindedness of the Soviet dictatorship, which is channeling far more of its technological research into military applications than into improvements in the comforts of life for Soviet citizens.

Edwin Hewitt of the University of Washington at Seattle, another distinguished mathematician, echoes Dr. Wirszup's concern. "When," he asked after reading Dr. Wirszup's reports on Soviet attainments, "will the United States wake up to the fact that the Soviets, despite their frightful political system, are overtaking us in education and technology?"

A chilling irony in this state of affairs is that a collapse of American education that may threaten the very survival of our republic has come about even as centralized government control over the schooling of the young has reached unprecedented heights. Interested more in social tinkering than in teaching the basics, educationists today are compelling nearly everything except study of the truly vital subjects.

A good place to begin correcting weaknesses in the preparation of our youth for life in a dangerous world is here in Virginia. State Board of Education member Henry W. Tulloch has asked his colleagues to consider the warnings of the Wirszup report in light of the state's math and science standards (only a year of each is required of Virginia high school students). The large lag behind our enemies in crucial training is no trivial matter. The Board of Education ought to act promptly to give proper priority to imparting to our young knowledge on which their lives may depend.

Newsday

Long Island, N. Y., November 13, 1980

It wasn't so long ago that a skilled backyard mechanic, with an occasional cussword and a barked knuckle or two, could make almost any auto repair short of a complete overhaul.

But now if such amateur mechanics set out to fix the carburetor, they may well discover that there is none; in its place is a fuel injection system, controlled by an on-board computer.

Their skills aren't the only ones that technology is outdating.

The typing pool is giving way to the word-processing center, where video screens have replaced typewriters. Production-line workers are being supplanted by computer-controlled machines programed to perform a series of tasks.

And all this is happening with bewildering speed.

That's why the nation's school districts should be concerned about a recent presidential report's warning that the United States is in danger of becoming a nation of technological illiterates, unable to understand or make informed decisions about the evolving high-tech culture.

Alarms about education are often sounded; at about the same time, another study commission was lamenting the damage that results from Americans' indifference to foreign languages. And youth has a way of embracing what's new; electronic games are a hot item, and computers abound in secondary schools.

Still, the report's two major points are well-taken.

First, it says, too many students leave high school without knowing much about technology and without the science background to help them acquire more knowledge later. Second, too little relevant material is available in colleges and adult education programs.

In the past decade or so, the self-fulfillment ethic and the frustration of rapid change have undercut the notion that all students should, or could, take courses that would familiarize them with the important fields of scientific knowledge.

So technology is offered as an elective for its enthusiasts. And the rest, when they need to understand what the auto mechanic is doing to a cranky car or comprehend why lagging technology is inflating the dollar, are left out in the cold.

Having raised the issue, Washington can now respond by funding teaching positions and sponsoring technology textbooks. That could help. But only if school districts see the need, too, and offer—or even require—technology courses. Given the subject's increasing importance in making the world run, they should.

C.P. Houston. Houston Chronicle

Chicago Defender

Chicago, Ill., June 5, 1980

The United States is always in competition with the Soviet Union understandably. But the evidence is, according to Izaak Wirzup, mathematician at the University of Chicago, that Soviet students are far more knowledgeable in mathematics and science than their American counterparts. He believes that the Soviets will use that advantage to overtake the United States in scientific and technological advances.

Various math and science teachers' organizations agree that there are problems with the U.S. education system. These are problems independent of the integration question and affirmative action issues. The problem is financing of education in these areas so that this country may become really competitive with what the Soviet is doing.

Bill G. Aldridge, executive director of the National Science Teachers Association said, speaking as an individual, that science education in this land is a "mess." He put the blame on insuffficent funding by the federal government.

He said that classrooms and school laboratories do not have adequate equipment. He added that, in combination with curriculum cutbacks by financially squeezed local school districts and a serious shortage of physics and chemistry teachers constitutes a decisive lag for American education. He fears "we'll wind up with physical education teachers and social science teachers teaching science."

It is a discouraging picture.

The Hartford Courant

Hartford, Conn., November 28, 1980

Sputnik prompted a vigorous competitive response in the American space program. It subjected a generation of American school children to a barrage of new approaches to math and science. Dick and Jane had to keep up with Ivan and Natasha.

Twenty-three years later a new presidential commission has concluded that Americans are again becoming scientifically illiterate. There is a vanguard acquiring a top education in the sciences, according to the White House study by the Department of Education and the National Science Foundation. "Those who are the best seem to be learning about as much as they ever did, while the majority of students learn less and less."

Arguably, not everyone in a technological society has to understand physics, computer science and biochemistry. But as jobs become more technical, the gap between the educated and the illiterate becomes more pronounced.

Technology itself may have contributed to the technical illiteracy of its own children. A calculator in every pocket and a computer game in every basement do not imply a youth who understands their inner workings and logic. In this, the Soviets once again may have surpassed us because technology is a tool, not a toy. A Soviet youth is probably more likely to comprehend the mathematics and mechanics of a computer than to have one.

The recommendations of the commission, although it was appointed by President Carter, should be heeded by President-elect Ronald Reagan, who has pledged to restore America's competitive edge with the Soviets. The commission recommends more courses in science and mathematics, like the catch-up programs after the Russians launched their Sputnik space program.

An immediate solution for parents might be to discourage their children from using calculators until they understand the basic principles of mathematics.

Scientific creativity stems from curiosity, not machinery. That's one quality America still should possess in abundance over the Soviets.

THE BLADE

Toledo, Ohio, February 14, 1980

THE Blade commented earlier this week on the Ohio board of regents' disturbing survey of remedial education among freshmen who entered state-supported schools in 1978. A subsequent report of students enrolled in remedial courses, broken down by school districts, prompts some further observations.

The report showed that 48 per cent of the 735 graduates of Toledo public high schools who entered state schools in the summer or fall of 1978 were required to take remedial math courses. And nearly 29 per cent had to take remedial work in English to meet college standards which, one might note, are none too high.

On the face of it, the Toledo schools' showing was worse than those of neighboring districts in Lucas and Wood counties. But even such highly touted school districts as Ottawa Hills and Sylvania had high percentages of graduates required to take remedial math courses — about 38 per cent in both cases. Adjacent districts apparently did a better job of preparing students insofar as English requirements were concerned.

One must be cautious in making generalizations because there are many wild cards in these statistics. For one thing, state colleges and universities have widely varying definitions of remedial education and approaches to carrying it out. Moreover, the "mix" of students attending state schools from Toledo and suburban districts varies considerably, which may have an effect on students' performances.

There may be a cause-and-effect relationship, however, between the minimal state and local district requirements in mathematics and the relatively poor showing of many students when it comes to tests of their knowledge of that subject. Despite the fact that arithmetic is a basic tool for living as well as for learning, the state of Ohio requires only one year of high school mathematics — a requirement which many students complete as early as the ninth grade.

And while the Toledo system requires three years of high school English — again the minimum specified by state law — many students in the past have been able to satisfy this requirement in part by taking literature courses that do not provide instruction in the basics of grammar and composition.

The response from the Toledo public school administration to the report was mixed. Dr. Rosie Doughty, assistant superintendent for curriculum and instruction, rather casually it seemed, blamed the high-remedial education figures on open-enrollment policies of the state universities and professed not to be alarmed by the results. However, Superintendent Donald Steele said that curriculum changes to take effect in the fall will require that students take more course work in basic math and English.

School officials have expressed the hope that Toledo public school graduates eventually will have a minimum of two years of mathematics, which is at least a step in the right direction.

The taxpaying public and students themselves are already paying a high price for the rampant permissiveness that has marked the attitudes of all too many educators, especially at the high-school level.

Although Johnny and Jane may not like it, more of their school time is going to have to be devoted to basic reading, writing, and arithmetic skills — "forward to the basics," as one Toledo educator put it. The same prescription no doubt is needed in many other Ohio school districts.

The Times-Picayune
The States-Item

New Orleans, La., August 29, 1981

Young people contemplating the best career opportunities in the next few years would be well advised to study mathematics and science. With the nation and the world in the midst of a technological revolution, there is in the United States a distressing shortage of people who are competent in mathematics and the various sciences.

In announcing $1.1 million in grants to schools to help overcome "acute" deficiencies in mathematics among minority students, the Ford Foundation observes that "signs are mounting that Americans are falling behind in mathematics and science training.

"In that respect, the nation might almost be called underdeveloped, especially in comparison to Japan and the Soviet Union."

In the age of the computer, the demands of business and industry are resulting in a critical shortage of mathematics teachers. The best teachers are quitting in droves to take much higher paying jobs in industry.

The growing math crisis "is a nationwide problem that is worrying leaders in the federal government, business and industry, universities and the professions," says Ford Foundation president Franklin A. Thomas.

Skill in mathematics is indispenable to advanced learning in the sciences. It also is essential for individuals to take full advantage of career opportunities in such fields as engineering, pharmacy, medicine, chemistry, the high-technology computer and communications industries, architecture, business management and teaching.

The National Council of Teachers of Mathematics also noted recently that mathematics training is not keeping pace with the revolution in technology.

Not surprisingly, the decline of competency in mathematics parallels the recent period of experimentation and permissiveness in the nation's educational systems. Mathematics is a study that requires discipline for mastery, and discipline has not been much in favor in the nation's schools in recent years. Fortunately, there appears to be a new appreciation of the fundamentals in education circles. It is coming none too soon, because the nation must have a wealth of scientists, engineers and technicians if it is to retain its position of supremacy in global affairs.

Chicago Tribune

Chicago, Ill., June 1, 1981

Education is a deplorably neglected aspect of the rivalry between the United States and the Soviet Union. According to Izaak Wirszup, professor of mathemetics at the University of Chicago, "The weaknesses of the American educational system have become a national malady that gnaws at our economic strength, our competitive edge in technology and production, and our ability to defend ourselves." The quotation is from testimony that Prof. Wirszup recently gave to the Senate Subcommittee on Science, Technology, and Space.

Prof. Wirszup's concern is shared by others who have compared the schools of the two superpowers. Soviet schools can of course be criticized for their high propaganda content and their centralized regimentation. But there is no doubt that the Soviet regime has put substantial content into the curricula of its elementary and secondary schools. A boy or girl emerges from a Soviet schooling with a far better grasp of mathematics, sciences, and foreign languages than an American boy or girl.

According to Prof. Wirszup, Soviet education "ensures a large pool of trained, well educated labor for Soviet military-industrial production." The compulsory Soviet program calls for years of study of mathematics, physics, chemistry, biology, geography, and a foreign language (especially English). But in the United States mathematics for most never gets far beyond repetitious drill of a few simple processes.

In our high schools, sciences and foreign languages are electives studied only briefly by a minority of pupils. Only 16 per cent take a year of chemistry, less than 10 a year of physics, less than 4 more than two years of a foreign language. We are short not only of pupils but of qualified teachers. The result, as Prof. Wirszup puts it, is that "military training manuals formerly written at an 11th-grade level are now being written at 6th-grade level."

Military needs hardly make an ideal basis for appealing for some substance and rigor in American schooling. There are other reasons for children to learn something useful and fulfilling during their years in school. But the military argument is a valid one, and for many it will be the most effective one.

Prof. Wirszup dates current Soviet educational policy from 1966. Before that, he says, "the Soviets were content with the creation of a small scientific elite, while the majority of their students remained at low levels of academic performance." It is dangerous for a totalitarian society to stimulate the mind, even in technical matters. But in the past 15 years the Kremlin has taken whatever risks to itself there are in a wide diffusion of basic knowledge of mathematics and the sciences.

At the same time, the United States has incurred the grave risks of producing graduates of colleges, as well as high schools, from whom little has been expected in the way of systematic learning of any kind. Most graduate and professional schools and university faculties still have high standards. But, aside from a small elite at the top, American education has leveled down rather than up as the average time in school has become prolonged.

It is ironic indeed that informed students of comparative education should conclude that education has become an advantage not of the free world but of the unfree world. For reasons of defense and for other reasons, we in the United States need to raise our expectations of both teachers and pupils. We could, if we would, teach our children no less than the Soviet Union is teaching theirs.

THE INDIANAPOLIS NEWS

Indianapolis, Ind., January 17, 1981

Little more than 23 years ago the U.S. was caught off guard when the Soviet Union launched Sputnik into the heavens.

But after a mad national scramble to catch up to the Soviets' space program, this country went beyond the achievements of the U.S.S.R. — put the first man on the moon and many more after that.

Today there is no Sputnik-like event to launch the U.S. into another technological flurry, but some quieter alarms have gone off to warn that America may be falling behind once more, particularly in the classroom.

To begin with, figures show that 98 percent of all Soviet children graduate today from high school, compared to a 75 percent graduation rate for all U.S. students. By the time they finish high school, almost all of the Soviet students will have had five years of physics, four years of chemistry, five years of biology, five years of geography, three years of mechanical drawing, 10 years of workshop training, one year of astronomy, five years of algebra, 10 years of elementary geometry and two years of calculus.

The high school science training for Americans is far less. A study by the National Science Foundation found that the few students who do take science courses take only a year of study in each area of science: Only 9.1 percent took physics, 16.1 percent studied chemistry, 45 percent had biology and 17.3 percent studied general science. Considering that some students took chemistry, physics and biology, the percentage of graduating seniors who studied science drops further.

These figures do not mean that Soviet schools are better than the U.S. counterparts, but it does mean they provide more opportunities for scientific training. American schools still produce some outstanding students, yet overall math and science scores are dropping.

Graduating all these students with heavy training in science has not been entirely a blessing for the Soviets. After drastically mobilizing the school system to provide all this training, the country now has more of these students than it can handle. Less than 10 percent can attend college. Many of those who cannot get more education are dissatisfied with the opportunities available.

Such is not the case in the U.S. where people are needed to run sophisticated military equipment, take top computer science jobs and help struggling American businesses and industry come up with innovative ideas to beat their foreign competitors.

The science shortage cannot be solved overnight. Many different steps must be taken. Each school should evaluate its own science program to see if the classes offered meet the needs of the students. Pay must be increased, particularly in the military, if people are to be attracted to science careers and retained to serve on technologically difficult jobs.

On a national level, a task force is preparing a report for President-elect Reagan that will recommend steps the government can take. The panel is expected to suggest that tax incentives be offered to industry to encourage new scientific research; that regulations discouraging technological development be scrapped and that new priorities be developed for distributing government research and development money (the Department of Defense and NASA are expected to receive more of such funding).

Izaak Wirszup, an expert on Soviet math and science at the University of Chicago, warns, "The recent Soviet educational mobilization, although not as spectacular as the launching of the first Sputnik, poses a formidable challenge to the national security of the United States, one that is far more threatening than any in the past and one that will be much more difficult to meet."

Obviously, the United States cannot let the challenge pass by. That is one of the big tasks of the '80s.

The Detroit News

Detroit, Mich., May 29, 1981

More and more high schools across the country are requiring would-be graduates to balance a checkbook. But mastery of this rudimentary skill, along with other minimum proficiency tests, is being challenged in some quarters as somehow unfair to the pupils.

Meanwhile, what must Soviet students master prior to *their* graduation from equivalent schools?

In the Soviet school system, according to Prof. Izaac Wirszup of the University of Chicago, basic arithmetic is completed in the third grade, and the teaching of algebra starts in the fourth grade. Also in the fourth grade, specialists begin teaching mathematics as a separate discipline. Calculus, too, is compulsory. And the standard Russian curriculum calls for 10 years of geometry.

In recent congressional testimony, Mr. Wirszup observed that only one-half of the U.S. student population receives a one-year course in plane geometry. Less than one-tenth of American high school students take a one-year physics course. Indeed, there are only 10,000 physics teachers for more than 17,000 U.S. school districts.

In short, Mr. Wirszup said, "the overwhelming majority of our population lives in a state of debilitating scientific illiteracy." Since it takes 20 years to produce a qualified scientist or engineer, he added, "if our public schools cannot attract and hold students to the sciences, a generation of future scientists will be lost forever."

Learning in the physical sciences and mathematics is cumulative — today's unlearned lessons will impede an understanding of tomorrow's concepts. Tests can't be fudged; facility of expression can't disguise a lack of comprehension. The study of science and mathematics requires a large measure of discipline.

As you know, discipline is in short supply in the American public schools. And the result could well be a failure to compete in an age of increasing technological complexity.

Those who would "protect" students from the rigors of balancing a checkbook will have much to answer for, should Mr. Wirszup's grim prophecy come to pass.

The Des Moines Register

Des Moines, Iowa, May 13, 1981

President Reagan might consider former Vice President Walter Mondale's words to the National Education Association the other day. Mondale castigated the administration for its proposals to reduce spending on education by 25 percent, and said, "If you believe in a strong defense for America, you ought to begin with good schools in America."

Mondale was not just mouthing traditional liberal rhetoric. In engineering education alone, low salaries for professors and high salaries for engineers in industry have created a situation in which professors (and those with engineering degrees at the bachelor's level) are siphoned off by industry. The University of Iowa and Iowa State University have trouble filling openings for engineering professors.

The Soviet firing of Sputnik, the first Earth satellite, in 1957 led to the 1958 passage of the National Defense Education Act. The program emphasized science, mathematics and foreign languages.

The Soviet Union, as part of a decade-long drive to enrich students' backgrounds in science and math, requires high-school students to complete five years of physics, four years of chemistry, one year of astronomy, two years of calculus and 5½ years of biology. The typical Iowa high school requires one year of science and one of mathematics for graduation.

Reagan's competitive attitude toward the Soviets clearly has not led him to the same education-policy conclusion as was reached by the Soviets and President Eisenhower.

Federal money covers only 7 or 8 percent of spending on elementary and secondary education, but the cutbacks in other federal programs for cities and states could pull municipal and state tax money away from schools to plug holes elsewhere.

If schools and colleges are gutted through ill-considered cuts in aid, a void will be left in national expertise. The harm will be done not only to foreign and military policy. An educated U.S. population is a resource. Just as economic health is necessary for national security, so is a sound educational system.

Science Education Troubles Spread Across the Nation

President Ronald Reagan's complaint that the schools were failing to turn out properly trained students was echoed across the nation by the press and business leaders. Student achievement was not the only area in decline; standards of teacher preparation also were lower than in previous years. Each school district had its own horror story of ill-prepared students and unqualified personnel. America was unfavorably compared to other countries, where the school year was longer and where there were many more required subjects. Reagan warned that the situation threatened U.S. military security as well as economic development, since technologically backward troops would not be able to handle sophisticated military hardware. He placed the burden of improving the school system primarily on the states. In his "new federalism" program, Reagan sought to abolish the Education Department and substitute a "Foundation for Education Assistance," whose scope for aiding local school systems would be greatly reduced. Aid programs for all subject areas were slated to be reduced to $10 billion in 1983 from $14 billion in 1982.

Detroit Free Press

Detroit, Mich., May 17, 1982

THERE IS some irony in President Reagan's complaint to a conference of science educators and business leaders about the deplorable job our schools are doing in teaching science and math. It is his administration that last year eliminated most of the funds for the primary and secondary school science education programs of the National Science Foundation. The administration also proposes to cut funds for Title I of the Elementary and Secondary Education Act, which subsidizes compensatory and enrichment programs, from $3.1 billion in 1981 to $1.9 billion in 1983.

By 1983, schools in Detroit will lose 37 percent of their federal funding. At the same time, more students will be forced to drop out of college as support for educational loans dries up. The next generation will be lucky if it knows enough to balance a checkbook or follow a route on a map — much less master word-processing systems or chemical compounds.

Secretary of Education T.H. Bell suggested the answer to the spreading scientific ignorance is for states to consider paying hard-to-find science teachers more than they pay other teachers. "Where is the funding going to come from to do all this?" asked Sarah Klein, president of the National Science Teachers Association. No one ever got around to answering her.

It is not, however, a problem the federal government can afford to brush aside or leave up to states and local school districts with widely differing capacities to respond. The U.S. risks losing its lead in basic science and endangering the efforts to strengthen our defense. The level of literacy among the armed forces has dipped alarmingly low; we could end up with a military technology too sophisticated for our own forces, unless we start pouring more resources, not fewer, into basic research and school programs.

And we stand to lose even more than military might or scientific supremacy. Our hard-won knowledge of the universe, its mysteries and marvels, is in danger as well. What will awe, inspire and shove man along on his quest for life's meaning, if not physics and zoology, calculus and astronomy?

Without the sense of wonder science spawns by its study of nature's ceaseless variety and overriding harmony, man becomes a pitiable creature, lacking any sense of his place in the scheme of things. That would be, for a species priding itself on eons of evolution, a kind of death in life.

The Morning News

Wilmington, Del., May 18, 1982

ARMS CONTROL, nuclear power, hazardous waste disposal and genetic engineering — these are complex issues on which the American people have to reach political decisions now and in the future. Do they have the know-how to do so?

Sadly, in most instances, the answer is no, because for many there has been far too little instruction in the sciences and mathematics. Quantity and quality of education in these crucial fields have been, and still are, inadequate, because as a nation we have not required that these courses be part of students' curriculum continuously until high school graduation. English is a required course year after year in public and private schools. Why not mathematics, including instruction in computers? Why not science, including life sciences as well as the basics in chemistry and physics?

There are several reasons. One is that traditionally these subjects have not been required for high school graduation. But they were not required because one could function as an educated citizen without mastering the basics of science and mathematics. Today, such functioning is no longer feasible in most workplaces, nor can one be a responsible member of the body politic without a grasp of these subjects.

Another reason for the current feeble status of schooling in mathematics and science is shortage of qualified teachers. With some exceptions, as at the University of Delaware, teacher education includes too little substantive instruction in these fields. At the same time, persons with good skills in mathematics and science find it more lucrative and challenging to go into industry and business than into teaching. That is true not only for high school teachers but also for those qualified to teach at the university level, where pay scales are also below those in industry. So top people are more often than not drawn to a profession other than teaching.

For a brief period after the Soviets sent Sputnik into orbit in 1957, the federal government increased its support of education in science and mathematics. Over the years, however, that support has dwindled, and the budget President Reagan proposed included only $15 million for science and engineering programs, as compared to $112 million in former President Carter's last budget.

Paradoxically, at the same time he is putting so little money toward education in science and mathematics, the president in remarks to science educators deplored the low state of instruction in science and mathematics and said that this threatens U.S. military and economic security. But neither he nor his spokesmen propose to spend more in this area.

That is unfortunate. For the gaps in science and mathematics instruction are nationwide, and require not just national recognition, such as they are now getting, but also national support to close them.

The Washington Post

Washington, D. C., May 29, 1982

PRELIMINARY TESTS in the classrooms of the city's public schools indicate what many parents and students already know about the state of science instruction. It's weak. In some classrooms, science is a joke, a series of dreary memorization exercises, assigned with little or no attention to the learning of analytic skills.

Science was downgraded when "back-to-basics" necessarily emphasized reading and math, and sharp cuts were made in the number of science teachers. This has meant that students of the highest and lowest abilities are lumped in the same junior high classes.

If this news is depressing, the reaction of school board members and the superintendent is not. Their focus is on improving science instruction. They mean to do it without rejecting back-to-basics math and reading or junking the competency-based curriculum procedures aimed at improving and monitoring the teaching of these subjects.

Reading and math remain fundamental skills whose mastery is essential to other academic pursuits. Still, just as foreign language study has been found to complement an English requirement, so can a strong science program strengthen students' abilities to draw inferences and think analytically. Many educators agree that the processes of observation, comparison and interpretation of evidence should be experienced early, in the elementary school years.

Superintendent Floretta McKenzie has been moving to improve instruction in subjects that were downgraded before—while continuing the emphasis on reading and math. Last year, the schools began requiring students to take two years of science, not just one, to graduate from high school. Even with budget constraints on the number of skilled science teachers, at least there can be increased staff training in science for those regular elementary school teachers who have had to assume responsibility for science instruction—and who, according to Mary Harbeck, supervisor of the system's science department, may suffer "science anxiety."

Supervisor Harbeck also says she thinks classroom materials may be a factor. She's right. Poor lab materials and illegible assignments cranked out on broken-down ditto machines (an unfortunate trademark of the entire D.C. school system) surely contribute to the weak showing of students.

So one preliminary test should not trigger educational panic or the dropping of everything else in a mad dash to the science lab. Solid assistance to all teachers of science—along with thorough evaluations of their instruction techniques—can begin to improve the skills, confidence and futures of Washington's youngsters.

THE BLADE

Toledo, Ohio, May 2, 1982

PERIODIC warnings that the quality of science and mathematics education are declining in the nation's public schools are unquestionably well-founded and are a cause for public concern.

A report of the National Science Foundation's Commission on Pre-College Education in Mathematics, Science, and Technology reveals that math and science achievement scores are declining. As many as 50 per cent of high school teachers in these subjects are uncertified, far too many good teachers leave their classrooms for higher-paying jobs in industry, and the curricula of U.S. public schools are too permissive for the pupils' and the nation's own good.

American scientific and mathematics instruction is woefully deficient. The accomplishments of Japanese and West German schoolchildren are surpassing those of American students, in part because for most U.S. children science or math instruction, such as it is, ceases by the 10th grade. The competition from these friendly nations often is overlooked because of a tendency to concentrate on what the Russians are doing.

Many warnings about American academic shortcomings refer to Russian curriculum reforms that threaten to outstrip this country's technological prowess. There is reason to be concerned when one looks at the efforts being made by our principal world adversary. For instance, it is reported that 5 million Russian schoolchildren take calculus as compared with just over 105,000 American children. It is not known how good Russian instruction in calculus is; there is a tendency even in the face of Russian ineptness in many technological areas to paint the Soviet education system in rosier hues than is warranted.

The problems outlined in the commission report cannot be solved simply by requiring more course hours in these fields, although that would help to some extent. The commission's report makes it all too clear that even if more hours are mandated, the quality of instruction simply will not be adequate in many school systems.

This scientific illiteracy does more than undermine the vitality of the American economy. It also is producing a generation of Americans unable to make intelligent public-policy decisions in technological areas of national concern, such as the role of nuclear energy in the generation of power or exploration of ways to provide rational environmental controls, to say nothing of the overarching question of finding ways to limit mankind's drift toward nuclear warfare.

Both the economic and intellectual health of the American people will depend upon whether this country assigns a much higher priority to education in these vital areas. To do so will require some tradeoffs — perhaps a willingness to tax ourselves enough to educate and encourage qualified science and math teachers to stay in the classroom or to eliminate some courses and expenditures that are less essential. At the moment, though, the outlook for such enlightened action is not promising.

The News and Courier

Charleston, S. C., April 5, 1982

For some reason, the argument over whether the public schools in South Carolina are good or bad tends to confine itself to lowest common denominators. That may be natural in a state where a great many children are having difficulty mastering fundamentals like reading, writing and arithmetic. But it suggests South Carolinians don't expect as much of their schools as they should. There ought to be far more discussion, for example, of what schools are doing in the increasingly important aspects of curriculum which embrace mathematics, physics and chemistry, the so-called "hard courses" — as opposed to the fluff of music, drama, social sciences and basketweaving.

It is not only that the hard courses give students what they need to cope in an increasingly scientific and technological age that makes them important. It is also the fact that they are usually the only courses which challenge children to analytical thinking and permit performance measured by accurate, unyielding standards. There is only one right answer to a problem in trigonometry, only one way to write the formula which provides a desired result in chemistry. The answer in the fluff course is, by contrast, whatever teacher says it is.

The Greenville High School teacher who testified before the state Senate Education Committee a few days ago was understating his case, if anything, when he said that by comparison with other places, South Carolina schools are lacking in concentration on math and science. Enrollment in those courses actually tends to be scandalously small, sometimes involving fewer than one percent of the students in a big school district.

Several reasons were given why that situation exists and is getting worse. They include low salaries, lack of federal aid and other familiar scapegoats. The big reason, we suspect, is that nobody pushes that kind of teaching. It should be pushed. A school system that isn't heavy in math and sciences is, as one senator said, fooling the people when it ought to be preparing them for life.

The Kansas City Times

Kansas City, Mo., June 2, 1982

In comparison with other countries, young America could be called scientifically and mathematically illiterate. Not only have Scholastic Aptitude Test scores declined — the mean dropped from 502 in 1963 to 466 in 1980 — but the disciplines traditionally get a stingy share of classroom time.

Proficiency tests as graduation requirements validate the sorry facts. Last week 81 Kansas City, Kan., seniors still had not passed that district's proficiency test in fundamental math, one covering basic skills teachers expect seventh graders to know. They had 11 chances to try. A daily specialized instruction period in mathematics and science for eighth graders in the Kansas City School District has been added to the curriculum as part of the five-year strategic plan. Already, achievement scores, particularly in math, have shown noticeable improvement. But tests won't solve the problem and neither will the enlightened efforts of individual districts; nothing short of a national crusade similar to the undertaking after the Russian Sputnik launch in 1957 will.

Blame administrators content to baby sit and school boards unwilling or unable to demand more; blame patrons who don't hold them accountable. It's all part of the scientific ho-hums affecting the U.S. where elementary students do well to get an hour of science and four hours of math a week. In the Soviet Union, East Germany, China and Japan, by the sixth grade, children have specialized study in mathematics, biology, chemistry, physics and geography; they'll spend triple the time on those subjects given by even the most science-oriented student here. In addition their school day and week are longer and vacation far shorter.

While this country is already paying a price in a shortage of classroom teachers — half the high school math and science instructors this year were unqualified, hired with emergency certificates — other countries recognize the importance of these disciplines in meeting future economic and social challenges. National policies promote them, thus encouraging students and eliciting citizen support.

At a recent conference at the National Academy of Sciences, the secretary of education lamented while the secretary of defense worried about the future shortage of scientists and engineers. President Reagan sent a message expressing concern that the problem could endanger America's industrial strength in the future. With such a consensus, what are we waiting for? Another dramatic Soviet breakthrough with which we can try to catch up? Again?

THE ARIZONA REPUBLIC

Phoenix, Ariz., March 26, 1982

THE U.S. may have learned something from Sputnik — but the lesson, it appears, was only a temporary one.

Having run sufficiently hard against the Soviet technological advances of the '50s to put several men on the moon in the '70s, America once again appears headed downhill toward what leading scientists call a pit of scientific illiteracy — a pit where economic and defensive dangers lie.

When Moscow sent up an earth satellite in 1957, the Soviets alone were superior to the U.S. in any meaningful phase of technology.

Today, there are threats from more than our cold-war nemesis. In the Soviet Union, Japan and West Germany, many students have surpassed America's brightest in science and math.

And, says President Reagan's science adviser, George Keyworth, those countries are demanding scientific educations while the American educational system has turned to flab.

In the years since Neil Armstrong set foot on the moon, there has been a nationwide trend toward reducing high school graduation requirements in mathematics and science.

As a result, Keyworth told last month's gathering of the American Association for the Advancement of Science, "We may find ourselves as a nation making poor decisions, bogged down in indecision or losing much of our freedom as decisions are made for us by others or by default."

The facts are grim:

- The average American high school student takes one year of geometry compared with the Soviet student's 10-year geometric curriculum.
- Only one-sixth of U.S. high school students take even a single year of chemistry. All Soviet students take four years of it.
- Among 14-year-olds in 19 leading countries, Japanese students ranked first overall in science. The American students came in 15th.

And so on.

As for those who might be teaching them, salaries in private industry are too attractive for schools to compete.

The Soviets' educational surge may not be as great as the one that sent Sputnik aloft, but between them and our allies — and rivals on the industrial front — the threat is there.

Because things haven't reached the late-'50s crisis stage, the U.S. has the opportunity for introspection; for a good, hard look at what its schools are producing — and why they're not producing more of the technology that made this nation great.

Phoenix, Ariz., May 27, 1982

DURING the two-day meeting of the National Academy of Science in Washington, Prof. Paul DeHart Hurd of Stanford University put his finger on the reason for the growing scientific and technological illiteracy of the American people.

Schools are playpens compared with schools in other countries, notably Soviet Russia, East Germany, China and Japan.

In those countries, he said, the school year is 240 days and absences are rare because parents are held accountable for them.

In the United States, the school year typically is 180 days but, because of absences, the average student attends classes only 160 days.

Their schools have a five-and-a-half day school week and a six- to eight-hour school day. Our children attend classses four or five hours a day, five days a week.

Their vacations are short and scattered over the year to avoid breaks in learning.

Our children get a three-month vacation from books during the summer.

Mandatory courses in mathematics, biology, chemistry, physics and geography begin in those countries in the sixth grade, and extend over four to six weeks.

Teachers are specially-trained. In Russia, a science teacher must carry out a research program in his field before teaching in a high school.

In the U.S., only 34 percent of the students who graduate from high school have completed as many as three years of math.

Only 8 percent have completed a course in calculus — only 31 percent of the high schools even teach calculus.

Most graduates have taken a biology course. Only a third have taken a course in chemistry; as few as 10 percent a course in physics.

Hurd noted that schools in the four countries he cited do not emphasize mathematics, science and technology at the expense of social sciences, humanities and languages.

On the contrary, he said, "foreign language study, usually English, is encouraged to make it possible for students to tap the world's largest resource of scientific and technical information — ours.

"There are more students and adults learning English in China than there are English-speaking people in the United States."

The picture Hurd painted was frightening because national security and economic progress increasingly depend on a population with scientific and technological know-how.

Other nations, including economic competitors, like Japan, and potential enemies, like Soviet Russia and East Germany, are developing such populations.

Some day, we may have to pay dearly for our failure to do the same.

Science Teacher Shortage Adds to Education Problems

The deterioration of mathematics and science performance in U.S. schools was made worse by a growing shortage of trained teachers. Experts who studied the teaching of math and science at secondary-school and college levels found a disturbing trend toward fewer teachers and fewer science and math students who chose teaching careers. In testimony April 15, 1982 before the Senate Labor and Human Resources Committee, F. James Rutherford, chief education officer of the American Association for the Advancement of Science, asserted that there was "a dramatic and disturbing decrease in the quality of science and mathematics teachers. . . . [M]any of the best chemistry, physics and mathematics teachers are leaving teaching for jobs in industry. Their positions in the classroom are . . . filled . . . more often than not by teachers who have little or no training in science and mathematics. . . . [T]he system is simply winding down." Rutherford blamed low salaries and poor working conditions for discouraging good science and math students from entering the teaching field. On the college level, the problem was the same: "B.S. engineer graduates are going directly into business and industry rather than entering doctoral programs. Our supply of researchers in these fields and of professors who can teach the next generation of engineers and computer scientists is rapidly depleting. Institutions . . . are also depending more and more on foreign instructors and foreign graduate students in their advanced programs."

The Seattle Times

Seattle, Wash., November 24, 1980

JAPAN, with half the population of the United States, graduates as many engineers, according to a study by the U.S. Department of Education and the National Science Foundation.

The report expressed justifiable alarm that American schools were relaxing mathematics and science requirements, whereas Japan, West Germany and the Soviet Union were putting heavy emphasis on those areas.

In notable understatement, the report observed that "the number of young people who graduate from high school and college with only the most rudimentary notions of science, mathematics and technology portends trouble in the decades ahead."

What is more, because the job market for engineering graduates is now so hot, students with four-year degrees are discouraged from attending graduate school.

This has created a shortage of faculty to train future engineers, threatening a further loss of America's competitive edge in technology in years to come.

In this state, engineering education has reached what officials at the University of Washington and Washington State University call a crisis stage.

Accreditation of some departments is threatened because of crowding, outdated equipment, and other problems. Says Dr. John B. Slaughter, W.S.U. academic vice president and provost:

"Both the U.W. and W.S.U. simply cannot now accommodate the number of students who want an engineering education . . . This is a high-technology state, and we're simply not educating as many engineers as we need."

While the lack of engineers critically affects The Boeing Co., the shortage also is felt in the nuclear-power industry, road construction, architectural engineering, and in this state's growing electronics industry.

The issue of inadequate engineering education in Washington is expected to be a hot topic in the coming session of the Legislature. And well it should be.

Statistics show that Washington is the third-highest state in the country in the number of engineers employed per capita, but well below the national average in the number of engineering students it educates.

The Washington Star

and Daily News

Washington, D. C., November 10, 1980

If a nation's technological future may be measured by the number and skills of its scientists and engineers, the United States appears to be taking significant risks. So suggests a recent National Science Foundation study, which reviewed the status of U.S. science, engineering and mathematics instruction — and found it wanting.

Consider this: The country now faces shortages of trained computer professionals and most types of engineers — as well as teacher shortages and falling educational standards. There are personnel shortages in some areas of physics and biological science. Outmoded facilities mean that many graduates are not properly trained in the latest techniques. Meanwhile, Germany, Japan and the Soviet Union, unlike the U.S., consistently pursue "more extensive and rigorous education in science and mathematics."

The news is not uniformly bad. Between 1973-79, for example, the number of "active" science and engineering doctorates increased by 42 per cent. In 1979, U.S. engineering colleges produced a record 53,000 graduates with bachelor's degrees. But, as a paper prepared by the American Society of Mechanical Engineers notes, "this is at or near the capacity of present facilities, but not nearly enough to meet the demand which is currently estimated by the Department of Labor to be 70,000 graduates per year."

There is nothing novel about the report. Between 1959 and 1962, the presidential Science Advisory Committee issued three separate reports recommending a larger federal role in support of science teaching, and more followed. The National Defense Education Act in 1958, after the Soviet launch of Sputnik, helped create new science-education programs. It also changed the statutory authority of NSF so that it could support science, mathematics and engineering programs at all levels.

But politics and fashion have often interfered with consistency and purpose — in science as in other areas. The NSF report notes that the lunar landing in 1969 "convinced many . . . that the nation's supremacy in science and technology could thenceforth be taken as a given. Meanwhile, the nation turned more attention to the social goals of providing equal access to education . . . while the NSF expanded its responsibilities to provide support to individual minority, women, and physically handicapped science and engineering students . . ." After 1968, real-dollar federal investments in research and development and science education began to decline, although they picked up again in 1975.

The concerns of a nation surely change from decade to decade; and a number of the social goals pursued in the '60s and '70s were worthy ones. But what happens when the goal of scientific excellence is put aside? The risks are obvious, and quickly multiply. A shortage of students in one decade is likely to produce a shortage of teachers in the next — no matter how many reports decrying the situation.

The "two cultures" gap described by the late C.P. Snow has never been so clear. Although what the NSF study calls "scientific and technical literacy" is increasingly needed, too many students are leaving school as scientific illiterates. "More students than ever," the report says, "are dropping out of science and mathematics courses after the 10th grade, and this trend shows no signs of abating."

For now, the NSF urges federal support for university-level programs that would permit qualified undergraduates to transfer to fields where shortages exist — or are anticipated. But the study also recognizes that far more complicated and far-reaching programs are needed; and that the workings of the marketplace cannot be entirely relied upon to find the solutions.

The report puts the peril of complacency plainly: "The economic well-being, security and health and safety of Americans during the remaining two decades of this century, and beyond, will depend increasingly on our ability as a nation to strengthen our technological and scientific enterprise."

The Dispatch

Columbus, Ohio, January 9, 1982

IT WOULD have seemed impossible a mere 15 years ago, but American universities today are in danger of running out of science professors. Lucrative salaries in the private sector are draining talent from campuses across the country. If the exodus continues, universities within the next 15 years will be hardpressed to prepare students for scientific careers.

Members of the American Association for the Advancement of Science (AAAS) met in Washington this week to discuss issues important to them. They had much to talk about. The heady days of the post-Sputnik, catch-up-with-the-Russians era are long gone. Research and development money is hard to come by. U.S. dominance of things scientific has been eroded by advances beyond our borders. The space program is a shadow of its former self. Other concerns have replaced the passion to explore mysteries.

All of these factors have contributed to the situation the scientific community now faces. The boom in the 1960s attracted students to the sciences, many of whom went on to work with the space program and related endeavors. The success of the space program provided countless scientific spin-offs with applications in the private sector. As interest in space started to wane and new technological industries expanded, more and more properly trained people were needed in industry. Top undergraduates decided to forego advanced degrees and to head into private positions, lured by attractive starting salaries and bright career opportunities.

Professors also started to leave the campuses, filling a need in the private sector for people capable of guiding the college graduates in their new work.

The situation leaves universities with fewer faculty members to undertake research projects, fewer graduate students to sustain the research and a reduced ability to attract public and private research funding.

The New York Times recently reported that while undergradate engineering enrollment was growing by 47 percent in the last decade, the number of doctorates declined by one-third. In 1980, an estimated 15 percent of the nation's engineering faculty positions was unfilled. In 1979, the number of positions open in industry and academia for "computer-related Ph.D.s" was 1,300, but the number of new doctorates was only 326.

This is a problem that affects not only the academic community but also private industry and the nation as a whole. If the universities cannot produce the trained people the private sector needs, the nation will soon fall behind in its scientific and technological expertise. The social and economic implications could be significant.

The AAAS, working with the federal government and representatives from the private sector, must address this problem and devise ways to remedy it.

The Oregonian

Portland, Ore., February 12, 1982

Basic research in science, the lifeblood of a technical nation, is in deep trouble because of federal budget cuts and state support for higher education that is taking a beating in places like Oregon and Michigan.

It takes a long time to train a scientist — up to nine years for a doctorate. It can take even longer to assemble research groups and get projects under way. Disrupting this process, even if only for a single fiscal year, can have a profound impact on research projects.

The costs of these losses can be translated directly into tax dollars going for national defense and space projects. Recently a National Aeronautics and Space Administration official blamed that agency's cost overruns in part on delays due to a shortage of technical workers and research teams. Colleges and universities are losing technical faculty members to industrial concerns, compounding the problem of producing more technical graduates in an era of reduced budgets.

The administration, in coming down hard in its proposed budget cuts for graduate education, has struck at the very heart of scientific development. In addition to raising loan costs for graduate students, the 1983 budget proposals would cut 1.9 million college students off various federal financial assistance programs.

Funding levels for National Science Foundation programs and engineering education would be cut nearly 80 percent below fiscal 1981 levels in President Reagan's 1983 budget proposals. Funding for basic research at the federal level is primarily directed at defense, space exploration, health and energy. But these categories are also being cut.

Clearly, America's scientific pre-eminence is being challenged from abroad, and at a time when the nation is preparing to neglect a high level of support for higher education. So, Congress ought to consider increases rather than cuts in some graduate fields because many states, caught in a sliding economy, have had to reduce outlays for higher education.

The Des Moines Register

Des Moines, Iowa, April 5, 1982

People have become so accustomed to stories about experienced teachers' being laid off because of declining enrollments, and about top-quality graduates' searching in vain for teaching jobs, that it's easy to forget that there are acute shortages of teachers in some subjects.

Science and mathematics are the fields with teachers in shortest supply. Only three of the 2,700 Iowans who graduated from colleges with teaching certificates last year qualified for a certificate in physics.

The number of students in teacher-training programs who are prepared to teach mathematics is 77 percent below what it was 10 years ago. Iowa surveys have revealed a similar decline among students at Iowa colleges. The decline in preparation to teach science is about the same.

Perhaps the most disturbing data come from surveys of high-school principals during the past two years. Asked to evaluate their new science and mathematics teachers, the principals rated almost half of them unqualified. Another survey indicates that veteran teachers may have fallen behind new developments in their subjects, such as the use of computers.

Recent studies revealing how badly U.S. high schools trail those of other countries in the quantity of mathematics and science courses take on a more ominous note in light of the science-teacher surveys.

•

Better salaries would help meet the competition from business and industry, but these would be hard to sell to taxpayers during a time when the general teacher supply is abundant and when resistance to tax increases has been stiffened by the recession.

Higher salaries for teachers in fields in which schools compete with business and industry were recommended by several groups at a recent hearing conducted by the National Commission for Excellence in Education. Some speakers acknowledged that pay differentials would be "traumatic" for less-essential teachers, but they saw no other course. Houston, Texas, offers up to $3,500 more for teachers in fields in high demand. However, this idea is not likely to be accepted by teacher organizations in states such as Iowa, where there is collective bargaining over salaries.

Iowa State University's College of Education is seeking funds for a program under which it would try to get more high-school students to prepare to teach math and science by offering a year's tuition for each year spent teaching in an Iowa school after graduation.

An approach that could produce more immediate results is developing summer institutes to fill the gaps for science or math teachers who are marginally qualified. A financial incentive might be needed to attract participants.

With federal education aid being slashed, the U.S. Department of Education is not a likely source of money for such efforts. Industry might be persuaded to contribute. It is questionable whether a Legislature on an economy kick would be interested.

This much is certain: The problem will not be solved merely by complaining about the poor quality of instruction being offered in too many classrooms and laboratories.

SYRACUSE HERALD-JOURNAL

Syracuse, N. Y., April 8, 1982

Visitors to the Discovery Center Science Fair at Nottingham High School this past weekend had a rare opportunity to view some extraordinary work by science students from schools around the area.

But this display, while underlining some outstanding individual achievements, belies what is a curious — and potentially dangerous — irony at work in America's schools.

With teachers being laid off in great numbers due to declining school enrollment, you would think there would be no problem in filling faculty vacancies, right?

Wrong.

School officials around the country say there is an alarming teacher shortage developing in the areas of science and mathematics.

The reasons for this shortage are several, and the possible consequences of it are grave.

▽ ▽

The most obvious factor is money. Why should a new college graduate with a degree in math or science take an entry-level teaching position at $13,000 a year when he or she can earn up to twice as much in private industry.

Another reason is the job itself. Aside from a few "born teachers," most young mathematicians and scientists find it much easier to deal with computers or test tubes than a roomfull of teen-agers.

This teacher shortage is a self-perpetuating trend.

As the number of good science and math teachers declines, fewer high school graduates head for college with an interest — as well as a cultivated aptitude — in these areas.

▽ ▽

The bottom line is a steadily declining number of scientists and mathematicians both teaching and in industry — a situation which in a couple of generations could turn the shortage from inconvenient to crippling and make the United States technologically a second-rate power.

How can more bright young science and math majors be persuaded to choose teaching as a career?

There's no way that schools can be expected to offer salaries competitive with private industry. And, unlike many of its chief competitors in the technology race, the U.S. government cannot dictate who among its young citizens is going to pursue what career.

▽ ▽

But the people offering the money that's luring teachers away have a vested interest in reversing this trend. If there are no science and math teachers now, there will be no bright-eyed young grads to recruit in the next generation.

Private industry, the chief ultimate beneficiary of good school science and math programs, must take an active role with the educational establishment in addressing this problem.

In an age when our economic and military security is predicated upon the use of highly advanced technology, we cannot afford to take our time.

The Houston Post

Houston, Texas, March 26, 1982

The United States is facing a shortage of top flight engineers, not because we lack students willing to enter the profession but because we lack enough university faculty members to do the teaching. This is only one item in a long list of educational deficiencies that could let the United States sag into mediocrity after so many years as a world leader.

Military and economic power have their place in the rating of nations. Both these strengths have been natural outgrowths in a nation that prized and encouraged excellence in science, technology, inventiveness and creativity. All cost money and require a willingness to recognize priorities. The United States reached a peak in World War II when, coming from behind, this country surged ahead to create the finest industrial complex the world has ever known and to surpass both Germany and Japan in the sophistication and power of the military instrument. It was an exciting time. The nation felt young, unified, confident.

In the postwar eagerness to return to life as it used to be, Americans relaxed in mind and spirit. We turned to the creature comforts and the amusements of a leisure bought by postwar affluence. Sputnik in 1957 jarred us out of the hammock. We discovered that we had been living the fable of the tortoise and the hare. While we rested on World War II laurels, the Soviet Union had beat us to outer space. Though Sputnik was scarcely larger than a basketball, it was there, over our heads in outer space, orbiting this, our Earth. Math returned in force to the curriculum. Science became glamorous. We put our money into public schools, colleges, universities and graduate schools.

But now? If the amount of money we are willing to pay is an indicator, then the nation values sports figures, rock singers and movie stars more than it values teachers, professors, scientists, inventors. Only 16 percent of our high school students take so much as a one-year course in chemistry; all Soviet students must study chemistry four years. When 14-year-olds from 19 countries were tested, the Japanese ranked first in science; Americans ranked 15th.

Every state has a shortage of math teachers, and in many high schools teaching is geared to the slowest students, not to the potential mathematicians. Twenty percent of our high school students drop out before graduation; 2 percent drop out in the Soviet Union. And though the United States will need 85,000 new engineers by the middle 1990s, many engineering schools have had to limit enrollment for lack of faculty.

The United States cannot go on like this and maintain the level of efficiency and expertise that made this nation great. This is the road to the decline of a unique moment in history known as the American civilization. It is a road we must not take.

The Dispatch

Columbus, Ohio, May 17, 1982

IF U.S. EDUCATION Secretary T.H. Bell meant to call attention to the drain of math and science teachers by proposing that they be paid more than other public school teachers, he has accomplished his purpose. Whether his proposal is the best way to solve the problem will certainly be debated by school boards across the land for quite a while.

Bell offered his suggestion Wednesday at a meeting of the National Academy of Sciences called to examine the weaknesses in math and science teaching in elementary and secondary schools.

There are many forces contributing to the problem. Perhaps the most significant is the fact that the expanding computer, communications and technology industries are pulling qualified people out of schools into better-paying private sector jobs. Universities were the first to feel the drain as their top people resigned professorships for positions in industry. Some schools are having a hard time finding faculty members — a problem that threatens future generations of math and science teachers.

In some cases, universities are competing with industry for the best and the brightest teachers at the high school level.

This competition is taking its toll. Stanford Professor Paul DeHart Hurd told the audience at last week's meeting that 50 percent of those hired to teach high school math or science in the current academic year "were qualified and are now teaching with emergency certification."

At a different forum, Ohio State University Professor Kevin Ryan recently noted that only one physics teacher was certified in Ohio in 1981 and that 25 percent of math teacher jobs go unfilled.

For his part, Bell would apply the Reagan administration's free enterprise, supply-and-demand principles to the problem. If you are short a needed commodity and the demand is there, the price paid for the commodity should be raised to improve supply. In other words, Bell says that school systems should abandon their single salary schedules and offer higher pay to math and science teachers.

The proposal would complicate teacher contract negotiations, of course, and it's doubtful that any public school system could keep pace with the salaries offered by private industries. But Bell's suggestion is one that should not be ignored and it could serve as a springboard for finding a solution to this growing problem. It should be considered.

THE DAILY OKLAHOMAN

Oklahoma City, Okla., March 15, 1982

THE TWO prime requisites for a great university are high degrees of excellence in both the quality of its teaching faculty and the product of its research activities.

But actual classroom teaching and research projects often are mutually exclusive endeavors. Put another way, an excellent teacher and personal communicator, a person who can develop a rapport with and motivate the minds of young people, does not necessarily have to produce any significant research work to be of immense value.

The converse also is true. Many innovative and gifted researchers in the physical and social sciences are simply incapable of performing equally well in the classroom.

What prompts these observations is a recent article in The Oklahoma Daily, the OU student newspaper, in which the state of the perennial "publish or perish" dilemma at OU was explored in interviews with several faculty members.

As you might expect, there is no consensus on what might constitute a proper balance between teaching and research, or how much of each should be considered vital in evaluating faculty performance for tenure purposes. But there was a disturbing undertone, reflected in the comments of some professors, that appears to downgrade the importance of teaching ability compared to what an instructor can compile in the way of published material in professional journals.

It was summarized by Russell Buhite, chairman of the OU history department, who said, "I can't imagine anyone going into a university environment and considering himself just a teacher."

Considering some of the useless drivel that is churned out annually by some college researchers, perhaps the emphasis on research at the expense of teaching is misplaced. In any event, the principal products of any university are the men and women it turns out, and if the best faculty minds are more occupied with research than teaching, the student body is being short-changed.

FORT WORTH STAR-TELEGRAM

Fort Worth, Texas, January 22, 1982

These are low times for higher education and the forecast doesn't seem to be very good.

A recent report by *U.S. News & World Report* on graduate schools at American universities shows that doctoral degrees are declining. And that is cause for concern for the future of intellectual America.

The reasons are several: A tight market for doctoral graduates, primarilly in teaching, has turned students away from advanced degree work; federal grant money has been reduced, and the general economic condition has caused states to tighten funding for higher education.

Low teacher pay, in addition to the dwindling number of jobs in higher education, is causing students to go into the business world after completing a bachelor degree and causing those in teaching to leave for higher paying jobs in business and industry.

Some schools have closed graduate departments. Others are cutting back. The University of Michigan is closing its geography department. The State University of New York dropped two doctoral programs in French literature and several schools in that state system have dropped doctoral programs in physics and philosophy.

Doctoral degrees in physics and astronomy declined 43 percent from 1971 to 1980 in the United States. Chemistry doctorates dropped 31 percent from 1970 to 1980, engineering declined 29 percent from 1972 to 1980 and mathematics and computer science degrees dropped 25 percent from 1972 to 1980. Those are overall figures that include foreign students. In some areas of study, as many as 50 percent of the doctoral graduates are foreign students who will take their knowledge back to their home country.

Those figures should tell us that unless something is done to reverse the trend, the United States will lack the scholars to teach, will lag in research and generally will decline intellecually. It would be shortsighted indeed if this nation were to neglect the quest for knowledge because of current hard times. Knowledge is an investment.

President John F. Kennedy in an address at Rice University in 1962 said: "If ... history ... teaches us anything, it is that man, in his quest for knowledge and progress, is determined and cannot be deterred." We must not be deterred.

Charles Darwin said that material progress of all kinds depends upon the work of the intellectual.

Some attempts to solve the graduate program problem are being made. The Exxon Foundation is providing grants for doctoral work in engineering and is helping to increase the salaries of engineering faculty. Carnegie-Mellon University is working up a loan program for students in which part of the loan can be canceled if the graduate spends some time teaching.

But more is needed. As the industrial nations of the world are gaining on and surpassing the United States in research and development, this country cannot afford to cut its investment in knowledge. And government — meaning the taxpayers — has to provide its fair share of the investment.

The Cincinnati Post

TIMES STAR

Cincinnati, Ohio, June 7, 1982

About four of every 10 engineering graduate students in the United States are foreigners, according to a report by the National Science Foundation.

The same ratio is true of engineering school faculties which, at the same time, have at least 2000 teaching vacancies.

The reasons for the faculty shortage are not hard to find. Industry has been luring engineering teachers away with salaries far higher than the colleges can pay. Even engineering graduates with freshly minted bachelor's degrees can earn more in industry than their teachers, without having to take the extra five years or so of study required for a doctorate.

Realizing that the situation will only hurt them in the long run, some of the country's largest corporations are exploring ways, such as student stipends and faculty salary subsidies, to encourage more engineering students to go into teaching and more engineering teachers to stay in.

That's good. But what we wonder about is why, with such potentially high salaries, there is also a shortage of high school graduates going into engineering in the first place.

The shortage of engineers is so acute in the electronics field —where in some cases 30 to 40 percent of the talent comes from foreign countries—that companies are alarmed at a proposed revision of the immigration laws that would require all foreign students to return home for at least two years after they graduate.

Here, some other data from the National Academy of Sciences may be enlightening.

A third of U.S. public schools, for instance, don't offer enough math courses to get a student into an engineering school. Many offer nothing beyond algebra. Worse, in 1981, says the academy, half of all newly hired high school math and science teachers were unqualified in those subjects.

Industry might do itself, and the nation, a better turn if it led an effort to attack the engineer shortage where it really begins—with the woeful state of math and science teaching in the elementary and secondary schools.

Otherwise, Americans could wake up one day to find that their vaunted technological prowess had gone the way of so many other things that are no longer "Made in the U.S.A."

Foreign Language Study Called Important for U.S.

At first glance, foreign languages seem to have little to do with America's technological position in the world. However, as these editorials show, the U.S. is directly affected by its ability to communicate internationally. A 1979 presidential commission reported that the situation of foreign language study in the U.S. was "scandalous." "At a time when the resurgent forces of nationalism and of ethnic and linguistic consciousness so directly affect global realities," it said, "the United States requires far more reliable capacities to communicate with its allies. . . ." Familiarity with foreign languages is considered important not only for national security reasons but also for commercial reasons. Sales of U.S. products abroad often are hurt by the inability of U.S. salesmen to understand the local language. With Japan, West Germany and other nations rapidly catching up to the U.S. in science and technology, America stands to lose valuable time and knowledge if researchers cannot read foreign-language articles or communicate with foreign colleagues.

Chicago Tribune

Chicago, Ill., January 5, 1980

The report of the President's Commission on Foreign Language and International Studies has been issued. The Council on Basic Education and others have commented on the commission's report. As the subject of cosmopolitanism in education is inexhaustible, here are some further remarks on the subject.

The commission viewed with alarm, asserting shrilly that "nothing less is at issue than the nation's security." As therapy, the commission proposes a wide range of undertakings, for the most part administered by the federal Department of Education, with a price tag of a "total immediate expenditure of new federal funds" of $180 million.

The Council on Basic Education, ever since its organization in 1956 a strong advocate of foreign language study, suggests that there is a reason "even more important than the national interest" to be served by making at least two years of foreign language study in high school "part of the education of all public school graduates." That reason? Freedom from confinement within one's native language and culture.

Commissions and councils can recommend and sometimes even inspire appropriations. But to shake Americans out of our complacent linguistic and cultural isolationism something much more is necessary. Motivation is the key. [And even bilingual Canada, with its strong reasons for becoming a nation of bilingual Canadians, has not solved its motivation problem.] What would conceivably motivate the generality of Americans to apply themselves to foreign language study?

The only convincing answer is a quality of teaching so high and so widespread that pupils would become more enthusiastic about foreign language studies than their predecessors ever were.

Among pupils, driver education and sex education will continue to be the glamor subjects. School administrators will remain hung up on desegregation body counts and the financial problems of reconciling falling enrollments, continuing inflation, increasingly militant employes, and ever more resistant taxpayers. If bilingual programs grow further, they may continue to isolate pupils from English more than they make people genuinely bilingual. There is not a chance that contemporary America will insist on two years of foreign language study for all high school graduates, and no guarantee that such a requirement would be worthwhile, considering the quality of the teaching.

If such pessimism is justified, is "the nation's security" imperiled thereby? Hardly. The national security requires a few experts in other languages and cultures rather than many briefly exposed, without contagion. The liberal education argument for widespread foreign language study is better than the national security argument, but until every public high school graduate is literate in the English language and the language of arithmetic, teaching everyone a foreign language will have to sit on the back burner. Good as it would be for the United States to be a nation of cosmopolites, we must give first priority to wiping out functional illiteracy in our own language and educating an intellectual elite to advanced levels.

The Dispatch

Columbus, Ohio, January 12, 1980

A PRESIDENTIAL commission reminds us again of what many Americans have perceived for quite some time — that this country's incompetence in foreign languages "is nothing short of scandalous." It is worse, unfortunately, a silent danger to the nation's strength in an unsettled world.

What are some of the more obvious consequences of this incompetence? The President's Commission on Foreign Language and International Studies says first it "diminishes our capabilities in diplomacy, in foreign trade, and in citizen comprehension or the world in which we live and compete."

Earlier reports had already disclosed that the majority of U. S. foreign service officers could neither speak nor understand the language of the country to which they were assigned.

In the foreign trade arena, this latest report notes, for example, that Japan has 10,000 English-speaking business representatives in the United States. There are 900 American counterparts in Japan, and "only a handful" can speak Japanese.

One can only surmise the unsettling implications of this ignorance to the U. S. throughout the rest of the world, including the more remote, yet strategic, nations of the Third World.

In the meantime, at least 175 other countries have been sending their young people — and some advanced scholars — to American and European universities by the tens of thousands each year. They come, knowing the required foreign language in advance, or they learn it on arrival before embarking on their chosen course of study.

All the while, the 34 percent of American universities which required foreign language for admission in 1966 regressed to only 8 percent by early 1979. Only 15 percent of U. S. high school students today study a foreign language beyond two years.

It was not always this way with Americans on the move. But somehow we have lost our way in languages and international studies since the 1957 post-Sputnik surge. We have also sadly corrupted our own English, and as the president's commission urges, it is imperative we find our way back into the real communicating world — without delay.

THE ANN ARBOR NEWS

Ann Arbor, Mich., January 18, 1980

AT A TIME when even an illiterate person could scarcely fail to notice the need for better communication among nations, it's interesting to notice the status of foreign language instruction in the Ann Arbor schools.

At one time, a proposal to require 7th graders to study a foreign language at least for a year was included in the intermediate school curriculum proposals the Board of Education expects to vote on soon. This idea has been dropped.

* * *

SEVENTH GRADE is far from the ideal time for providing youngsters their first opportunity to begin understanding, as a personal reality, that the world is full of people for whom English is a foreign language. This learning will come more easily if it starts in the early elementary grades. That's when youngsters in Ontario and Canada's other chiefly English-speaking provinces start learning French, in conformity with their country's bilingual policy. And that's when European youngsters start learning English and each other's languages.

Still, a seventh grade foreign language requirement would represent an advance over the local schools' practice of not requiring any foreign language instruction.

In the form now under discussion, Ann Arbor's proposed intermediate school program lists beginning classes in French, German, Spanish and Latin as electives, equated in educational value to four music classes, a performing arts class, and non-remedial reading and writing classes, from which students must make two choices.

If a 7th grader needed and wisely elected a remedial reading course, he or she would be excluded by this proposal from taking a beginning foreign language course or any of the other electives. This is because the remedial reading class would meet daily, and count as two electives, leaving no time for scheduling any of the foreign language classes or other electives, which would meet every other day. Yet, the proposal finds room for moving mandatory health instruction classes from 8th to 7th grade, and to build in personal conduct counseling at the cost of Social Studies classes, in government, history and geography.

Skepticism as to whether this makes sense was voiced, as it should have been, by several Board of Education members during Wednesday's discussion. But not all focused on the low status being assigned to language instruction.

Several voiced dissatisfaction with the prospect that 7th graders with reading problems, but determined to take a music class, would not have their reading needs and musical interest simultaneously accommodated.

* * *

DISSATISFACTION that a considerably larger number of students – all of them – would still encounter no required foreign language study was voiced Wednesday only by Trustee Wendy Barhydt, who pointed out that foreign language study "can help clarify English" for many students, and by Trustee Patti Cerny, who said she is disappointed that the proposal for some required foreign language instruction has been dropped.

That it was dropped, and apparently won't be reconsidered, is a move very much in line with national practice. Less than 1 percent of all U.S. elementary school youngsters and only about 15 percent of students in all secondary school grades are studying foreign languages, according to a survey the Smithsonian Institution issued last fall.

It's no wonder that so many foreign observers who lambast this country as a declining, arrogant giant have learned to do so in English. They want to be clearly understood.

The Honolulu Advertiser

Honolulu, Ha., January 2, 1980

If Americans over the past few years entertained hopes that the country could quietly withdraw from international affairs, the taking of hostages at the American Embassy in Iran and the sacking of other embassies has effectively ended such dreams.

Rather than withdraw into post-Vietnam insularity, our country has been forced by economic and political conditions to continue and to expand its contacts with other nations.

IN MANY INSTANCES, we have been poorly prepared for the increased responsibilities. And perhaps in no single area have Americans been so ill-equipped to deal with other peoples and cultures as in foreign language ability.

This point has been made for some years, and is again emphasized in the report of the President's Commission on Foreign Language and International Studies.

The commission, which visited Hawaii in August, recommends a $178 million program of increased foreign language requirements in schools, improved teacher training and exchange programs, and more assistance for schools sponsoring language courses.

It is an ambitious program, though not surprising given the interests of the commission's members. Dr. Priscilla Ching-Chung, a visiting professor at the University of Hawaii Manoa, is a commission member. Her comments are presented on the facing page.

ASIDE FROM the obvious intellectual and academic benefits of learning another language, the fact is today one's professional development is enhanced by study of another language.

Trade is rapidly expanding with Asian markets, and here the tourism industry places a premium on second- and third-language ability. As a language program coordinator told The Christian Science Monitor, "English works fine as long as you're buying products, but you can sell much more effectively in Brazil if you're speaking Portuguese, or in Japan if you're speaking Japanese."

In issuing its report, the language commission has placed the ball in the schools' courts. It will be our secondary schools and universities that will have to implement many of these recommendations.

A GOOD PLACE to start might be in our public secondary schools, where foreign language ability is not a prerequisite for graduation. Or the University of Hawaii might consider requiring a second language as an admission requirement, thus placing pressure on the public schools to tailor their language programs.

However, in either case, the language commission's report may serve as a catalyst to get us from the "foreign language study is good for you" stage to actually doing something to improve programs and increase study opportunities.

THE MILWAUKEE JOURNAL

Milwaukee, Wisc., February 11, 1979

Travel to the Orient, Europe or Latin America and you'll find lots of people who speak English, from store clerks to business executives to government officials. But look around you in the US. How many people do you know who are proficient in other languages? Not many, we'll bet!

Foreign languages just aren't popular in this country. Part of the problem is the great size of our nation. Most people don't leave the country and don't see much use in being able to communicate in another tongue. Another part of the problem is an ethnocentric arrogance. Too many Americans traveling overseas expect everyone to speak English. Moreover, there is a lack of emphasis on language training for Americans when it counts, during the early years of schooling. Today only 15% of the nation's high school students take foreign languages and only 2% pursue such studies beyond the second year of training. Right now foreign language training in America is at an all-time low.

Yet ironically the US is entering a period when the need for language proficiency is likely to be more pressing. America is becoming much more dependent economically upon the outside world. The largest minority in the country soon will be people of Spanish speaking origin. We have to learn to communicate better, both inside and outside our nation.

Creating interest in foreign languages is not going to be easy, especially at a time when school budget cutters regard such studies as "frills." It will take a concerted effort ranging from the local school district to the federal government.

Washington should take the lead. It has the power to give foreign language training and education high visibility and financial support. Such leadership is needed soon, for the nation may find itself in the near future woefully short of Americans trained to deal with other peoples of the globe in their native tongues.

THE BLADE

Toledo, Ohio, June 4, 1979

ONE of the side benefits of the post-Sputnik rush to expand federal aid to education was the increased emphasis given to the study of foreign languages, a benefit which has now been largely discarded at the very time when such studies could be of great diplomatic and commercial importance to the United States.

The melancholy statistics are that one-fifth of the country's public high schools offer no foreign-language instruction and that only 14.3 per cent of American college freshmen have knowledge of a language other than English. A recent news article in The Blade mentioned that while there are 10,000 English-speaking Japanese businessmen in America, there are but 1,000 American businessmen in Japan and only a few of them speak Japanese.

To be sure, English has become the international language of technology, business, communications, and even diplomacy. But to a considerable extent, American businessmen abroad have to depend upon others to fill in the linguistic gaps during their negotiations. "They're under a significant disadvantage not knowing what is going on around them in a bargaining situation," according to a spokesman for the President's Commission on Foreign Language and International Studies.

Toledo area high school students have for more than a dozen years had access to courses in Chinese and Russian language, history, and literature at the Chinese-Russian Study Center at DeVilbiss. While enrollment has not been large, it has been growing and more than 50 students have indicated interest in the language courses next year. The Toledo public schools have continued to support the program since federal funding was discontinued in 1970. Last month, Dr. Donald Steele, superintendent of schools, and Jude Aubry, member of the board of education, met with Rep. Thomas Ashley in Washington to discuss, among other things, federal aid to the language program in light of the increased interest in improved trade and diplomatic relations with the Soviet Union and China.

It should not be supposed, however, that the only reasons for studying foreign languages is their commercial or political utility. Command of foreign languages has always been considered the mark of an educated person, particularly in parts of the world other than our own continent — and even the framers of our Constitution in many instances were fluent in French or other important languages of the time. Moreover, study of other languages makes a person more aware of the nuances and subtleties of his native tongue.

There are, of course, no quick solutions to the long-term decline in foreign-language instruction, which is due in large part to the abandonment of foreign languages in the curriculum requirements of many colleges and universities. The solutions suggested by the presidential commission include federal funding for advanced language studies, reinstatement of foreign-language requirements, and development of magnet schools in which foreign-language studies play an important role.

A good argument can be made for continuing to support the Chinese-Russian program entirely with local funds inasmuch as it has brought the local schools a measure of well-deserved recognition. But federal aid to expand a program already in place and working well would certainly not be amiss.

The Washington Post

Washington, D. C., October 15, 1979

IN A CROWDED third-floor classroom, the junior high school's sole surviving teacher of French is packing in the students, 39 at a time, and still has to turn dozens away. Throughout the country, the stories are similar: classes in Latin, Spanish and other languages are being cut from school programs—reversing a trend in the 1960s toward more language studies. It is not that students dislike language studies; the cuts are mostly in response to budget problems and to a general lowering of foreign language requirements of colleges.

Unfortunately, too many financially pressed school boards and administrators, charged with mandates to revive "the basics," have not made the connection between the study of Latin or other languages and improvement in English. As a result, the foreign language class becomes a "frill," lumped with woodworking, music, art and other worthy subjects that necessarily take a back seat to the three Rs. In some urban school districts, authorities argue further that their students need to learn standard English before they tackle any other vocabulary and grammar.

Perhaps. But still other inner-city educators claim that imaginative classes in Latin, for example, have been enormously helpful in improving the English-language skills of even some of their poorest students. Latin studies also have helped Spanish-speaking students learn English. And a recent report in The Wall Street Journal noted that in Philadelphia, Latin programs have been expanding for more than a decade to include elementary school children. One study there showed that fifth- and sixth-grade students who had studied Latin had performed at a full grade higher than their counterparts who had not.

There are other reasons for preserving and expanding foreign language studies, not least the value of this kind of knowledge in a world of increased travel and growing economic interdependence. The rigidity of budgets may continue to take its toll on language instruction, but America's failure to expose its young to foreign tongues is a shortsighted and unfortunate educational policy.

HOUSTON CHRONICLE

Houston, Texas, August 10, 1981

The controversy surrounding bilingual education and the need for immigrants to learn English has overshadowed another need: In a shrinking world of increased commercial competition and threats of military conflict, Americans need to be able to understand and communicate in foreign languages.

During the 1960s, student protests against college curricula had one principal casualty: Many universities canceled their foreign language requirements. Now, Americans' lack of foreign language proficiency is taking its toll.

U.S. corporations are facing increased competition in foreign markets, where knowledge of the native language can make a crucial difference. A classic example, one American car sold poorly in Latin America because its name, "Nova," told Spanish speakers that it doesn't go, *no va.*

On a less humorous note, ignorance of foreign language can handicap American foreign and defense policy. Fewer and fewer U.S. diplomats can speak the language of the country in which they are stationed, while the Army cannot fill many intelligence and liaison positions because of the lack of trained linguists.

However, a Modern Language Association survey indicates that the trend away from foreign language study has been reversed. Of the colleges and universities surveyed, 18 percent have retained or reintroduced foreign language requirements for admission, and more than half require some foreign language study before graduation.

The Houston Independent School District's foreign language program mixes traditional courses with the innovative, although none of the courses are required. High schools offer French, Spanish, German and Latin, while several magnet programs allow students to study a language continuously from kindergarten through the 12th grade. In addition, Bellaire High School has an academy that offers classes in Hebrew, Russian and two Chinese dialects.

About 8 percent of HISD students at any given time are enrolled in a foreign language class, although a larger percentage of students study some foreign language before being graduated.

Increased interest in foreign language study, and the attendant insights into world affairs and other cultures that such study brings, should be encouraged in both primary and higher education.

The demands of international competition and the threat of world conflict demand it, for America's future may not depend solely on mom and apple pie, but on *die mutter und die apfelpastete, madre y pastel de manzana,* and *la mere et la tourte aux pommes.*

Roanoke Times & World-News

Roanoke, Va., January 28, 1981

There are more teachers of English in the Soviet Union than there are students of Russian in the United States. Such has been told to, and reported by, the Council for Basic Education in its January bulletin. Of the foreign languages traditionally taught in American college, only Spanish is holding its own. French and German enrollments have dropped drastically.

There are some 10,000 English-speaking Japanese businessmen in the United States — but of 900 American businessmen in Japan, only a handful have a working knowledge of the Japanese language.

These figures are shocking. They have nothing to do with the issue of requiring the public schools, in the name of equality, to teach the foreign language to the foreign-born public school student who cannot speak English. They have to do with Americans who think they are becoming educated, who seek to become, or who have been labeled as "educated."

Commissions have been appointed and they have viewed with alarm but there has been, in the words of the bulletin, "no answering cry." Have Americans become egocentric or just lazy? Do the people, and not university scholars, decide what makes a liberal arts education? Is it possible that somewhere in the land an M.A. degree can be obtained without a working knowledge of one foreign language, and a Ph. D. without two?

For once, we have not the smidgen of a possible answer. Perhaps we should not even be surprised at the questions. When students are not taught to read, write and spell correctly the English language, it may be forlorn to wish they would also learn another language. The only recourse is to hope.

The Boston Herald American

Boston, Mass., August 19, 1981

The United States now has more visitors from foreign countries than Americans traveling out of this land. And it should be getting embarrassing for us.

Whole groups of visitors from other countries do not understand a word of instructions given them on an airplane. They don't know what in the world the park ranger is saying to them.

They can't read the signs at Disneyland. They can't read the menus in San Francisco.

They can't order room service in their hotels. And they have a hard time finding the restrooms.

This is because we are such a stubborn, one-language country for the most part.

You can find English spoken or printed to a large extent in most of the rest of the free world. It's about all you can find in this country, except for a little Spanish along the border states.

Are we smug about language in the United States, or just backward?

THE SAGINAW NEWS

Saginaw, Mich., July 25, 1981

In the widespread public concern about basic education, which The News has shared, maybe we've all been missing something: An adequate definition of what's "basic."

The point was brought home by a survey in the Chesaning School District that reportedly found most residents willing to pay for "the basics," but ready to sacrifice music, drama, athletics, shop and foreign languages.

The last item, especially, jars us. It should bother most citizens.

Unquestionably, the so-called three R's — reading, (w)riting and 'rithmetic — are fundamentals. No student can cope in today's world without the ability to read, write and perform calculations at minimal levels of competency. Yet studies indicate that millions of Americans are functionally illiterate, meaning they have trouble, for instance, filling out a job application. That suggests grave faults in basic education.

But today's world, in our view, demands more from educated persons.

It's a world in which a meeting in Switzerland decides how much we pay in Saginaw for a gallon of gasoline. Elections in Germany and France affect the defense of America, and how much tax money Americans pay for defense. Talks in Tokyo may lead to the recall of workers to Saginaw auto plants. Negotiations in Mexico City play a role in the price Saginaw Valley farmers get for their bean crops.

By any definition, those decisions in far-off places make basic differences in the lives and livings of Saginaw-area citizens. If the major role of education is to enable us to understand, and deal with, our own world, it must include an understanding of the world around us.

Foreign language courses help provide that understanding in a way difficult for other disciplines to match. Passing high-school knowledge of Spanish, French or German may not, by itself, make us competent globe-trotters. But the process of language study also takes students beyond words and grammar, into awareness of the way other people think, live and behave, at home and toward "foreigners" such as Americans.

When we know better why the rest of the world is the way it is, we know better what's happening, and why, in the world immediately around us. Certainly this is "basic" education.

Good cases undoubtedly can be made also for music, drama, athletics, shop and other programs that are often labeled "frills." But if choices have to be made — as they must be not only in Chesaning, which will try again for millage approval July 16, but in all public schools — we urge voters and school boards alike to realize that the language of basic understanding is not English alone.

Student Loan Cuts: Another Worry for Education

President Ronald Reagan's budget plans for 1983 included tighter restrictions on loans for college students. Eligibility requirements would be stiffened, and the interest rates on government-backed loans would be raised. Students would have to demonstrate a measure of financial need instead of being eligible automatically for the guaranteed loans, and the "origination fee" would be raised. That fee was an advance amount a student had to pay on the interest charge of the loan, and the Reagan administration proposed raising it to 10% of the interest charge instead of 5%. Students wishing to pursue graduate or professional education after college would not be eligible for guaranteed loans from the government but would have to apply for loans whose interest rates were set by the banks (which generally charged higher rates than the government). The program of Pell grants for low-income undergraduate and graduate students also would be reduced, under Reagan's budget plan. The total amount of money for the program would be $1.4 billion in the 1983 fiscal year (which would begin Oct. 1), instead of the current level of $2.2 billion. The maximum individual grant would drop to $1,600 from $1,670, and the number of eligible students would be lowered to 1.8 million from 2.2 million.

The Hartford Courant

Hartford, Conn., March 13, 1982

President Reagan's view that education should be almost entirely a concern of state and local governments has led him to propose drastic cuts in student aid programs, cuts that threaten the national well-being.

In Mr. Reagan's 1983 budget, total federal spending for education would fall by 23 percent, to $10 billion.

Even more shocking is the 45 percent cut in funds for several loan programs and for Pell Grants for low-income students. In Connecticut, the number of people who would be eligible for the Guaranteed Student Loan Program, the major source of education financing for middle-class students, would be reduced by as much as 15 percent.

In making these recommendations on the heels of last year's education cuts, Mr. Reagan is reneging on a national commitment, established more than 20 years ago, to expand educational opportunity in America. That commitment — an investment really — was rooted in civil rights concerns and in an appreciation for education's role in making the United States a world power both economically and militarily in the first half of the 20th century.

It is ironic, then, that an administration claiming that the United States has slipped from the pinnacle of technological and military power in recent years would advocate a policy that would limit educational opportunities. The modernization of national defense is an administration priority; but Mr. Reagan's education policies will shrink the pool of scientists and technicians needed to develop and operate the highly technical weapons systems the military desires.

Similarly, a highly trained work force will make it easier for the United States to compete with other industrialized nations, particularly in high technology fields. Yet, Mr. Reagan's cuts in student aid and loan programs threaten to create a manpower gap in technology and commerce that will make economic recovery more difficult.

In addition, there is a serious social concern inherent in Mr. Reagan's proposed student aid cuts. While student aid for the very poorest of potential college students would remain available, the group that would be hardest hit are those from families with incomes between $12,000 and $20,000 a year. Young people from this income level, many of them minorities, would be without the means for a college education. The resulting social stratification and the animosity along racial and class lines that could develop would be a dangerous step backward for American society.

There have been abuses in the student aid programs in recent years, primarily by the more affluent recipients of aid. Reductions in federal aid are no doubt possible. But a national investment in broad educational opportunity is no less important today than it was 20 years ago. Mr. Reagan's aid cuts seriously threaten that investment and, for the national good, they should be tempered.

THE COMMERCIAL APPEAL

Memphis, Tenn., March 9, 1982

COLLEGE PRESIDENTS have been telling Congress the administration's plans to reduce financial aid to education programs would hurt students who depend on federal loans.

They undoubtedly are sincere in their concern for students. But they are concerned just as much about financing their institutions in future years.

If students drop out because they can't get federal loans or other assistance, there will be less money to pay faculty salaries and expenses, to amortize the debts accumulated through campus expansions during the boom education years, and to buy the land, new buildings and equipment they would like to have.

Institutions of higher learning have been mushrooming since Congress enacted the GI Bill after World War II and sent thousands of veterans back to school. They were given another big assist when the Soviet Union sent Sputnik into orbit and the nation became scared that the Russians were overtaking the United States in scientific development. The benefits of those programs have been many and proved they were wise.

Because that proved wise, the federal government established the Guaranteed Student Loan program to help youths to enroll in colleges even though they didn't have the necessary funds.

IT ESTABLISHED the principle that higher education is the right of every citizen. And the college administrators who have been speaking up in Washington now are warning of the consequences if those loan funds are taken away.

But some educators also are becoming concerned about what this program is doing to the system of higher education, as well as what it is doing for it.

They see colleges and universities eagerly reaching for students in order to get the revenue those enrollments represent without regard for the ability of those students to cope with the demands of a college education. They accept high school graduates who cannot read beyond the fifth-grade level, don't even have the ability to write a simple sentence, have no sense of history or government and little sense of values. They make passing efforts to provide remedial education for such students and when many of them lack either the ability or the desire to absorb even that teaching they finally dismiss them or let them drop out and seek yet another crop of such incompetents.

Not all institutions do that, of course, and not all students enrolled in colleges are such incompetents. But there are enough of them around to raise some legitimate questions about a program that doles out loans simply on the basis of financial inability.

A GROWING number of private businesses, realizing the value of a true higher education, have been encouraging their employes to take courses, especially in fields that are job-related.

Those employers offer to pay the cost of such education, but many tie their offers of assistance to the level of educational achievement. A student gets 100 per cent of his tuition and fees paid by the employer if the grade received is an A. The student gets less if the grade is a B, still less if it is a C, and so on.

That sort of correlation makes sense and it may be that Congress should change the student loan program to make such a link between guaranteed loans and grades. It might even go further and require youths seeking loans to demonstrate a greater capacity for learning than the colleges now demand of them.

THE INDIANAPOLIS NEWS

Indianapolis, Ind., March 22, 1982

Abuses in the college loan and grant programs, viewed by themselves, are dismaying: Doctors refusing to pay back their loans; parents in collusion with their children to deposit loan money in the banks; students and parents misrepresenting their needs — and more.

These cases, however, represent the minority of "bad apples" — and they are in every barrel. Most of the students receiving loans and grants have advanced themselves through their studies, and they are paying more taxes today because they have qualified for better jobs.

In instigating the cuts, the President looked at the bad apples and at the alarming increase in numbers: 3.5 million students receiving loans last year and 2.6 million disadvantaged students receiving grants. Truly, it is a shocking increase since the plan was introduced in 1958.

The President has gone far beyond his usual formula, however, of reducing the rate of growth. He wants to cut Federal aid in half by 1984 — from $14.7 billion in fiscal 1981 to $7.7 billion. He would also eliminate professional and graduate student participation in the Guaranteed Student Loan program directed mostly to middle-class families. Across the board, eligibility requirements would be tightened, interest rates increased and loan fees boosted.

The case is strong for such tightening and a downward trend in numbers, but the President has used a broad ax. The American Council on Education estimates that almost a million students would be forced to drop college plans if all the Reagan cutbacks are adopted.

The million students who do not attend college would pay $156 billion less in taxes over the next 20 years. Reliable studies show that for every aid dollar allocated to a deserving and needy student, the government can expect to get back $4,300 in tax revenues over 20 years.

A college diploma is not a guarantee of a job or of certain success, but over a lifetime a college graduate earns more money and has much more security, on the average, than a person without that diploma.

Families sufficiently aware of the value of education may find new channels to funds in spite of hardship, as indeed they should. But in times of inflation and recession, some families have no avenues for new finances. Their college age children will simply stay home, joining the growing ranks of the unemployed.

No one can estimate how many potential members of the "scarce" professions, such as engineering and the physical sciences, will be shunted out of the educational stream. No one knows how many potential doctors, dentists and nurses will be diverted to the ranks of unskilled labor. Whatever the number, it will be too many.

No surveys are needed to prove that the nation's economic future lies in the "knowledge" industries, not in assembly lines. The economic future of the whole nation will be determined by how many educated persons will be available to meet the needs of a "service" society.

One of the reasons Indiana has the third highest unemployment rate in the nation is that it built its economic future on assembly lines. This is a dead-end strategy.

Actually, this state does not have an adequate pool of educated personnel to match its economic hopes and objectives to the realities of the 1980s.

Facing such a future, the nation — and this state — can afford whatever it takes to provide educational opportunities for its youth. To ignore the deserving youth of this society is to live dangerously.

The President can logically slow the rate of government subsidy in many lines, even education. But to slash it as he has done, or is about to do, is extremely shortsighted. He should retrace his steps before it is too late.

The Knickerbocker News

Albany, N. Y., March 11, 1982

"In a word, drastic actions have a high probability of failure when there is no allowance for a period of adjustment," says Walter A. Fallon in criticizing the Reagan administration's proposed cuts in education funds.

Mr. Fallon isn't a father who must scrimp to send his children to college and who, with justification, might accuse the Reagan administration of trying to narrow budget gaps on the backs of the hard-pressed. Neither is he a Democratic congressman quick to label Reaganomics nothing more than old fashioned Republican "trickle down" theory that helps the wealthy long before benefits seep into the ranks of the less fortunate.

No. Mr. Fallon is chairman of Eastman Kodak Co. — the kind of corporate leader who might be expected to support the president's goal of reducing student aid as part of an overall plan to reduce federal spending and get the government off the people's back.

But Mr. Fallon knows better. If the Reagan education cuts take effect, he estimates, nearly one million students across the nation will have needed funds reduced or eliminated. Some student aid programs may be slashed by as much as 46 percent.

When you run a technologically-oriented company like Kodak, you know how short-sighted it is to aim the ax at the future work force. Less student aid simply means fewer graduates in the future, at a time when the nation will need more skilled persons in order to remain competitive in a technological and complex marketplace. It's a fast-paced marketplace, too, as Mr. Fallon can attest: the majority of Kodak products on the market today have been there less than five years. Obviously companies must have a supply of qualified workers and managers to survive in such a world.

Neither Kodak nor American business in general is looking for a free ride in educating its future employees. In 1981, for example, Kodak gave $4.9 million in support of higher education, part of $1.1 billion that American corporations donated to colleges and universities.

Given these impressive totals, Mr. Reagan would have a hard time telling businessmen to "put up or shut up," as he told congressional critics a short while ago. Corporate leaders have indeed been putting their money where their mouths are — and they should not stop talking about the wrongheadedness of butchering student aid programs until the president pays heed.

Newsday

Long Island, N. Y., March 15, 1982

When President Reagan said he wanted to downplay the federal role in education, he wasn't kidding: First he cut school spending in this budget year by $1.9 billion. Now he proposes to lop off $2.7 billion more in fiscal 1983, and carry out his campaign pledge to eliminate the Department of Education.

Getting rid of the Education Department is all right with us; its functions can reasonably be shifted to other agencies, at a saving of more than $50 million. But reducing federal education expenditures by more than 30 per cent in just two years will put school districts, colleges, college students, and state governments under immense new pressures.

We understand the urgency of holding down federal spending and deficits. But the overall reduction in domestic spending is only about 3.6 per cent; the education sector is being chopped disproportionately and far too abruptly.

Reagan's proposals would lump many existing public school aid programs into several block grants, to reduce overhead and give state education departments more flexibility in allocating funds. That's okay. But the possible savings will hardly make up for cuts of more than $1 billion in aid for the disadvantaged and handicapped and funds for adult and vocational education.

Those reductions will hit particularly hard in the Northeast, where there are big groups of disadvantaged pupils and little tax growth.

Reagan also proposes cutting college tuition aid programs by 28 per cent, or almost $1.7 billion, by reducing total grants, tightening eligibility and lowering loan subsidies. Graduate students would be shifted to a more costly loan program.

Denying federal loans to wealthy students is good policy, but cutting back so massively on aid to low- and middle-income households is not. The result is likely to be a damaging shift from private to state colleges. Besides, discouraging young people from higher education, particularly advanced technical education, threatens to stunt the country's future economic growth.

Congress is already wary of Reagan's education proposals, as it should be. The nation would be better served if school and student programs were cut more equitably and more gradually.

THE ANN ARBOR NEWS

Ann Arbor, Mich., February 23, 1982

THE TREND of the '80s toward self-reliance (and given impetus by President Reagan's "less is better" approach to government) is hitting higher education square amidships.

Michigan educators warn that thousands of moderate-income students may be forced to drop out of school if Reagan's plan to slash loan programs is approved.

The president's fiscal year 1983 budget lops a whopping $1 billion from the current $3.4 billion federally guaranteed student loan program.

Speaking to this cut, U-M President Harold Shapiro said, "I'm sure there would be students who would have to leave college because of this, most of whom probably come from lower-and middle-income families."

About 60 per cent of U-M students receive some financial assistance. At MSU, the percentage is slightly higher.

Cut, too, are the EOGs – the educational opportunity grants, called Pell grants. These grants, many of which go to minority students, would be cut 39 per cent.

THE SIGNALS President Reagan is sending are unmistakable. Education is playing second fiddle (and a very poor one at that) to Titan missiles and tanks.

Higher education is already facing a host of recession-ordained budget slashes. Tuition increases are making it tough for students to pay their way through school even with outside help.

How high is up? Well, the average tuition at Michigan's 15 four-year public colleges is the third highest in the nation, according to the Chronicle of Higher Education.

Tuition rose an average of 13 per cent this year, largely because of state cuts in college support. As for the smaller, private colleges, phased-back student aid is devastating because a very high percentage of their students rely on federally guaranteed loans.

Reagan's secretary of education, Terrell Bell, defends the cuts by saying the federal government can no longer afford "posh student aid."

But a candid appraisal of student need would not conclude that either the loans or their recipients are "posh."

Here is an issue, finally, on which faculty and parents, students and administration find common ground. If these forces combine to protest student aid cuts, the political impact could be formidable.

WRITING IN The New York Times, Anthony Lewis says "education is a crucial way for individuals to escape from a background of poverty. Federal loans now enable a young man or woman without family resources to become a professional..."

And again: "For the Reagan administration to make such proposals is peculiar in another sense. The central theme of the president's domestic policy is that the U.S. must become more productive, modernize its economy, compete more vigorously in the world.

"Are we going to do that while reducing our people's opportunity for education?"

It used to be said that the truly determined kid would get a college education, regardless. Reagan is saying that the person with incentive can get his/her degree without government assistance.

In these times, at today's inflation and in a recession economy?

A CONTINUING investment in education will insure wise, informed leadership in the years ahead.

And in a super-sophisticated world of micro-circuitry, tangled geopolitics and complex human relationships, let's not pretend that anything less than a strong, all-out commitment to educational excellence and opportunity will be sufficient.

THE CHRISTIAN SCIENCE MONITOR

Boston, Mass., March 12, 1982

The Reagan administration is correct in arguing that the United States must increase the proportion of financial assets going into endeavors that lead to new production and wealth, such as capital development, research and investment, and new technology. It is precisely because such "reinvestment" is necessary that it is difficult to fathom the rationale for the proposed new budget cuts in federal loans and grant programs for college students.

For in the long run the most important investment the US can make is the education and schooling of its young people. The ultimate "payoff" – in terms not only of greater national prosperity but public enlightenment – is incalculable.

The proposed fiscal year 1983 budget cuts come on top of reductions already made in the 1982 budget, as well as a persistent recession that has forced many young people to rethink their future college plans. This week the Congressional Budget Office said the cutbacks already enacted or sought by President Reagan would cut federal aid to college students almost in half by 1984 – from $14.7 billion in fiscal 1981 to $7.7 billion. This would go far beyond simply cutting *growth* in federal spending.

For example, the administration would limit Pell Grants, the basic grant program for disadvantaged students, to $1,600 for families earning less than $18,000 a year. Up until now somewhat larger grants have been available to families earning up to $27,000. Another program targeted at needy students, Supplemental Educational Opportunity Grants, would be scrapped.

The administration would also eliminate entirely professional and graduate student participation in the Guaranteed Student Loan (GSL) program geared essentially to middle-class families; tighten eligibility requirements and double the fees paid at the time student loans are made; boost interest rates; and require that loans be made only upon a demonstration of true financial need.

In defense of its proposed cuts, the administration argues that if reforms are not made federal costs in the GSL program will rise from $1.6 billion in 1980 to $3.4 billion by the next decade. And that, since most of these loans go to middle-class families who have access to private bank financing, the government is justified in seeking significant reductions.

Unfortunately, there have been abuses in the student loan programs. Moreover, federal loan and grant programs have grown enormously since the government began providing direct college aid to students back in 1958. Last year alone some 3.5 million students received student loans, while 2.6 million disadvantaged students received direct grants.

Tightening eligibility, however, is quite different from ending loan availability altogether for thousands of students who might not be able to attend college without it. For that reason, many lawmakers are echoing the position of New Jersey Republican Congresswoman Marge Roukema, who said recently that "we've gone about as far as we can go" in slashing the student loan program.

Indeed, Congress should consider any further student loan or direct grant cutbacks only with the utmost care. The nation's graduate schools, for instance, are now experiencing declines in enrollments in such fields as the physical sciences and engineering. Yet how is the US to maintain its industrial prowess without a growing body of educated engineers and scientists?

In the meantime, colleges and the US industrial-business community would seem to have a responsibility to develop ways of aiding college-bound youth besides just directing them to federal aid offices. That means increasing scholarship endowment programs; underwriting bank loans; and establishing jobs programs for students.

Admittedly, establishing such programs will be difficult for financially pinched colleges and business communities. But it is crucial to the nation's future that every qualified young person who aspires to a college education receive that education.

Stretching financial resources today will assure sufficient human resources tomorrow.

Detroit Free Press

Detroit, Mich., February 17, 1982

BY NOW it is clear what the short-term effects of cuts and reductions in the growth rate of aid to education will be: the elimination of some programs for the disadvantaged and handicapped and the sharp curtailment of others. But the long-term effects of the Reagan administration's educational austerity program may be more devastating. If carried to completion, such policies could make America a rigidly class-divided society at odds with both our democratic rhetoric and humanitarian ideals.

Already, some of the nation's top universities are reviewing their student aid policies in light of proposed cuts in the federal college loan program. Wesleyan University is ending an admissions policy that did not consider student ability to pay and will now reject some students who cannot afford the full tuition. Other colleges reviewing their admissions policies include Columbia, Barnard, Harvard, Boston College, Cornell University, Princeton University and Wellesley College.

Does this mean bright students from impoverished backgrounds will no longer be able to attend Harvard? Not necessarily, particularly if they are very bright. College officials will no doubt devise alternative strategies, in some cases perhaps subsidizing a genius at the expense of the merely gifted or giving needy students only partial aid. But with even the best schools forced to restrict funds for student aid and scramble for research grants and declining state aid, the quality of American higher education is in jeopardy.

That puts American productivity, growth and, yes, national security in jeopardy as well. American science and technology, along with its popular culture, remain the most exportable U.S. commodities. We cannot expect to build stronger armies without skilled servicemen and women and innovative scientists and managers. Nor can we rebuild our economy without technically skilled workers.

In the short run, the administration's educational policies will save money, or, more accurately, transfer money from education and social services to defense. In the long run, with such policies, we all lose.

THE DENVER POST

Denver, Colo., February 22, 1982

THE BUDGET proposed by President Reagan for the 1983 federal fiscal year beginning Oct. 1 aims a neutron bomb at American higher education. If adopted, it will leave the ivy campuses intact but decimate their students.

The prospective budget would slash student aid programs in half from 1981 levels. Yet the savings, $1.7 billion, would barely buy a half-dozen of the proposed B-1 bombers.

This newspaper didn't protest the initial round of cuts in student aid in the current 1982 fiscal year. Skyrocketing commercial interest rates and sloppy administration had riddled the program with "arbitrage." Simply put, it was possible and common for students from affluent families who didn't need aid to borrow up to $10,000 interest-free from the government anyway. The surplus cash could be invested in money market funds for a tidy windfall. Trimming such abuses allowed the Reagan administration to cut aid budgets 10 percent in the present budget without any real harm.

So when it came time to look at the new budget proposals, there we were, staring contentedly out the window at the cascade of fiscal bathwater thrown out by Budget Director David Stockman. And what to our wondering eyes should appear? The baby!

Actually, it's a whole nursery school plummeting toward the pavement. The health professionals loans, supplemental educational opportunity program, state student incentive grants and direct student loans would be eliminated. Pell grants, another form of aid, would take a 40 percent slash and the work-study program a 27 percent whack.

Finally, as if to guarantee the U.S.A. will never rebuild its economic and military strength, the administration wants to bar graduate students from the guaranteed student loan program.

Gutting education produces only short-run savings, just as eating seed corn fills a belly for only one winter.

This is the administration that wants to restore our military strength? Modern war isn't a matter of bashing away with clubs, it's a high-technology duel. The next generation of weapons systems will be designed by people who are now in graduate schools. Well, in Soviet graduate schools, anyway. We're dropping out of the race.

On the economic front, the next generation of products dominating international markets will be designed by people now in graduate schools. Well, in Japanese graduate schools, anyway.

Of course, not all college students are in technical fields. Yes, some people with doctorates in philosophy are driving taxicabs. Is it bad to have learned people in humble occupations? We remember a cobbler named Socrates and there was this Hebrew carpenter . .

Ultimately, a society that sacrifices learning to build economic and military strength is like a man who gives up sex because he'd rather have children. We hope Congress understands cause and effect better than Stockman and restores the cuts in education. The Colorado Legislature should, meanwhile, increase state aid programs to offset part of the damage.

The human mind is not a frill.

THE TENNESSEAN

Nashville, Tenn., March 7, 1982

PRESIDENT Reagan's proposal to trim the budget by cutting back on federal student aid might prove to be a delusion which could end up costing the nation more than it saves.

Tennessee higher education officials estimate that the President's proposal would affect 25% of this state's college students. This includes public college students as well as those who have become accustomed to receiving aid to attend private institutions. State officials say the budget cuts could cause a shift from private to public colleges and universities, threatening the existence of some of these institutions and increasing the burden on the taxpayers in supporting public colleges.

The cutbacks are also likely to sharply reduce black enrollment, since many black students come from low income families and receive federal assistance. Education officials say this would tend to reduce desegregation of colleges in Tennessee at a time when education officials and the courts are attempting to achieve an integrated system of higher education.

The Reagan administration appears totally insensitive to the needs of education and unappreciative of the role of education in a progressive, thriving economy.

The administration seems to think the nation can prosper by devoting the major share of its resources to national defense without due consideration for other areas of American life. But educated people are needed even for supporting the national defense and operating modern weapons of war. If the U.S. shuts out millions of its young people from obtaining a higher education in order to develop more weapons, the national defense will be weakened rather than strengthened.

It is hard to see what the administration expects to gain by an elitist policy of education for the privileged few and an end to opportunity for the underprivileged many.

This not only will undermine the national defense but it will also cripple the economy in the long run. People without education and training are not qualified to handle self-supporting jobs in a technical society such as this one.

But possibly there is a reason why this doesn't worry the administration. As long as Mr. Reagan and the GOP are running things it doesn't seem there will be many jobs anyway.

The News and Courier

Charleston, S. C., February 23, 1982

Those heart wrenching moans that grow louder each day are from liberals decrying the Reagan budget proposal for cutting back on federal tuition loans and outright grants to college students. One such moaner even went so far as to recommend building three fewer B-1s, stopping the demothballing of two battleships or bringing back 25,000 troops from Germany as the quid pro quo for restoring the program's funding. (Of course, his neat solution would extend the funding for one year only. Perhaps he would repeat the defense cuts to fund the following years?)

In fact, the program desperately needs a complete overhaul. If it's done right, there will still be funds enough for those in financial need who are seriously interested in a post secondary education.

Currently there are no uniform federal policies regarding grade standards. It is indicative of the clout of educational lobbies that every effort to include standards has been defeated.

An official of the General Accounting Office testified recently before Congress and cited numerous cases of student aid abuse. He described students attending college for four years and never making a passing grade. One such collected over $13,000 in federal handouts. Another received aid for 58 credit hours but completed only three. A third, who received $5,000 in aid while seeking a degree in business administration, failed an accounting course four times and a math course four times before finally scoring a D.

In a study of 20 colleges, the GAO found that 19.9 percent of students with federal grants and 23.1 of Social Security students had grade averages below C. In those same colleges, the student default rate on loans was 16 percent.

At the very heart of such abuse is the lack of uniform grading standards. Individual colleges are allowed to set their own yardsticks. Some apparently lower the standards for federal aid recipients while others ignore standards completely. Regrettably, it appears that some institutions chase dollars and enrollment statistics more than academic quality.

It seems a reasonable demand that students attending college on taxpayers' money, make respectable grades or have the funding withdrawn. Establishing credible, uniform standards would ensure that sincere young people who need the assistance get a fair shake.

WORCESTER TELEGRAM.

Worcester, Mass., March 7, 1982

For generations, Americans have been raised to believe that a college education leads to a well-paid job. But over the past decade, that hasn't been the case for thousands of able graduates. Instead, they find themselves in service jobs or doing clerical or blue-collar work and wondering why they shelled out all that tuition money.

That is the disturbing picture painted by U.S. Labor Department statistics and by a variety of private employment experts in a recent article in The Wall Street Journal. One economist calls it "a further tattering of the American dream."

What seems to have happened is this: Colleges churned out so many graduates in the 1960s and 1970s that members of the work force with degrees nearly doubled, to 17.6 percent in 1979. Meanwhile, there was more growth in service jobs than professional and managerial positions. This mismatch has been made worse by the back-to-back recessions of 1980 and 1981-82. The upshot is that large numbers of Americans are now working at jobs for which they are overqualified.

Richard Freeman, a Harvard University economist and author of a book called "The Over-Educated American," notes that even college people lucky enough to hold college-level jobs are generally paid a smaller "premium" for their education nowadays. He says male college graduates going into industry in 1969 averaged 24-percent better pay than the work force average. By 1979, that premium had shrunk to 5 percent.

It's a disturbing trend, all right, and not just for college students and graduates. We Americans like to think that education is the key to success, that there is nothing we can do for our children more important than encouraging in them a love of knowledge and a determination to acquire it.

But before we conclude that Americans have been wrong about the importance of education, a number of matters deserve consideration.

The statistics cited by all those experts are averages. They do not take into consideration that certain professional skills continue to be in demand. Scientists of many kinds and engineers of most kinds still have little trouble finding jobs that pay well. Here in Massachusetts, even in the middle of a recession, the high-technology industry continues to complain that it can't find enough electrical engineers and computer technicians.

This suggests that something is wrong in high school career guidance. Perhaps something is wrong, too, in the thinking of young people who insist on education for a particular line of work, without reference to kinds of graduates society now needs and is willing to pay well for.

There is also the thought often set forth by advocates of the liberal arts — that higher education should not be just vocational education, that a person also benefits by becoming a better-informed citizen and by a heightened ability to contribute in many ways and to enjoy reading, the arts and the world's passing parade.

This era of overqualified waiters, typists and sales clerks is a good time for Americans to rethink their attitudes toward higher education. Before asking it to prepare people for jobs, shouldn't we try to find out what sort of jobs will be available? And before writing off the non-economic advantages of college, shouldn't we consider that education broadens everybody, clerks and production-line workers along with dentists and business executives?

The Chattanooga Times

Chattanooga, Tenn., March 5, 1982

Looking for ways to cut government spending, the Reagan administration has settled on federal aid to college students, proposing to cut the general student loan program by nearly half. The administration is correct, as Mr. Reagan asserted at his last press conference, that the program has been abused. The government has had difficulty collecting some loans. And as Mr. Reagan pointed out, some students obtain low-interest education loans from the government and invest the money in high interest accounts. But Congress can, and should, prevent those abuses without devastating a valuable program merely to gain negligible savings.

It's important to keep in mind education's enormous value. Federal loans for college educations have enabled children from lower- and middle-income families to improve their ability to compete in an increasingly technological society. Equal educational opportunity, after all, is a crucial complement to equal economic opportunity. Denial of the former would do more than just prevent thousands of children the chance for a better life. It would ultimately penalize the nation's future.

The administration's proposal would fall especially hard on graduate students because it seeks to bar them from the general student loan program. Under that plan students borrow money at 9 percent and repay it after graduation. But the administration wants to put graduate students in the auxiliary student loan program, which charges borrowers 14 percent interest and requires repayment while the student is still in school. Under that plan, therefore, a student borrowing the maximum of $40,000 over five years would have to begin repaying nearly $500 monthly.

The administration is not entirely wrong is trying to cut back the GSL program. It increased dramatically between 1977 and 1981, and is budgeted at $3 billion this year. The administration wants to require students from all income levels to prove their need in order to become eligible for a subsidized loan. Such a requirement might prevent some of the abuses Mr Reagan mentioned.

But it is counterproductive in the long run for the nation to force graduate and professional students out of the GSL when the auxiliary student loan program is unavailable in some states. True, future doctors or lawyers might be able to obtain loans without too much trouble because their post-graduation income would ensure repayment. But what about students interested in other fields — the nation's future economists, teachers, mathematicians, natural scientists, engineers, physicists?

The American Council on Education says that congressional cuts last year meant that nearly a fourth of the 3.6 million students holding education loans were declared ineligible. And the Action Committee for Higher Education now says that cuts proposed by the administration — approximately $2.7 billion out of the loan programs' total of $5.7 billion — will force several hundred thousand students to forget about a college education. The impact will also be felt by the schools, many of which are already struggling to stay afloat.

It is foolish for the administration to pretend that encouraging the devastation of higher education will not have far-reaching adverse consequences, or to assume that those consequences will not have an impact on such national goals as increased productivity, technological advancement — and national security. It is an incredibly shortsighted proposal which Congress should correct.

THE ATLANTA CONSTITUTION

Atlanta, Ga., March 9, 1982

American society is becoming more and more complex, with so many of its components rapidly moving toward total computerization and high technology. The age of science and communication is here for many people; just ask any person who is unemployed and lacks the skills to enter today's job market.

It's unfortunate that the Reagan administration is not aware of, or does not care about, the suffering many people endure because they have no skills and no place to learn them. If the administration succeeds in making all of the scheduled cuts in education — particularly in student-loan programs — more and more people will become useless to the new society. The day may even come when this country will be forced to depend on foreign minds and know-how to make it all work.

U.S. Education Secretary Terrel H. Bell, in Atlanta over the weekend, said while he regrets cuts in education programs for disadvantaged children, he has no misgivings about cutting college student loans because he believes higher education is a privilege and not a right. Bell apparently is confused about where this country is heading and the role the government has played in bringing society to this point.

The responsibility for education does not change between elementary school and college. The chance for a disadvantaged child to make something of himself or herself should not change, either. To say that higher education is a privilege and not a right is to say that opportunity in this country is limited to wealthy families who can afford the luxury of sending their children to college. Ridiculous.

In the long run, society will be worse off if cuts in education become a reality. The American dream would vanish for many, and thousands of people with contributions yet to be made would become useless, a burden rather than a boon to society.

Study Reports Japanese IQs Surpass U.S., European Scores

As if Japan's economic triumph weren't punishment enough for the U.S., a Northern Ireland researcher concluded that Japanese pupils were smarter than their American counterparts. Richard Lynn, a psychologist at the New University of Ulster, wrote in the May 20, 1982 issue of *Nature* that Japanese schoolchildren scored an average of 111 on the American Wechsler scale of intelligence, at least 10 points higher than the mean score for U.S. and European pupils. "Since intelligence is a determinant of economic success, as it is of success in many other fields," he wrote, "the Japanese IQ advantage may have been a significant factor in Japan's outstandingly high rate of economic growth." Lynn concluded that 10% of the Japanese population had IQs of over 130, considered "genius" level, as opposed to only 2% of the European and U.S. populations. Lynn said the results came from testing 6-year-olds, which meant that factors other than superior education were involved. He cited "largely environmental improvements," such as better diet and health care and the migration of peasants into Japanese cities since the end of World War II. *Nature* editor Alun Anderson cautioned, "Whether the difference in IQ represents a real difference in 'intelligence' or simply implies that the Japanese are better at IQ tests remains open to question." However, he observed that "the future implication for societies which are becoming increasingly dependent upon technological innovation may be profound."

Arkansas Democrat

Little Rock, Ark., May 23, 1982

Claims by a North Irish psychiatrist that the Japanese have higher IQs than Europeans (including Americans of European descent) won't please many of us. We all glory in our "smarts" and general knowhow. However, the Irishman's way of accounting for the Japanese superiority doesn't sound very scientific to us.

He shows the Japanese turning what was a small IQ edge over us less than half a century ago into a huge one today. Those Japanese now between 36 and 46 years of age are only 4 IQ points ahead of us – but those in their early 20s, outsmart us by a full 15. How could that happen? It's unheard of.

"Environmental improvements" yonder since World War II, says our Irish expert. No Japanese genetic changes could have occurred in such a short space of time. Right, they couldn't – but "environment" can't work such a change either.

Though the environmentalist IQ school in this country is loud in its claims that IQ differences (especially between races) are solely a product of social and economic advantages and disadvantages, there's been no such dramatic IQ gains here as in Japan – and our gains in living standards since World War II certainly exceed hers.

So why have we lost IQ ground to the Japanese so dramatically in the past few years? Were they latently that much more intelligent than us all along? We think that there's a better answer than either that or environment. It lies in the theory of IQ itself.

That theory is that IQ is less a measure of raw intelligence than of reading ability, knowledge and memory – and that its main use is to predict scholastic success.

That sounds like a self-fulfilling prophecy. So what we may well be looking at in the case of the Japanese is ever-better and continually-improving schooling. That explanation doesn't take a thing away from the Japanese IQ – but if it's the right one it certainly says something pretty piercing about the state of American education.

The Boston Herald American

Boston, Mass., May 25, 1982

You will have trouble reading this unless you are Japanese. But if you sound out the words and touch each one with your right — no, your right — forefinger, you will get the idea. Eventually.

That is because Americans are not as smart as Japanese.

The mean IQ of Americans is 100. The mean IQ of Japanese is 111. That is not mean as in nasty, but as in average. That is what Richard Lynn, a psychologist in Northern Ireland says he found out.

How did he find that out. Jane? How did he find that out, Dick?

He compared standardized intelligence tests, he said.

He said 10 percent of the Japanese have IQs of 130 and up. He said only two percent of Americans have such big IQs.

He has made the National Education Association angry. It is angry. It said, "balderdash." (Ball-duh-dash.)

The psychologist also said that ever since World War II, Japanese have been getting smarter and smarter.

How did he find that out?

Maybe he looked at the balance of payments of Japan.

Maybe he looked at the balance of payments of America.

Maybe he compared them and found out.

If he is so smart, how come he didn't think up Pac-Man?

Detroit Free Press

Detroit, Mich., May 22, 1982

NOW WE know one more reason the Japanese are wiping us off the map. They're smarter, and they're getting smarter all the time.

According to a recent study, the Japanese score higher than any other nationality in the world on standardized intelligence tests. Whereas the mean American IQ is 100, the Japanese score 111. More astoundingly, over 10 percent of the Japanese score over 130, compared with only two percent in the U.S.

The high scores can't be attributed just to superior genes. The Japanese have made steady IQ gains since World War II. Japanese born between 1936 and 1946 have a mean score of 104, but those born in 1960 score 115.

The psychologist who conducted this study, Dr. Richard Lynn of the New University of Ulster in Northern Ireland, credits the IQ rise to improved nutrition and general quality of life. He also believes Japan's unusually large pool of bright citizens has contributed significantly to its phenomenal economic growth.

So maybe amid our frantic studies of Japanese labor relations and capital investment rates and robotics for the solution to our current economic woes, we ought to ask what we're willing to invest in our own citizens. Good food and good schools lack the trendy gloss of high tech schemes. But if we continue to shift our public resources from basic human services to military hardware, we not only put our money where it brings the least economic return, we also cripple our raw human product. We can't refashion our economy to compete internationally on a foundation of protein deficiencies and illiteracy.

Part IV: Space

Space exploration gave America its first experience of overcoming technological inferiority. In the 22 years between the launching of the Soviet Sputnik satellite in 1957 and the first U.S. moon landing in 1969, Americans proved to themselves that once challenged, they could win the race. Achievement in space exploration and by extension, in science and technology, became a matter of national pride.

This sense of commitment, though intense, was short-lived. The sixth and last moon walk took place in December 1972, less than three years after the first. Subsequent U.S. manned missions were sent to space stations, but they, too, ended after less than three years in 1975. Six years elapsed before the next U.S. manned venture, the shuttle *Columbia,* whose maiden voyage was in 1981. Unmanned flights fell victim to the some waning of energy. The Mariner 2 probe of Venus in 1962 began a series of explorations that sent back information on planets at least every 12 months for almost 20 years. After Voyager 2 completed its tour of Saturn in 1981, however, a five-year silence descended upon transmissions from space. It will be broken, if all goes as planned, only in 1986, when Voyager 2 is scheduled to pass by Uranus.

The problems facing the U.S. space program are not short-term. The National Aeronautics and Space Administration cannot be "shelved" until public interest and government funding revive. The enforced inactivity will erode the personnel and programs that nourish NASA's future. One participant at the annual Lunar and Planetary Science Conference in March 1981 remarked, "In three years, 40% of these people will be out of the business." (The speaker, Jeffrey Warner, confirmed his prophecy by leaving NASA the next year.) Higher salaries in private industry, government cutbacks that curtail research and the emergence of new fields of scientific study are producing a "looming planetary brain-drain," in the words of another conference participant. Budget cuts and staff reductions hurt in other ways. Nancy Evans, a Voyager program scientist, remarked the following year that only 10% of the data received from the Viking probes of Mars had been properly analyzed. "Mariner 10 data [from Venus and Mercury] was never adequately looked at, because there wasn't enough money," she added. Commenting on the two unmanned Soviet landings on Venus in March 1982, Eugene Levy, chairman of the National Academy of Science's Committee on Planetary and Lunar Exploration, said, "I know that we all look forward to the advances in human knowledge that soon will be appearing in Soviet scientific journals. . . ." Levy left his audience with a clear impression that he did not expect any such advances to appear in American scientific journals.

Carter Cuts NASA Budget, But Congress Restores Funding

In January 1978, President Jimmy Carter requested a total of $4.3 billion for the National Aeronautics and Space Administration for the fiscal year 1979 (which would begin in October 1978). The amount was $288 million more than NASA's 1978 budget, but NASA Administrator Robert Frosch said it only "holds us even" after taking inflation into account. The budget was held down by cutting out funds for a fifth space shuttle, which had been planned by the previous administration of President Gerald Ford. By the time the budget bill was passed by Congress in September, however, NASA had received an extra $100 million. Congress restored funds for the fifth shuttle, making the shuttle program the largest single item in NASA's budget at $1.443 billion. Other items were: research and program management ($914 million), space flight ($316 million), tracking and data acquisition ($305.4 million), physics and astronomy ($285.5 million), aeronautical research and development ($275 million), lunar and planetary exploration ($187 million), construction of facilities ($150 million) and life sciences ($42.6 million). Despite public and congressional pressure for cutting federal spending, the space program remained an American favorite.

The Times-Picayune The States-Item

New Orleans, La., October 15, 1978

Few can argue against the Carter administration's new space policy, and some will applaud it as a return to reason. The only quibble we have with it is in its apparent inclination toward penuriousness, but that is not entirely controlled by the White House, and specific budgets can be judged on their merits as they are proposed.

The new policy offers three basic guidelines:

— Space projects "will reflect a balanced strategy of applications, science and technology development."

— They will be pursued only "when it appears that national objectives can be most effectively met."

— "It is neither feasible nor necessary at this time to commit the United States to a high-challenge space engineering initiative comparable to Apollo."

The latter responds to arguments in and out of the space program that "spectaculars" — particularly manned spectaculars — waste resources best used for greater overall gain by other means. It means we are not going to Mars anytime soon, but there is no reason why we should.

Such adventure was appropriate to the heady days when we were just getting our feet wet, as it were, in space, but now we are in the slower-paced phase of consolidating our knowledge and turning our experience to practical uses. The space shuttle is the project for this, and though it carries heavy front-end costs, they should be fully justified by the life and returns of the program.

But the key to space program budgeting lies in assuring that it does not dip below the level necessary to maintain a core of space specialists — scientists, engineers, designers, even astronauts — whose loss could not quickly be made up. This is the time for long-range planning and excution, not expensive improvisations, and continuity of staff and stability of funding are essential to success.

The Boston Globe

Boston, Mass., October 14, 1978

The Carter Administration, cutting its style to the perceived mood of the public, has set a modest series of goals for expenditures on the nation's space programs. The modesty of these goals is highlighted by the spectacular character of the moon program that preceded it but America should not lose sight of the space-effort's aim of giving mankind a better understanding of the solar system and all of outer space.

Our National Aeronautics and Space Administration, custodian of the space program, is little more than a shadow of its former dramatic self. NASA's budget in the current fiscal year is about $3.3 billion, far below the $6.8 billion spent in fiscal 1966, even before allowing for the ravages of inflation.

The Apollo-dominated space program of the 1960s, climaxed by man's arrival on the moon in 1969, was very flashy and very expensive. It would be difficult to justify such a program today.

But NASA operates a variety of basic scientific projects that have been highly useful both in terms of basic science and applications for general use. The space shuttle, which uses about 40 percent of NASA's budget and will begin operations next year, is expected to open up a large number of low-orbit, earth-oriented observations, technology, potential manufacturing and satellite-repair projects. In addition to federally funded experiments and exploration, the shuttle program will also allow commercial and academic customers, including international organizations, to buy places aboard the shuttle for experiments.

While the shuttle is expected to increase the range of space activities and to reduce costs, work in space is still very expensive. NASA right now is struggling with a malfunction aboard the Seasat-A satellite, only three and a half months into a year-long project for a highly detailed mapping of the world's oceans. Early data from Seasat have been enormously rewarding to NSA scientists but if the satellite fails, there will be a difficult choice about funding another $75 million to replace it.

The danger is not that the Administration or NASA fails to appreciate the need for the slow, steady, unspectacular kind of effort that goes into support of basic technological programs like Seasat or more scientifically oriented space probes like the upcoming solar-orbit satellite that will pass over the sun's poles. The danger is that lack of the circus atmosphere created by an Apollo program will lead Congress away from support of such projects — as nearly happened with a recent Jupiter space shot.

The Seattle Times

Seattle, Wash., October 4, 1978

IT WAS just like old times. Standing amid the futuristic paraphernalia of Cape Canaveral, the President of the United States pledged this week to maintain the nation's leadership in space.

Nineteen-sixties nostalgia being popular these days, President Carter's Canaveral visit was a timely reminder that the decade of the Vietnam debacle, student riots, and inner-city turmoil was also the decade in which America put a man on the moon.

Yet the tendency to think of space feats solely in nostalgic terms poses a potential problem for the West. Russia, whose pioneer space achievements were so decisively eclipsed after the U.S. recovered from the sputnik shocks of the 1950s, is still aiming at being No. 1.

Every so often, the Russians hoist a couple more cosmonauts into orbit, and the Western world simply yawns. But the Russians never have lost sight of the likelihood that control of near-earth space will be as decisive in the military balance of the future as air power is today.

There was reassurance in Mr. Carter's pledge at Canaveral that "we will not give up the leadership of the United States in space."

Even more reassuring was congressional action last month in providing final authorization for $4,401,600,000 in fiscal-1979 funding for the National Aeronautics and Space Administration. That is a modest $30 million more than the amount requested by the administration.

The President rightly observed in his Canaveral speech (marking NASA'S 20th anniversary) that funding for space programs must compete with the nation's other needs.

The days of the 1960s-type crash programs are over. No one doubts that. In today's dollars, a NASA budget of under $4.5 billion does not provide for a crash operation but rather for steady progress in maintaining space leadership.

The present program, built around the reusable space shuttle, rightly aims at a judicious balance between further scientific exploration and numerous practical benefits.

One of those benefits, the President made clear — and why not? — is maintaining the vital photo-reconnaissance capability over the Soviet Union.

Houston Chronicle

Houston, Texas, August 9, 1978

Sen. William F. Proxmire's narrow-mindedness about scientific research projects, including those involving the National Aeronautics and Space Administration, might save the taxpayers a little money right now, but the long-term results could be damaging.

A two- or three-sentence summary of any research project might make the whole endeavor sound ridiculous or pointless, but it would be advisable in most instances to withhold the laughter until a few more facts are presented.

Proxmire was behind the move that resulted Monday in Senate deletion of $5 million for research and development in the NASA budget. That money would have continued funds for research at several laboratories across the country on the lunar samples brought back during the Apollo missions.

Without the NASA budget funds, the research undoubtedly will suffer since other sources of funding will be less generous and probably less reliable.

Paying a scientists to study moon rocks isn't a waste of money. The analysis is a necessary part of the space program and a loss of momentum in the research now could mean a considerable loss in the Apollo investment.

Proxmire also was instrumental earlier in getting cut from the NASA budget a $600,000 item that would have paid for equipment to survey the sky for possible radio signals from outer space. Although funding the study of the moon rocks is easier to justify, the listening post isn't really any more far-fetched than sending satellites to see if there's life on other planets.

The House had previously approved the $5 million lunar research appropriation but had cut back funding for other important NASA projects. There is no predicting what the final bill will include, but Congress apparently is going to saddle NASA with some tough decisions about how its funds will be used and where the corners will be cut.

THE BLADE

Toledo, Ohio, October 16, 1978

GIVEN the Administration's current efforts to get a handle on inflation, President Carter's decision to tighten up the nation's space budget makes sense as long as it does not emasculate the program.

There can be little doubt, certainly, that the space effort over the past two decades has brought immeasurable benefits in many areas, some completely unrelated to space. But, as was noted by Frank Press, the President's science adviser and chairman of a space review committee, the economic health of the country and the mood of the people are quite different today from what they were during the drive to put a man on the moon.

So a White House statement on space policy is on sound ground in spelling out priorities for the next 10 years in which costly and spectacular missions are rejected in favor of unmanned scientific exploration and practical application of existing technology. In the same vein, military and civilian projects will be consolidated where possible in order to save money and increase efficiency.

It should be possible to conduct a respectable space program on something less than the $4.3 billion that is currently going to the National Aeronautics and Space Administration, and the statement indicated this would shrink significantly as spending for the orbiter shuttles winds down.

At the same time, however, the stakes in a continuing, effective space program should not be lost sight of. The benefits already realized are myriad, ranging from satellite surveillance and medical applications to improved weather forecasting and mineral-resource exploration. With the investment the United States already has in space technology, it would be short-sighted to cut the program to the bone because the ultimate benefits are almost certain to outweigh the cost of operating it.

After 20 years of intensive space exploration and accumulation of knowledge, the time may well have arrived when this country should cut back a bit, assess and make use of what has been learned, and set pragmatic priorities for the future. To the extent that Mr. Carter's newly enunciated policy is aimed in that direction, as it appears to be, it can be accepted as recognizing the realities of present economic conditions in the United States.

The Hartford Courant

Hartford, Conn., October 16, 1978

President Carter's latest space policy places a sensible emphasis on less spending. Of necessity, it is also less exciting.

Plans for the next decade do not include any new spectacular, costly missions, according to Frank Press, the president's science adviser. Instead, scientific explorations will be stressed, including more probes of Jupiter, Saturn and other planets. Another major project will send two spacecraft into orbit around the sun in 1983 to take various scientific measurements.

Mr. Press pointed to the economic health of the nation and the mood of the people as the reason for lowering America's space sights. "Nobody in Congress or the federal government or the public has put forward a case for a U. S. manned Mars mission," he commented. "And if the Soviets decide to spend $70 billion to land men on Mars in five years, we say: God bless them."

This is a decided shift from previous administrations. The late President Kennedy committed the nation to the successful $24 billion Apollo project to land men on the moon. Former President Nixon directed the development of the reusable Space Shuttle at a cost of $10 billion.

The current budget of the National Aeronautics and Space Administration is $4.3 billion, $288 million higher than last year. That figure is slated to decline over the next decade as spending for the Space Shuttle tails off.

The Space Shuttle is designed to make numerous flights to and from orbit. In 1980, it will ferry into space the Skylab, a manned laboratory. If successful, the project will be an engineering triumph, though lacking the novelty and excitement of May 5, 1961, when Alan B. Shepard Jr. became the first American to orbit in space.

Seventeen years later, the country has become accustomed to space accomplishments, and increasingly aware of their high cost. That does not diminish their significance. The spinoff of space technology has been invaluable in other fields, particularly medicine. The development of spy satellites is crucial to the success of monitoring international agreements to limit arms. Though now restricted in scope, the space effort still deserves a high priority.

The Virginian-Pilot

Norfolk, Va., August 12, 1978

Since astronauts picked up the first ones nine summers ago, moon rocks have come a long way down. Chauffered to earth in a billion-dollar spacecraft, some of them subsequently were lost in the mail. Little pieces were given to 51 governors and to 157 heads of state, and were displayed in museums, exhibitions, and Las Vegas hotels.

Undeterred by youths throwing earth rocks, Spiro Agnew presented the first moon sample to Ferdinand Marcos shortly before he became dictator of the Philippines. The second sample Mr. Agnew gave to President Thieu of South Vietnam.

Scientists soon were allowed access to the samples, and their work was lavishly funded for a while. But the Johnson Space Center's public relations office continued to have the largest share of the material, and of the 84 pounds of rocks and dust brought back to earth by six Apollo missions, 74 pounds remained sealed away.

Research on the rocks has, among other things, given scientists clues to the origin of the solar system and indicated that the moon would be better for mining than farming. Thirty universities and nine laboratories, seven operated by the Government and two by industry, continue working with lunar analysis under NASA grants. But the Senate has just passed an appropriations bill to eliminate all such NASA grants.

The House had approved the full $5.7 million for lunar-sample analysis, cutting instead parts of NASA's proposed solar and Jupiter investigations. The final version of the bill will be decided later this month by a joint conference of the House and Senate.

A year ago NASA chose rather than cut the grants to stop monitoring transmissions from the millions of dollars worth of equipment that five Apollo missions set up on the moon. These atomic-powered modules continue to send back data on temperature, solar wind, and seismic shocks, but since last September nobody has been listening.

Much that could be learned from the modules and rocks has by now been learned. In some respects, more is now known about lunar than terrestrial geology. It may well be time to come down to earth. However, several hundred scientists whose work will be cut short could make a case that, after spending $30 billion in getting the moon samples, the Senate is wasteful to cut off research on them in order to spend more money on new excursions. After all, only if the scientists can further man's useful knowledge of the universe will the billions prove to have been well spent. Without that new knowledge, all we got for our money was entertainment and some mighty expensive souvenirs.

The Morning Union

Springfield, Mass., October 15, 1978

In the context of other needs for federal funds, President Carter's decision to pare expenditures on space to a much lower level than in preceding administrations is all to the good. Outer space travel will be left to the Soviet Union, if indeed the Kremlin is inclined to duplicate the U.S. lunar landings — or even go to Mars.

The exploration of the solar system does not lend itself to impulse, or even to a timetable. Considering that the earth is a speck of dust in the universe, any program of manned flight among the planets would require unbounded time — even if funding were unlimited. It may be, in fact, that mankind will never achieve the technical competence to put men on Mars or any other planet.

Nobody has said for sure, but there have been "hints" that the Soviet manned space flights now being conducted are part of a Soviet plan for a manned mission to Mars, the nearest planet to earth. It is expected that such a mission would be preceded by thorough exploration of the moon, with a view to using the moon's surface as the launching site for the excursion to Mars.

This country's Mariner and Viking unmanned photo missions to Mars revealed little about that planet that made it much more interesting than the moon. Viking landers carried soil - testing equipment but the analyses have been inconclusive. Photographs transmitted to earth indicate terrain features that must have been formed by flowing water, though there is presently no water on the planet.

There remains a fascination about Mars — if only the possibility that it resembles in some degree what the earth may look like a few million years from now. But for practical reasons its deepest mysteries probably will survive many more generations of mankind.

The Houston Post

Houston, Texas, September 14, 1978

We had to play catchup after the Soviet Union ushered in the space age with the launching of its Sputnik satellite in 1957. We had to play catchup again when it put the first man in orbit in 1961. In both instances, American scientific and technological superiority closed the gap in a relatively short time. We went on to outstrip the Soviets with such feats as the moon walks and the Skylab missions.

But no American has been in space since the linkup of our Apollo and the Soviet Soyuz spacecraft in 1975 when detente was in flower. Interest in the U.S. manned space program has visibly lagged in the administration and in Congress. Meanwhile, the pace of Soviet earth orbital missions has quickened to the point that flights to and from the Russian Salyut space stations have become almost routine. In the past few months, three Soyuz craft, two of them manned jointly by Russians and cosmonauts from other Eastern European countries, have docked with the Salyut 6 space station.

American astronauts will not return to orbit until the new space shuttle plane is launched late next year at the earliest. We will depend on periodic shuttle flights to maintain our presence in near-earth orbit for several years to come. The veteran astronaut, Thomas P. Stafford, pointed out recently that the Russians are developing the capacity to keep their cosmonauts in virtually continuous orbit. We, on the other hand, have opted for a major technological advance—the reusable shuttle that can greatly reduce the cost of orbital flights. Lt. Gen. Stafford, deputy chief of Air Force research and development, said the Russians told him five years ago that they intended to put a permanent space station in orbit.

The Soviets, as Stafford notes, are gathering valuable scientific and technological information from the current series of space missions, as well as practical experience in living and working in space. But the Kremlin's bid to put Russians in orbit on a more or less permanent basis has an ominous implication. A continuously manned space station orbiting the earth could pose a military threat of disturbing proportions. The Soviets have developed an operational hunter-killer satellite capable of crippling or destroying our reconnaissance and communications satellites. They have experimented with orbital nuclear bombs. Such weapons carried aboard a space station could prove to be more effective and less vulnerable to counterattack than similar weaponry launched from earth.

The Kremlin has cooperated with us in space ventures—up to a point. But much of the Soviet space program is still shrouded in secrecy. Where our national security is concerned, we cannot afford to play catchup with the U.S.S.R. as we did in the early years of the space age. We might not have a chance to close the gap a third time.

WORCESTER TELEGRAM.

Worcester, Mass., October 16, 1978

It is, quite obviously, over.

The whole race in space, the fierce competition over which country would be first to land on the moon, on Mars, is over.

Spectacular space missions are a thing of the past, according to a policy statement released by the White House. Instead, the new emphasis will be on simpler — and cheaper — programs.

Over the past 20 years of the National Aeronautics and Space Administration's lifetime, there had been criticisms that the space program was too costly, that the results were not worth the investment. It was argued that the money could be spent more productively, more humanely, on earth.

But the mood of the nation, the catch-up spirit inspired by Sputnik, was different back in the '50s when President John F. Kennedy committed the country to the $24 billion Apollo project to land men on the moon. President Richard M. Nixon committed us to develop the reusable shuttle-spacecraft program at a cost of almost $20 billion.

We wanted to be first on the moon, and we were.

But today, no one is overly concerned that the Russians may be aiming at a manned flight to Mars. Our priorities are elsewhere. President Carter has also rejected such high-cost projects as space colonization, space manufacturing and solar-power satellites.

Instead, the emphasis will be on unmanned scientific exploration and practical application of existing technology.

The country's newly conservative space policy, perhaps, is not just a matter of economics. In the winding down of the more spectacular space missions, people had become jaded. All the astronauts started to look alike and the extraordinary feats in space became routine. People began to prefer their space exploration in science fiction movies, where it had all started in pre-Apollo days.

The dwindling of the space program is tinged with sadness, of course.

But nothing can ever dim the glorious memories of "One small step for man . . ."

The San Diego Union

San Diego, Calif., October 16, 1978

The National Aeronautics and Space Administration entered its 21st year the other day facing, among other things, a diminished share of the federal budget.

As if to underscore the point, the White House last week released a lengthy statement of its space policy for the coming decade. A NASA official tactfully characterized the Carter administration's shrinkage of the space program as "a go-easy approach." Actually, the policy statement describes a space program pared to the bone.

Space exploration will continue to be de-emphasized. Only unmanned vehicles will be assigned exploratory missions and then only when there appears a good chance for direct collateral benefits. Projects designed to explore the feasibility of space colonization, manufacturing in space and solar-power satellites were specifically rejected. At NASA, the glory days may be over for quite some time.

The space agency's basic political problem is that its constituency is generally limited to its own employees, a few visionaries in the scientific community and some factions within the aviation and aerospace industries. Public support has proved transitory and NASA no longer has a friend in the White House. The popular fever to cut taxes and government spending compounds the space program's problems.

Declining public support for space programs can be attributed in part to a lack of public understanding of the real worth of these endeavors.

Weighed against its benefits, the cost of the space program has always represented a bargain. Since 1958, NASA has spent a total of just under $72 billion. Counting space program contributions from other federal agencies, the aggregate cost of America's two-decade effort in space is $100 billion.

Aside from the direct benefits of our immense new knowledge of space and space travel, the program's so-called technological spin-offs are truly impressive.

They include: the revolution in electronics made possible by miniaturization of components and circuitry, quantum leaps in computer science, much of the most advanced intensive-care equipment used in hospitals, a whole range of new fire-retardant fabrics and materials, space photography with all its geological, agricultural and military applications, a vast new knowledge of climate and weather patterns that has transformed the science of meteorology, and a similar revolution in communications and navigation due entirely to satellites.

The value of these developments is incalculable.

A failure to comprehend the central role the space program played in the evolution of these technologies is deplorable enough. But we wonder if waning support for space exploration might signify something far more disturbing and profound.

If exploration is an act of faith motivated by confidence in the future, what does the current apathy toward space programs tell us about the nation's vision?

The Department of Health, Education and Welfare now spends more in nine days than NASA spends in a year. Can that be the extent of this country's collective desire to push back the boundaries of man's physical limitations? We hope not.

Skylab Falls to Earth Ten Years After U.S. Moon Walk

The tenth anniversary of man's first flight to the moon was celebrated on an ironic note July 11, 1979 by the fall to earth of Skylab, America's 77-ton space station. Skylab had been in orbit since 1973 and had been home to three crews of U.S. astronauts. The $2.6-billion program had produced the greatest amount of scientific information of any U.S. manned space-flight series. The astronauts spent a total of 172 days in the craft and brought back thousands of pictures of the earth and the sun, plus miles of tape-recorded data. Skylab's orbit began to decay after the last mission in 1974, and the National Aeronautics and Space Administration periodically had maneuvered the craft from earth in an attempt to prolong its orbit. NASA hoped it could keep Skylab in flight long enough for a space shuttle to reach it and attach a rocket that would either lift Skylab into a higher orbit or aim it at a remote spot in the Pacific Ocean. However, the maneuvers could not counteract an unusual amount of solar activity that affected the earth's upper atmosphere, increasing friction on Skylab and slowing its orbital speed. Meanwhile, the shuttle that was to rescue Skylab fell behind schedule because of technical problems and budget cuts.

NASA was virtually helpless to control Skylab's fall and could not even predict where the craft would land. Skylab finally broke apart upon descent and scattered its pieces over Australia, after NASA had made an unsuccessful attempt to aim it at the Pacific. Fortunately, no populated areas were affected, and chunks of Skylab became the object of humorous treasure hunts. Skylab's fate marked an ignoble end to an era of high hopes and confidence in the American space program.

The Evening Gazette

Worcester, Mass., July 12, 1979

Ten years ago this month millions watched in awe as civilian engineer Neil A. Armstrong planted a human footprint on the moon. The landing was an incredible triumph of technology . . . and it also buoyed the American spirit.

But it's somewhat ironic that as we near the 10th anniversary of the lunar landing this week another space venture — Skylab — tumbled to earth in a shambles. Skylab's inglorious descent is symbolic of the direction of the American space program in the 1970s.

The NASA budget is about half what it was at the time of the moon walk. The program is, in fact, a mere shadow of what it was in those spectacular days of Apollo 11.

A Congressional Quarterly review of where the U.S. stands on the space program points out there is little interest in coughing up the enormous sums needed to restart a full-throttle space program.

Generally, the mood in Congress is to invest relatively modest amounts for space projects. President Carter, too, favors a low-gear effort.

The current NASA schedule includes a launch sometime next year of an orbiting space shuttle. Unlike Skylab, this machinery is reusable. It will be able both to launch and retrieve satellites.

The other half of the current program is the continuation of interplanetary exploration by unmanned probes.

Critics of a winding down of the space program argue that many in Congress and the president aren't paying attention to the tremendous scientific gains we could continue to reap from a strong space program.

And, they ask, why retrench now that we've seen the substantial benefits to life here on earth the space program has achieved?

They're right on that score, but we suspect most Americans agree with the go-slow approach.

With the mind-boggling list of domestic problems — energy, inflation and the possibility of recession — it's understandable that the public won't buy a return to the golden days of Apollo.

THE INDIANAPOLIS NEWS

Indianapolis, Ind., July 20, 1979

Ten years ago today the nation gathered 'round to watch a man walk on the moon.

It was a magical moment and a near-magical achievement. It was an accomplishment in which all Americans shared. On July 20, 1969, Americans were shooting for the moon — and they made it.

The sense of triumph has faded somewhat since then. Many would say the world has soured since then. But what of the space program? What of the program that accomplished the impossible in such a short time, providing an inspiration to all Americans?

It's been scaled back. "It is," according to the Carter administration, "neither feasible nor necessary at this time to commit the United States to a high-challenge space engineering initiative comparable to Apollo." In inflation-devalued dollars, the National Aeronautics and Space Administration now receives less than half the money it received in its peak year, 1965. In that year the agency employed 34,000 persons. It now employs 23,000.

Therefore, NASA has largely turned its eyes from space back to earth. Through satellites, NASA is helping America 1) assess and manage its natural resources 2) enhance its communications capabilities and 3) monitor arms control agreements. NASA engages in international cooperative efforts, exchanging personnel and information with other countries. But (Skylab aside), NASA doesn't hit the front pages much any more. Satellites aren't as interesting as manned flights. Work still goes on, though not at the previous breakneck speed.

Not all the romance is gone, however. Voyager II vied with Skylab for headlines this month. The spacecraft buzzed the planet Jupiter and sent back absolutely spectacular photos of its moons. Scientists were totally unprepared for what they saw. In the words of one, "We've discovered. . . how narrow our vision really is."

There is also NASA's next big venture upward: The space shuttle. NASA is building four space shuttles — powerless airplanes that perch atop rockets and glide back to earth. The shuttles are a tremendous bargain — they're reusable. Each should last about 100 flights. Private companies and foreign governments can rent space aboard the shuttles, partially defraying the cost.

In short, our space program may not soon provide us with many nights like the one 10 years ago today. But it is providing us with invaluable scientific knowledge and extremely useful technology. Our leadership in space must be maintained. The space program produces results. It should be continued and perhaps expanded. It is an excellent national investment.

The Birmingham News

Birmingham, Ala., July 19, 1979

Ten years ago tomorrow, the first men landed on the moon. The world watched in fascination as U. S. astronaut Neil Armstrong took the first tentative steps from the landing vehicle to the gritty surface of the earth's only moon. With him was Edwin "Buzz" Aldrin. The two men knew they were making history. The watching world knew it was seeing history made.

Beyond the desire to know more about our solar system, the motivation was the spirit of adventure, the response to a challenge and the keen pleasure of competing — qualities which have distinguished American history — to see who would be first on the moon, the United States or the Soviet Union.

But while the spirit of competition was certainly a factor in the successful moon landing, it required at the same time one of the greatest displays of cooperation known to human experience. The complexities of the venture and the corresponding demands on the spirit of cooperation are beyond description.

Well, the crazy Americans had done it again, Europeans remarked. But some Europeans were not sure what the moon landings were all about or whether they were worth the effort, much less the billions of dollars they sucked out of the world economy.

Americans, proud of their success, were a bit longer arriving at those questions — not until the Vietnam war became a stalemate and the Great Society staggered and lurched ahead amid growing factionalism.

Interest in space exploits dwindled rapidly. Today, a rocket-launching is greeted with a shrug and the promises implicit in the moon walk seem as dead as last year's New Year's resolutions. Even the adventures and engineering marvel of Skylab, its birth and demise, creates little excitement or interest except as a bizarre happening.

Now the focus is upon the environment, the energy problem and the "me generation." The sudden and rampant selfishness is a mystery to many. But certainly it had its antecedents in the campus riots, draft-card burnings and a variety of demonstrations which attempted to give the lie to high national purpose. If the me crowd is selfish, it surely had beautiful examples of how to focus on one's own needs and desires to the exclusion of broad national goals in no-growth groups, union cartels, racist minorities and professional groups.

But the moon is still out there and so is the technology that made the landings possible. And whether we approve or not, technology is the wave of the future. Desirable or not, it will be more and more a factor in the survival of humankind on earth.

And while inhospitable planets in our system may hold no prospects for colonization, they certainly do offer rich lodes of scarce vital minerals. Doubters should be reminded that more than 100 years passed between Columbus' first voyage and colonization in North America. And it has taken civilization some 6,000 years to have to resort to mining the inhospitable depths of the ocean.

It may be in the 22nd, 23rd or even 24th century, but if divine providence sees fit to preserve life on this planet, the future of that civilization may very well lie with the stars.

THE BLADE

Toledo, Ohio, August 24, 1979

THE Soviet Union's latest achievement in sending two men into space for a record 175 days in orbit around the earth is not likely to jolt Americans into a surge of scientific activity as did the flight of Sputnik I back in 1957. Still, the Russian success unquestionably does put the USSR ahead of the United States, at least in the area of longevity in space, and further along in that country's quest for a permanently manned orbital space station.

Although this country has not been exactly niggardly in its government appropriations for space ventures, these have been concentrated more on unmanned missions to other planets than on orbiting satellites such as the big Skylab vehicle that recently plunged to earth. The shuttle project, which has been designed to operate in conjunction with a space station, lagged as public interest waned. Now that program — while behind schedule and currently without a station with which to link up — is getting renewed attention. The shuttle will be operated by itself as an orbiting space-research station because it has the ability to return to earth and be used again.

The ultimate ramifications of establishing permanent orbiting space stations can hardly be ignored. They will include further advances in numerous fields such as weather, medicine, environment, and solar energy, not to mention military and security uses of which we are now seeing the first examples. It is an area of technology in which the United States can ill afford to fall behind, and it is one that, based on the record so far, promises to return far more in benefits to mankind than it requests in investment.

THE DENVER POST

Denver, Colo., July 15, 1979

WEDNESDAY'S FIERY crash of Skylab is only partially symbolic of the present state of America's space exploration effort. The National Aeronautics and Space Administration has grown used to a quieter disintegration of this nation's once-indomitable will to press on with the exploratory spirit upon which our civilization was founded.

The space agency's budget shows graphically the erosion of interest in solving the barely probed mysteries of the cosmos. In dollar terms, NASA's projected 1980 budget of $4.6 billion is the same as its previous high in 1966. But there is a difference. The more robust 1966 dollars fueled the Apollo project which led to the unforgettable moon landing on June 20, 1969. A decade later, the inflation-riddled dollarettes of the present budget are devoted to consolidating those early breakthroughs.

That consolidation, of course, is yielding great benefits. Space technology has affected our lives in such spectacular ways as weather forecasting and communications satellites, and in such mundane matters as pocket calculators and non-stick cookware. NASA's central effort now is to develop the space shuttle, a reusable "airliner" that will be able to carry as many as five satellites at once into space for perhaps $20 million a mission, compared to the $400 million for each Apollo mission.

The shuttle program will help orbit and maintain intelligence satellites. Without such reliable monitoring systems, strategic arms control efforts would be doomed.

The shuttle's ability to launch and recover multiple satellites will be a boon to science. A giant telescope and a space laboratory, both unmanned, will be orbited by the shuttle. The late lamented Skylab will be replaced with a manned laboratory called Space Lab, planned by the 10-nation European Space Agency.

Commercially, NASA is offering to rent space to private companies aboard its orbiters for $3,000 per 1.5 cubic feet. Experiments in zero gravity and absolute vacuum can help proliferate commercial applications of space technology.

But the shuttle, valuable though it is, is still puttering around in mankind's backyard. Interest has waned in more aggressive efforts to explore the infinite frontiers of space. Unmanned probes, the latest being the recent fly-by of Jupiter, continue to yield information. But there are no present plans to take the next obvious step — a manned exploration of Mars.

A special study group in 1970 headed by then-Vice President Spiro T. Agnew urged such an expedition by the century's end, but the proposal was rejected. The logic used against it was the same as then senator and now Vice President Walter Mondale used in 1972 trying to stop the shuttle program, which he labeled a "senseless extravaganza." He demanded that the funds be used instead to "solve such human problems as mass transit, housing, education and the environment."

Obviously, we are in a period of fiscal retrenchment, but the Carter administration went too far last October in a statement arguing that it is "neither feasible nor necessary at this time to commit the United States to a high-challenge space engineering initiative comparable to Apollo." The policy rejected such follow-on efforts as space colonization, orbital manufacturing programs and satellites to generate solar-electric power for use on Earth.

It is time to explore these projects, as well as a Mars landing, seriously. "High challenge" is associated with great reward. Had Queen Isabella in 1492 been a convert to the Mondale-Carter philosophy, she would have hocked her jewels to provide a few more hitching posts for the horses of Castille and perhaps given a dollop to the poor. But discovery of the vast resources and immense opportunity of the New World would have had to await a more imaginative leader.

THE CHRISTIAN SCIENCE MONITOR

Boston, Mass., July 12, 1979

Ten years ago this month the inhabitants of Earth watched in awe as the first human beings set foot on the surface of the planet's moon. Millions were caught up in the romanticism of exploring outer space. Others were dazzled by the sheer technological proficiency of the moon mission. Still others were inspired most by the new view of Earth, like American poet Archibald MacLeish, who commented that to see our planet from the moon was to "see ourselves as riders on the earth together, brothers on that bright loveliness in the eternal cold – brothers who know they are truly brothers."

The glow of that lunar walk a decade ago soon faded. Other earthbound events – the Vietnam war, political assassinations, social turmoil, Watergate, inflation, energy – sobered the public mood. For most, space exploration receded in interest and importance. Today technology itself seems to be under fire following such events as the near-disaster at the Three Mile Island nuclear plant and the grounding of the mighty DC-10. Even the premature (but gratefully safe) fall of Skylab because of miscalculations has sprouted headlines about the "tarnished image" of technology.

Is there actually a serious public backlash to science and technology? We doubt so. But if there is any temptation to join one it ought to be resisted. It would be ludicrous to weigh the technological "disasters" in the same scales as the achievements. Science and technology are an indispensable part of the fabric of everyday life and anyone who uses a copying machine, who flies across the ocean in a jet, who computes his tax on a pocket calculator, who plows up 500 acres on a modern tractor, or who telephones a friend in another country can only stop to wonder anew at the blessings of technology.

We are nonetheless living in a new time, a time of more sensitized questioning of the purposes and impact of technology. No longer can we ignore the social, economic and, most important, moral and spiritual challenges raised by scientific knowledge and technological progress. Natural and social scientists are asking: Is "big" technology always appropriate to achieve economic growth? Are specific technologies perilously destroying our planetary environment? Is some scientific investigation, such as DNA research, dangerous to life? Are some technologies not worth the cost? How do you forestall human error to ensure safety (and recent accidents that were caused largely by human errors)?

Surely the growing investigations of such questions point to a new intellectual maturity and a dawning moral sensitivity to the world around us. In every age humankind has had to learn to adapt wisely to scientific and technological developments and it is no different in our times. Perhaps only more difficult because of the extraordinary explosion of scientific knowledge and the breakneck pace of technological progress. But would we wish to stop the exciting and liberating march of scientific research – on Earth or in space? Who can fail to thrill from the delightfully colorful canvases of Jupiter relayed to Earth by the Voyager spacecraft from a mind-boggling 500 million miles away?

We would not be starry-eyed about space exploration, if readers will pardon us the pun. We in fact share the view of planetary physicist Richard Goody that the mere search for knowledge in space is an insufficient objective, despite such practical results as weather and communications satellites. Earthkind needs a more overarching goal, the goal perhaps of rational, humane exploitation of the solar system for man's benefit. If we develop a space shuttle system, for instance, what do we transport, where to, and why?

Our endeavors, in short, whether soaring in outer space or applying technology closer to home, must have moral and spiritual purpose. They must enhance life for Earth's "riders" and their beautiful environment, making us all more aware of the brotherhood of man. With such purpose, science and technology need never fail. The first moon walk – and the fall of Skylab – are both instructive benchmarks along the widening way.

THE ANN ARBOR NEWS

Ann Arbor, Mich., June 22, 1979

THERE WAS a time when technicians at the Space Center in Houston could justifiably expect an enthusiastic response when they announced a success. That time seems to have ended soon after Neil A. Armstrong climbed back into the Eagle and came home from the moon, 10 years ago next month.

The National Aeronautics and Space Administration has had an amazing comedown, in every sense.

* * *

ON WEDNESDAY, NASA technicians were able to announce in Houston that they've succeeded in turning Skylab sideways in relation to earth's equator. This, they say, improves odds that when the 80-ton satellite crashes next month, chunks won't tear into Europe or Asia, and might still be subject to some control before disintegration.

Instead of renewed respect for accomplishing this much with Skylab, NASA is receiving mostly skepticism for having gotten the world into the fix of having to wonder who or what Skylab's bigger parts will hit. That didn't just start; it's skepticism reflected in NASA's budget. When allowance is made for inflation, that budget is worth about half of the $4.6 million NASA was getting from Congress a decade ago.

When it comes to cutting the federal bureaucracy, NASA is out in front. The space agency had 34,000 employees in the period just before Armstrong's "giant leap for mankind." Now, there are 23,000.

Earth-orbiting satellites other than Skylab are the only NASA projects serving purposes clear enough to everyone to go unchallenged.

Spy satellites will be essential in enforcing the new strategic arms limitation treaty, if it's approved in the Senate. Others serve such wide-ranging functions as enabling the U.S. Geological Survey to keep track of the undulating land bulge in southern California's earthquake zone, providing daily weather pictures, and also pseudo-serious purposes, such as enabling Walter Cronkite to tell us live from Vienna he has nothing new to report about the Carter-Brezhnev meeting.

* * *

ONE PROSPECT for making space-related research again look useful and exciting to more members of Congress, and more taxpayers, is offered by NASA's $18 million evaluation of possibilities for using satellites to beam down a reliable flow of solar power for conversion to electricity.

A proposal to boost that portion of NASA's budget to $25 million was approved a year ago today in the House, by 267-96 (including a "yes" vote from Rep. Carl D. Pursell, R-Plymouth, a member of the House Science and Technology Committee). But that idea has been spurned in the Senate. It wasn't mentioned by President Carter as he was voicing his new enthusiasm for solar energy this week.

Plans must precede money.

Proposals to set specific earth-resources research and space exploration goals as far ahead as 2010 have been offered by Sen. Adlai E. Stevenson, D-Ill., who heads the Senate's science, technology and space subcommittee, and by Sen. Harrison Schmitt, R-N.M., a former astronaut and ranking GOP member on that subcommittee.

Their proposals don't call for funding. They call for commitments beyond the day-to-day view of major problems – and challenges – that seems to dominate both in Congress and the White House.

The Houston Post

Houston, Texas, July 5, 1979

A tall young American was walking along a narrow sidewalk in Paris two years ago when he came face to face with two smiling young Tunisians. He looked foreign to them. "Where are you from?" they asked. "Houston, Texas," he replied. They laughed happily. "Skylab!" they said, and walked on, their curiosity soothed. Like the two young Tunisians in Paris, people in other parts of the world who had never heard of Houston or Texas have been captured by the audacity of Skylab.

The threat of Skylab's falling has never seemed quite real to us. NASA has been too successful too long for anything to go so drastically wrong. Unbelieving, lacking any feel in our bones that this enormous thing really could come tumbling down, we have tended to treat the whole prospect with levity. Chicken Little has not had so much free publicity in all his long and honorable life in children's literature.

But the fact is that Skylab's failure is a tragedy. Even were it to land in some remote ocean depth, harming nobody and nothing, its loss is tragic. Skylab is the biggest and most useful craft ever put in orbit. Our allies view its failure not only with concern, but with a degree of indignation. The London Daily Telegraph editorialized: "Experts predicted a great increase in sunspot activity in 1978 and 1979 which would swell the earth's atmosphere and disrupt Skylab's orbit. NASA ignored their predictions. . . . How sad."

The United States has too much invested in outer space, too much of our spirit and confidence as well as time and money, to let this great loss weaken our drive toward the stars and beyond. We have gained too much in benefits to all humankind to be discouraged. Expensive though it has been, our space program has been inexpensive compared to any war we have ever fought. The research done for each war has always resulted in peacetime development, but only at the price of thousands upon thousands of lives. Our research for space is the first effort of comparable scale done not for war but to advance earth's people to new horizons, new frontiers. Skylab's fall must not be allowed to damage that aspiration.

The Oregonian

Portland, Ore., July 12, 1979

Teasing a flaming piece of space junk into a splashdown in the Indian Ocean near Australia is hardly comparable to the glorious days of America's Apollo program and all those marvelous footprints left on the moon. Drowned by all the splashing news of efforts to bring Skylab down in a "safe" ocean are the problems of the lagging U.S. space program.

NASA has been forced to occupy itself with saving abandoned junk while the Soviet Salyut 6 program has been setting new men-in-space records, leading to the establishment of a permanent space station.

One solid U.S. success, which caused only a blip in the attention span of Americans worried about getting hit by Skylab, was the fly-by of Jupiter by Voyager 2. It sent back over a billion miles spectacular pictures of the Jovian moons.

But while the Voyagers race on toward the giant planet Saturn, NASA has been trying to keep the Shuttle program on schedule, fearing that it will have to seek more funds if the Shuttle launch is delayed beyond March 1980.

The odds that NASA would be able to dump Skylab into an ocean, rather than having it spew its parts over some land area, were much better than are the chances NASA will be able to launch the Shuttle by March, which are down to 20 percent. Technical problems, including the need for denser heat protection tiles, have delayed the launch, once targeted for this fall, perhaps until next summer.

It was penny-wise austerity that prevented Skylab from being equipped with rockets that could have brought it down in a timely manner. The cuts in NASA funds in the 1970s have also caused the Shuttle delays.

Nursing a derelict space ship is hardly what flight controllers from the command center at Houston were trained to do. But had they failed in their jury-rigged mission center, permitting debris to strike populated areas, it would have triggered more congressional opposition to space expenditures.

So, Skylab, a 77-ton junk yard, momentarily has entertained and frightened the public, like a one-star horror movie. But its hot, spectacular plunge to earth should not hide the deplorable fact that the U.S. space program has been put on a cool back burner.

The Washington Star

Washington, D.C., July 7, 1979

It is one of those odd coincidences that the 10th anniversary of America's most extraordinary space venture, the first lunar landing, roughly coincides with our greatest space embarrassment, the decline and fall of Skylab.

It is difficult to consider one without the other. Indeed, Skylab was, in a very real sense, assembled from "off the shelf" Apollo hardware — made available when the last three moon landings were scrubbed.

Today, apart from bouts of nostalgia, the Apollo missions appear to have vanished from our collective imagination. Like relics from another age, spacecraft and models from the program sit on view in the National Air and Space Museum. What engages our attention now is the slightly alarming, slightly comic end of Skylab.

Skylab, of course, is about to fall; re-entry, as of this writing, is calculated to be on Wednesday. In the jargon of scientists, it will leave a 4,000-mile long, 100-mile wide "footprint" across the planet. Of its 77.5 tons, some 55,000 pounds are expected to survive the fall, in pieces ranging in size from a pound or so up to 5,000 pounds — this last an "airlock shroud."

One is quick to imagine a rain of lethal debris, although the odds are vastly against anyone being hit. Some months ago, a NASA spokesman suggested that most of the fragments will be like giant pieces of skin, "floating like leaves and that sort of thing."

Realistically, Skylab's descent is the equivalent of a tiny meteor shower, a man-made addition to the great mass of rock and metal which regularly, and quite naturally, come down upon us each year. Although no one has ever been reported killed by a meteor, there have been seven "incidents" recorded in the last 200 years. There have also been artificial "meteors." In 1975, for example, the 40-ton Saturn II, which launched Skylab, returned to earth, off the coast of Africa. There was a flash in the sky and it vanished without a trace. Skylab should behave in similar fashion.

Still, Skylab's fall remains one of the intriguing events of the season. Thoughts of it tend to bring on a kind of wild hysteria which, after all, may be the best defense against the idea of things falling from the heavens. Even if, as we believe, its 77 tons are unlikely to be noticed, we recognize the *hubris* in stating such beliefs.

In any event, it is ironic that the American space program has become so identified with this elaborate piece of human error, particularly as we remember the success of Apollo; or Viking and Voyager; or, indeed, as we recall the considerable scientific accomplishments of Skylab itself, which was a home-in-space to three crews of astronauts.

In a way, Skylab's tumble downward is a perfect counterpoint to the early space program, and in particular the short life of Apollo, which landed 12 men on the moon before being canceled. That decision to cancel after such an investment, as Carl Sagan has observed, was much like purchasing a Rolls Royce and then declining to drive it because of the cost of gasoline.

So Skylab is about to come crashing down, accompanied, as it were, by odds of 152-1 against someone being struck. But its danger as an object out of control is eclipsed by its symbolism, its reflection of a space program that once signalled a national purpose and is now, like Skylab, barely aloft.

Voyager's Saturn Fly-By Produces Spectacular Photos

Sagging morale in the space program was given a boost in November 1980 by the flight of Voyager 1. The unmanned spacecraft transmitted astonishing pictures of Saturn, revealing aspects of the ringed planet that had never before been noticed. Voyager 1 came within 78,000 miles of Saturn itself and 2,800 miles of Titan, the planet's largest moon. In addition to taking photographs of the planet, its mysterious rings and its moons, Voyager transmitted data about the atmospheres of Saturn and Titan, the only planetary satellite in the solar system known to have an atmosphere. Voyager 1 was launched Sept. 5, 1977 on a four-year mission to probe Jupiter and Saturn. Its sister satellite, Voyager 2, was launched two weeks before, but Voyager 1's flight path brought it to Jupiter and Saturn earlier than Voyager 2. The Voyager satellites cost more than $300 million but represented a reduced version of what the National Aeronautics and Space Administration originally intended to be a close look at all the outer planets in the solar system. Budget cuts had forced the program to limit itself to the two largest planets. NASA's budget for fiscal 1981 was set at $5.4 billion, a 9% increase from the previous fiscal year. Most of the funds were earmarked for the shuttle, whose cost kept rising because of technical problems. Only one other new program was scheduled, a satellite to measure gamma rays. In contrast, the Soviet space program was constantly making the news, with six flights in all to its Salyut 6 space station, including a world's record stay in space.

The Cincinnati Post

TIMES STAR

Cincinnati, Ohio, November 17, 1980

We could spill superlatives all over the page trying to do justice to the accomplishments of the Voyager space probes. Suffice it to repeat the assessment of a spokesman for the Jet Propulsion Laboratory, manager of the project:

Thanks to Voyager I, scientists learned more in a week about the rings and moons and planet of Saturn than in all previous history.

The same was true of the encounters of Voyager I and II with Jupiter and its system of moons last year. And if all goes well, it will be true when Voyager II reaches the planet Uranus in January 1986 and Neptune in August 1989.

Talk about getting the most from your buck! (In the Voyagers' case, about $460 million of them.)

So immense are the distances involved that children in kindergarten could be analyzing data sent back by the Voyagers as they reach the outer edges of the solar system in the 1990s.

The same time and distance factor, however, points up the fact that America's space program is really feeding on past commitments.

In real-dollar terms, NASA's budget is about half what it was in 1966 in the heyday of the Apollo moon landing program. Only about three unmanned planetary probes are planned for all of the 1980s. The stunning spacefaring capability created by the Apollo program has literally been thrown away.

Americans will once again go into space when the space shuttle finally begins its earth-orbital flights sometime this decade.

But the trouble with the shuttle, besides its myriad technical problems, is that it is not part of some longer-range concept.

For example, it could be used to assemble a permanent manned earth satellite, or to construct orbiting solar collectors to beam energy to earth, or to plant a scientific-industrial colony on the moon or to launch a manned mission to Mars.

Of course, this nation cannot possibly do all of those things, but it should have some goal. Without a larger vision of what the future of mankind should be in space, we are really indulging in mere stunts.

The Virginian-Pilot

Norfolk, Va., October 22, 1980

The Soviet Union has not yet put a man on the moon. But it could become the first nation to place a large, permanently manned space station into near-Earth orbit. That possibility alone should stir Congress to greater funding for the United States's lagging space program.

The U.S. space-technology lead is real. And it will be reaffirmed whenever the trouble-plagued American space shuttle settles gracefully to a landing after rocketing into orbit—an event now scheduled to occur next March. But the space shuttle's technological glories may less impress humankind than a relatively primitive Soviet space station spinning overhead.

Will space stations be used for peaceful enterprises or military advantage or both? The military is involved in each superpower's space ventures, so the strategic and tactical implications of space stations cannot be ignored.

Manned space stations, as science-fiction buffs have known all along, could be employed to intimidate and destroy. It would be to humankind's benefit, of course, if the superpowers' weapons rivalry were not extended to the space arena. *Star Wars* errors are entertaining on the screen, but you wouldn't want to live them.

Sticking to tried-and-true space machinery, the Soviet Union has logged roughly 46,000 manned-space-flight hours, more than twice the U.S. total of 22,493. Soviet cosmonauts hold the records for endurance in orbit. Leonid Popov and Valery Ryumin this month exceeded the previous Soviet achievement of 175 days in space. The American endurance record, 84 days in the three-man Skylab, was set in 1974.

Space fever in the U.S. died down after the Apollo moon-flight series. Despite new products spun off by the space program, the National Aeronautics and Space Administration's share of the federal budget gets smaller. The $5.2 billion allotted NASA this year is no improvement over last year.

The Soviet Union meanwhile is launching manned and unmanned spacecraft at a fast clip—87 launchings compared to 16 for the U.S. in 1979. The pattern continues. The shock of a Soviet space station probably would reverse NASA's fortunes overnight, as did the shock of Sputnik. But it would be well not to wait.

For NASA could lose valuable scientists, technicians, and know-how in the interim. Congress should contemplate now a steady strengthening of the space effort to assure continuity and growth.

The space program has been a boon to computer technology, electronics, and the study of natural forces. And U.S. industry would be even more at the mercy of foreign competitors without space-stimulated advances. We could improve our position with a rejuvenated space program.

Houston Chronicle

Houston, Texas, September 30, 1980

Author James Michener is a man who looks at life through the span of many generations. In his books, what one generation does affects the next generation and the next. The novels fascinate people worldwide.

Michener is a member of a council appointed to advise the National Aeronautics and Space Administration. He brings to that task the same grasp of history shown in his novels.

In Houston recently, he pointed out that what has been done in space so far has changed the course of history, and that what can be accomplished in the future could be even more dramatic.

Satellites now can make inventories of a region's resources in water, land and forests. Areas worthy of exploration for new energy can be defined. Technology developed for space now helps make possible drilling for oil offshore at greater depths. Oil spills can even be monitored. The revolution in communication is tremendous.

Beyond those sweeping changes there are more mundane but nevertheless important applications of space technology. NASA research helped streamline livestock trailers to save energy. Automobile manufacturers are turning to space-developed computer techniques in designing, engineering and production. A coating derived from space technology helps produce glass containers. Materials developed for the space program now are used in insulating blankets and thermal suits. The list could go on and on.

As Michener said, it is impossible to predict what the benefits of space activity might be. But it is possible to say that unless this country continues to be dedicated to space exploration, it will fall behind in technology and those now undreamed-of benefits will never ever come to pass.

Tulsa World

Tulsa, Okla., October 28, 1980

SOMETIME next March, if all goes well, the United States will be back in the space business, this time with the space shuttle.

The shuttle is expected to move the U. S. into the exploitation of space, since the program will involve much more productive activity than past programs which involved exploration of space.

The shuttle will be capable of a wide range of orbital activities, including the launching and servicing of satellites, recovering and returning them to earth, conducting on-board scientific and technical experiments and assembling huge space structures.

Between 1981 and 1985, 44 shuttle flights are planned in what has been called the most complicated piece of equipment ever built.

Unlike previous space shots, the shuttle will provide low thrust and an environment that scientists and technicians of ordinary physical conditioning can stand.

It has been more than 5 years since the United States has sent a man in space, so the start of the space shuttle should renew public interest in the program.

The massive effort to land a man on the moon perhaps was one the most exciting and significant achievements in history.

But the space shuttle is the start of the use of space, and future historians are likely to conclude that it is perhaps as important as the moonshots.

Lincoln Journal

Lincoln, Neb., October 21, 1980

The National Aeronautics and Space Administration budget for fiscal 1981 is $5.2 billion, about the same as 1980. And, because of inflation, it is actually significantly less than the $4 billion NASA spent in 1969, the year Apollo 11 landed on the moon.

That, of course, is one reason nothing as exciting as the moon landing has been pulled off lately by NASA. True, next month Voyager I is due to fly by Saturn, and next August its sister planetary probe, Voyager II, will reach the same neighborhood. Both will undoubtedly send back useful information. But both, also, were launched some time ago. They are not new feats by NASA.

Neither, in a sense, is the oft-delayed space shuttle, now scheduled for launching in March. It has been planned a long time.

The truth is that America's space program seems, as it were, adrift in space. And unless the shuttle can revive interest in as well as provide transportation for construction of permanent stations in orbit, the program may limp along indefinitely.

All this stands in contrast to the new Soviet endurance record for astronauts. Cosmonauts Leonid Popov and Valery Ryumin have just achieved 185 days in orbit. Moscow's space program in many ways has been not only less fruitful but also less dramatic than that of the United States. Yet the Russians keep plugging away. Their eye is on the long term, obviously.

Americans' waning enthusiasm for space exploits is understandable. As a society, our attention span tends to be short, our appetite for new fads considerable. And costly undertakings are not popular now.

Yet many technological advances our nation has made in recent years owe something to the space program. In addition to benefiting wide areas of human endeavor, space exploration has increased our knowledge of our world and universe.

What is to become of our space program receives almost no attention in this presidential election campaign. That may accurately reflect the mood of the electorate. But one wonders if nothing but a Sputnik-sized breakthrough by the Soviet Union can penetrate this apathy.

The Houston Post

Houston, Texas, December 26, 1980

Long before the first moon walk, Buckminster Fuller foresaw the strategic importance of outer space. The open cockpit planes of World War I, he reminded, foreshadowed World War II. The European war of the 1940s was won only when the Allies had gained complete control of the skies. World War II, he continued, foreshadowed the ultimate power struggle: Any nation gaining the mastery of outer space will be the nation with the power to control the Earth — for freedom and peace or for oppression and destruction.

In the 1960s the United States won the race for the moon hands down. And now, like the self-content hare, this nation is letting the Soviet tortoise plod its massive, tireless way to outstrip us in space exploitation. The Soviet Union has devoted money and research to the development of a killer satellite that can destroy the space satellites of other countries. But only belatedly did the United States notice the implicit danger.

While scientists, astronomers, leaders in the space industry and a whole series of television spectaculars have pleaded the need for the United States to return to full speed on space research and exploration, our achievements have been scattered and few. Sen. Harrison H. Schmitt, R-N.M., is urging Congress to get on with what may well be the most important task facing the nation. Schmitt, one of the last astronauts to walk on the moon, said that the United States should develop a permanent space station, that it should be "in place, operating and visible" by the end of this decade. Such a station would enable us to see if it is practical to build solar-power satellites to collect sunlight, convert it to electrical energy and beam the energy back to Earth.

Meanwhile, only four space shuttle craft have been approved for construction, with a fifth projected. Schmitt says the United States should have at least four shuttles for civilian use and another four for military. The shuttle is designed to be thrust into space by rocket engines for a mission of up to two weeks before gliding back to land. But the first shuttle launch is already three years behind schedule. The new launch date is set for March.

Fortunately, Sen. Schmitt is on the senior advisory staff of President-elect Ronald Reagan's transition team. It is to be hoped that he will be able to balance the influence of economic advisers who had lumped NASA programs among the first to cut from the budget. The scientific, medical, military, agricultural and commercial spinoffs from space research have contributed enormously to the national economy. Space research has filled at least some of the lacks created by sagging research and development in other fields. Future economic historians will be able to see and evaluate the contributions. But looking ahead, it is possible to say that American dedication to space exploration in the 1980s may make the difference between a world that is oppressed and a world that is free.

THE LOUISVILLE TIMES

Louisville, Ky., December 16, 1980

Exploring space seems to bring out a manic-depressive trait in the American personality.

The triumphs of technical genius give us a temporary high. But the euphoria is quickly dissipated by the cares and pressures of the workaday world, to which moon rocks and Jupiter's storms have little relevance.

Among scientists who study the cosmos, the descent from exhilaration to gloom is especially precipitous.

One moment they chirp that age-old mysteries have been unraveled by the latest planetary flyby. They rejoice over clues to the origin of the solar system. They are ecstatic about pictures of a remote satellite that may look the way earth did eons ago.

The next thing you know they are blubbering in their tea about public fickleness, congressional miserliness and an unenlightened nation that lacks enthusiasm for space spectaculars.

To the bewilderment of many in the space program, public interest in moon landings flagged rapidly after the first few. By 1979, according to a poll, more than half of us didn't think the effort was worth the cost. To make matters worse, NASA got bogged down in the space shuttle, which has the technical and financial problems of a military weapons program without the redeeming attribute of pizzazz.

Just weeks after the extraordinarily successful Voyager flight to Saturn, a Cornell professor, in an article on the page opposite this one, fussed about the "faltering American planetary program." Another probe, he said, isn't planned for five years.

He went on to complain that other countries may seize the initiative. The United States could end up like Spain, which made the early "discoveries" in the New World but was then upstaged by France and England.

The scientists who feel slighted seem to have lost all sense of time, space and human nature.

They would do well to reflect on the amount of public money invested in their projects. The Apollo program cost $24 billion. Billions more were spent to make those admittedly thrilling pictures of Saturn and Jupiter. It would be strange if taxpayers failed to ask about the tangible benefits.

Indeed, today's explorers should thank their lucky stars that television permits mass participation in space adventures. They are better off than poor Columbus, for instance, who couldn't even show pretty pictures when his patrons griped that his voyages served no useful purpose.

Another reason the public feels no sense of urgency is that the solar system will be out there for a long time. If all the big questions aren't answered in this century, they can be tackled in the next one. In the meantime, *Star Wars* sequels will keep us amused.

While our victory in the race to the moon is a source of national pride, most Americans surely don't insist on being first everywhere, especially at today's prices. The fact that Galileo saw Saturn's rings before a German or Swede spotted them is pretty trivial in the cosmic scheme of things. We should be delighted, if anything, that Europe and Japan will send spacecraft aloft to study Halley's Comet in 1986 while the U.S. watches from the sidelines.

Despair, in any case, is an inappropriate response to what will be at most a pause in space studies. Exploration is bound to proceed in fits and starts, according to the priorities of the moment. But so long as there are uncharted spheres out there, it will inevitably continue.

THE INDIANAPOLIS NEWS

Indianapolis, Ind., November 27, 1980

Voyager I was a marvelous project. Still, for all the billions of miles the space probe has traveled to deliver unprecedented peeks at Jupiter and Saturn, it has not gone far enough.

Actually, it is the U.S. space program that hasn't gone far enough.

After the memorable triumphs of the '60s when the U.S. raced to put the first men on the moon, interest in the program has faded like a child tired of a toy. Today's space program is only a glimmer of what it used to be. Americans wearily turned their attention to problems of the Earthbound, such as poverty, illness, war, racial conflicts and many other woes.

However, turning away from space exploration is just what the U.S. should not do. Harrison Schmitt, a senator from New Mexico and a former Apollo 17 astronaut, has spoken out against the demise of the space program, saying it can be a valuable tool for solving problems of the folks still on the ground.

"There still is a large misconception about the benefits and costs of the space program," he says. Those aren't the rantings of an astronaut out of job. He's seen what the space program can do with all the technological advances it has made and the resulting products that have been invaluable to medicine, food processing, transportation, education health services, communications and many other industries.

Even now, while the country is concerned about energy, scientists at the Kennedy Space Center are trying to develop a way to produce water-free alcohol efficiently for cars used on Earth. The project is in its early stages and could be killed easily if funding doesn't come through. But if the scientists are given the chance to try to solve problems in the process and if the process is expanded, "it could be possible that the United States could be close to fuel independence within a few years," one engineer told the *Christian Science Monitor*.

But there is another, more ominous reason for pushing ahead with the space program, according to Schmitt: "If the Soviet Union dominates space, they'll dominate the Earth, too. We need to replant the seed of individual freedom somewhere safer than the Earth."

That alone is reason enough to rededicate the country to a better-funded space program.

DAYTON DAILY NEWS

Dayton, Ohio, November 27, 1980

Do Americans want to sell off the investment in space technology that has paid off so well for this nation?

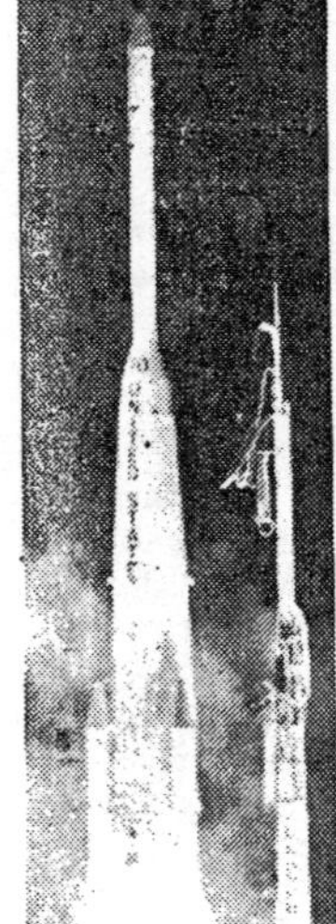

Voyager 1's incredible photographs of Saturn and its rings have reminded Americans that, yes, there still is a space program. But there may not be as great a program in the future if Washington and the public do not realize its importance.

The payoff in knowledge of the solar system is incalculable — but there is a more tangible payoff, too. The space program brought many of this nation's science and engineering disciplines together to make one technological breakthrough after another.

It is too glib to say this country ought to be helping its poor rather than flying men around the moon.

The fact is that human welfare has not been ignored in the era of U.S. space programs, and the space effort has produced developments ranging from miniature electronics to satellite broadcasting, to the benefit of the whole economy. Some of the technology used to transmit images from Voyager 1 may be used to improve deep-sea oil drilling.

The United States must maintain the space program to keep the momentum of scientists working purposefully together. And although Americans get tired of hearing about the Russians, this nation *is* competing with the Soviet Union — and with other industrialized countries.

The Soviet Union, technologically inferior to a vigorous United States, can get the edge on an indifferent United States. With long-range commitment to space work, the Soviet Union now has the only active manned space flights going, has twice as many communications satellites as the United States, twice as many photographic spy satellites, early warning satellites, ocean surveillance satellites and navigation satellites and is ahead in its killer satellite testing and orbital bomb testing.

The Soviet Union is superior yet, but its progress is a warning that the Russians can take over the lead if the United States lets its space program deteriorate, as it has in recent years.

The Charlotte Observer

Charlotte, N. C., December 5, 1980

The marvelous discoveries of the Voyager I spacecraft, along with public interest in shows like Carl Sagan's "Cosmos", come at the beginning of the end of U.S. space exploration as Americans have known it.

Kennedy

The space era came into full bloom shortly after President Kennedy announced that the United States was going to put a man on the moon. "No single space project will be more exciting," he said on May 5, 1961, "or more important for the long-range exploration of space."

For the next two decades, Americans were able to vicariously visit the Moon or another planet on the average of every six months. Students entering college today can't even remember when we weren't exploring space.

Many Projects Left Undone

Yet all that is about to change. The space spectaculars of the late 1960's and '70s — the unmanned missions to five planets, the Apollo trips to the Moon — were conceived in the early '60s. Projects envisioned in the last decade are, for the most part, still on the drawing board, victims of an increasingly unenthusiastic public and a government that has never funded the space program for the right reasons.

Historically, when the federal budget has had to be trimmed, the space program has been among the first to be cut. A Reagan administration dedicated to slashing federal spending may be tempted to follow this pattern. That would be both unfortunate and misguided.

Next year's $5 billion federal appropriation to NASA, the U.S. civilian space agency, represents less than a penny of each taxpayer's dollar. The NASA budget has been declining since 1966; cutting it more would barely make a dent in federal spending but could cripple the agency.

The history of the space shuttle — the next U.S. space effort — shows what can happen when funding shortages force too many corners to be cut. The shuttle is a combination airplane/spaceship designed to do just what the name implies — shuttle communication satellites, space telescopes and people — back and forth between earth and space. It departs from and lands on a runway and is re-usable, thus reducing the cost of space flight.

We're Two Years Behind

NASA estimated the shuttle costs in 1971 at $10 billion, but got about $6 billion. The agency took engineering shortcuts, deferred procurements, and ended up with a building program plagued with engine failures and other problems. If the shuttle is launched next year as planned, it will be two years behind schedule.

Lack of public interest in the shuttle's travails, or in space exploration in general, is understandable. When 500 billion people in the world don't get enough to eat every day, it's hard to argue that we need photographs of Jupiter's atmosphere.

But the real benefits of space travel are often overlooked, even by the U.S. government. They include little-known technological spinoffs like heart pacemakers, lightweight braces for paraplegics, and engineering on the Alaska pipeline, as well as the better-known communication, weather, and crop satellites. Space travel encourages interdependence between countries; it also forces human beings to ask important religious and philosophical questions.

Emotions Don't Persuade

Washington has never talked about the space program in these terms, however, preferring to emphasize the military necessity of competing favorably with the Soviet Union in a space race. Americans have been persuaded by emotional language, not logic, that rings hollow when those same citizens are faced with unemployment and high prices.

It is high time that Americans realized the real returns for every dollar spent exploring space, and start supporting the space program again, this time for the right reasons.

The Burlington Free Press

Burlington, Vt., November 16, 1980

Voyager 1 is speeding toward deep space after having flawlessly performed its spectacular mission to Jupiter and Saturn.

Its discoveries boggle the mind and the senses. Scientists at the Jet Propulsion Laboratory in California will be kept busy for years interpreting the millions of bits of data and the photographs they received as the tiny space probe flew closely by the two giants of our solar system.

Voyager ventured more than a billion miles over 38 months to reach the ringed planet, Saturn, after having swept closely by Jupiter in March 1979.

Was the expense worth it? Struggling with our everyday problems in an increasingly difficult world, we wonder why so much money is being spent and, in our practical way, wonder what's in it for us? We may also wonder if the money couldn't have been put to more practical use in feeding and clothing the poor or rehabilitating our decimated cities, or cleaning up our environment.

What use is it to us to know how many rings circle Saturn and how many moons orbit that gaseous giant or what chemicals they are composed of? What use is it to know that Io, one of Jupiter's moons, is volcanic still? We can't fill our bellies or warm our bodies with that information, so why spend so much money to send spacecraft to probe our solar system?

Some would say, with a simplicity that needs no explanation, because it is there. Scientists would say because there are things out there we need to know about to better understand the tiny planet third from our sun on which we live and love and work. There are as many reasons for going as there are scientists involved either directly or peripherally with the Voyager missions. (Stand by, folks, Voyager 2 will fly by Saturn next August.)

Perhaps the most important reason expressed by one Jet Propulsion Laboratory scientist is by studying the lifeless planets nearest us, we may learn just how much we can pollute our own planet with poisonous gases before we create an irreversible situation that eventually will lead to extinction of all life on earth.

Our planet is an insignificant sphere circling a sun that is but a luminous dot located three-fifths of the way from the center of a single galaxy containing 200 million or more suns. That galaxy is only one of perhaps as many as a million more spinning slowly in the vastness of the cosmos. There may well be other planets like ours, close enough to be warmed by their sun, that contain life as we know it. We should attempt to find them. Should we soon launch another Voyager beyond our solar system in search of life on some planet in the solar system closest to us, we would be long dead before it arrived. But generations yet to come will know that probe is on its way and will anxiously await the information.

No future space missions are planned in this country or anywhere else in the world. Our space program, wounded when manned flight was scrapped after the moon missions, is to languish and die unless we rise to find the answers we need to know about how and why we exist in the universe and to learn whether there is life out there like ours.

We must not let the space program die.

Reagan Seeks Cuts in NASA While Supporters Pass the Hat

Soon after taking office, President Ronald Reagan called for "moderately reduced" spending by the National Aeronautics and Space Administration. The cuts were part of Reagan's budget-reducing program for the upcoming fiscal year, which would begin in October. Former President Jimmy Carter had planned to increase NASA's budget to $6.6 billion in the 1982 fiscal year, but Reagan asked for a smaller increase, to $6.2 billion. "The space program has been and is important to America," Reagan acknowledged, but he added that "a reordering of priorities to focus on the most important and cost-effective NASA programs can result in a savings of a quarter of a billion dollars." News of the pending cuts had prompted thousands of people to send donations to NASA, mainly to enable the agency to continue receiving and processing information from the Viking1 landing craft on Mars.

As if to tantalize NASA in the face of the budget cuts, Voyager 2 sent back even more astounding pictures of Saturn than had Voyager 1. The craft passed within 63,000 miles of the planet in August, a closer distance than Voyager 1's route, and was able to take photographs of Saturn's rings from a vantage point that Voyager 1 had not enjoyed. The transmissions served only to whet astronomers' appetites for more, since they solved few of the mysteries Voyager 1 had uncovered. Current plans, however, called for no more flights near planets until 1986, when Voyager 2 would pass Uranus.

THE MILWAUKEE JOURNAL

Milwaukee, Wisc., September 14, 1981

You don't have to be a Star Trekkie to appreciate the mystery and splendor of the universe. Those pictures beamed back to Earth by Voyager 2 and its predecessors surely made even the most cynical pause to contemplate the awesome majesty of our neighbor planets.

But now the more mundane concerns of Earth have spelled an interruption — a "hiatus," if you will — in active space exploration by the US. Oh, Voyager 2 is still out there, speeding silently toward Neptune. And the Galileo journey to Jupiter is still precariously scheduled for launch in 1985. But their targets are five years and more in the future. In the meantime, the nation's space scientists and engineers, who helped take all of humankind where it has never gone before, will look for work elsewhere.

It is difficult to quarrel sharply with those who find space exploration a luxury in this era of austerity. The squeeze on the federal budget, continued national security threats from near and far, the critical need to see all Americans adequately fed and sheltered — these certainly must take center stage.

But let's not forget what has been gained from the space adventure. There have been the "spinoffs" in everyday things — such as Teflon on our frying pan and the microwave in our kitchen and the minicircuits in our television set. Remember, too, the advances in predicting the weather, and better coping with it. But, above all, there has been the enlarged knowledge of the universe — an intellectual expansion of inestimable value. What great mysteries might this enhanced knowledge help unlock in the years ahead? Perhaps the most important benefits of space exploration are yet to be derived.

Since the beginning of time, conquering new frontiers has been a passionate pursuit of humans. The admirable urge will never cease. And someday, when the nation's economic house is in better order, America can again take up the challenge put forth by e.e. cummings: "Listen. There's a hell of a universe next door. Let's go."

The Times-Picayune
The States-Item

New Orleans, La., January 10, 1981

We do not lack examples of citizens complaining about paying taxes so that the government can do things the complainers don't want it to do. But now we have an example of citizens voluntarily paying extra for what they *do* want the government to do.

In a ceremony at the National Air and Space Museum in Washington Wednesday the National Aeronautics and Space Administration was presented a check for $60,000, contributed to by 10,000 individuals, and a pledge of $40.000 more. It was to supplement NASA's thin budget so it could continue for at least two more months to receive, log in and study the signals still being sent fom the Viking I landing craft on Mars.

Viking I was the first spacecraft to undertake detailed studies on the surface of another planet. Its star turn was scooping up Martian dirt and processing it for signs of life, but every 37 days since its landing July 20, 1976, it has been automatically sending back photos and other data. It could continue to do so for the rest of this century, but NASA has had to stop listening because it can't stretch its budget that far. It has also shut down some moon probes that could still transmit.

The Viking messages are no longer full of surprises, and George E. Cranston, executive secretary of the American Astronautic Society, says, "We really don't expect some big discovery." We wouldn't expect drama from a weather station in Trapper's Toe, Wyo., either, but its routine data would add to what forecasters needed to serve the public. A science station on Mars will add far more effectively to our rudiementary knowledge of everyday conditions on that planet.

Tight budgets are a national imperative these days, but the space budget is more a long-term investment program than funding for an immediate service. A good deal of the difficulties with the space shuttle, currently three years behind schedule, has been such a small budget that there was no room to plan for problem-absorbing backups and spare parts. Perhaps the most crucial factor favoring substantial budgeting for the space program is the need to maintain an experienced, high-technology work force of engineers, scientists, technicians, analysts and others whose simple existence, not to mention the fruits of their labors, is a national asset of perhaps critical importance.

THE ATLANTA CONSTITUTION

Atlanta, Ga., December 9, 1981

The quote is an interesting one: "The space program was born of politics, has declined because of politics and will revive only when the space constituency becomes directly involved in politics." The statement was made by Thomas Frieling, 28, of Bainbridge, Ga., who is executive director of Campaign for Space Political Action Committee. And, come to think of it, Mr. Frieling is probably right.

The U.S. space program, for sure, didn't really take off until President Kennedy promised to put a man on the moon before the 1960s ended. Once underway, the space program continued to accelerate while there was extensive public and political support for it. And it has declined in recent years, in major part, because of a lack of that support — or, at least, public demonstrations of that support.

Despite the great success to date of the space shuttle, the Reagan administration and Congress are moving to cut back on expenditures for space exploration. There's little argument that, if the federal budget is ever to be balanced, nearly all aspects of government spending has to be slowed. But it's quite obviously foolhardy to cut back too much on the U.S. space program.

The pro-space argument can be made that the "future" is in space, and indeed it is. But a much more practical argument can be made — not only made, but demonstrated — that the space program already has contributed greatly to the improvement of life on Earth. Numerous examples of improved medicine, nutrition, communications, computers, aircraft, weather forecasting and appliances that had their genesis in space program developmental work.

Freiling predicts that the space program will revive when people see the space program "affecting their daily lives." It does now, of course, but perhaps not so obviously to many persons. But it's very obvious to persons with interest in space, and a number of pro-space lobbying groups have organized around the nation. These groups range from organizations headed by "unknown" space buffs, such as Freiling, to nationwide "celebrity" committees, made up of such persons as John Denver, Carl Sagan, Johnny Carson, Isaac Asimov and James Michener. These groups are increasingly beginning to "lobby" with senators, representatives and members of the Reagan administration for increased emphasis on space exploration, not only maintaining the current program but boosting it.

In Washington, there's a lobby group for almost all "special interests." But the space program is a special interest for all of us.

THE CHRISTIAN SCIENCE MONITOR

Boston, Mass., March 5, 1981

The United States is entering a new phase in the development of the space frontier. The Space Transportation System, to use the official name for the reusable space shuttle, will do more than launch satellites cheaply. It will open up new possibilities to explore the manufacture of special materials under weightless conditions. And it will reestablish the United States in the fast-developing field of manned space flight – a field in which the Soviet Union now is the world leader.

Thus the shuttle is very much a doorway to the future in space. Yet, as the series concluding in today's Monitor bears out, the United States seems reluctant to pass through it and uncertain of what it wants to do in the new world of space exploration and practical applications that lies beyond.

For a decade now, the space posture of the country that first put men on the moon has been impaired by lack of consistent policy. An undoubted capacity for achievement is underutilized. The major challenge from the Soviet Union is the persistence that has enabled it to carry on with a long-term program. This has given it a formidable space capacity, including what amounts to a prototype for an operational manned space station.

There is much concern that President Reagan's proposed cuts in the US space effort will emasculate vital programs. Certainly it would be unwise to cut back on planetary research to such an extent that the country loses its hard-won capability in this field. The Jet Propulsion Laboratory, which carried out many of those spectacular planetary missions, now feels its very existence threatened.

Certainly, too, it would be dishonorable for the US unilaterally to pull out of international agreements, especially when the other partners have already spent a great deal of money and rearranged their own priorities for the joint ventures. Yet this is what is threatened by proposals to drop the US half of the Solar-Polar Mission to send twin spacecraft over the poles of the sun and to restrict US participation in Spacelab missions. Spacelab, a capsule carried by the shuttle in which scientists can work in orbit, is Western Europe's contribution to the shuttle program.

Yet, while it is easy to criticize what appears to be thoughtless budget slashing, there is a larger issue. The proposed cuts are thoughtless precisely because there is no guiding policy – no long-term goal – to give budget planners an overall perspective.

The US never made up its mind what it wanted do after the stunning success of the Apollo moon program. There was talk of focusing on practical applications. In the early 1970s the National Aeronautics and Space Administration, together with the administration and the Congress, worked out a strategy for such development. This gave a central role to the shuttle as an efficient way to launch spacecraft. It included a commitment to constant budgeting, meaning the same level from year to year adjusted for inflation. Unfortunately, the concept did not even last a year. NASA has lived in a hand-to-mouth manner for a decade. On-again, off-again programs have proved wasteful in many areas. This has led to unilateral changes in international projects that have angered European partners long before the present cutbacks were threatened.

Thus the challenge facing the US as the Columbia prepares for its maiden voyage is to get its act together, set some meaningful goals which it can pursue consistently, and fund the effort at a sustained level. Important as it is to restrain federal spending, the administration should give very high priority to defining such goals. Meanwhile, it should be careful not to undercut the country's space capability with hasty budget trims.

The Morning News

Wilmingham, Del., January 12, 1981

The good news is that private citizens are donating their hard-earned dollars to help the National Aeronautics and Space Administration to carry on. The bad news is that they have to.

NASA is taking handouts like a waif while the U.S. Postal Service is advertising for a *distribution clerk, machine.* Applicants need have no experience, educational background or interests or anything else in particular except for some manual dexterity and adequate eyesight: basic starting salary, $18,282 per annum. That's 121,880 stamps at 15 cents each. Irrelevant?

On Wednesday, meanwhile, a check for $60,000 was handed over to Dr. Robert A. Frosch, NASA's administrator. The money came, as the Associated Press put it, in dimes and dollars donated to the Viking Fund, a San Francisco-based volunteer organization that has received contributions from more than 10,000 people nationwide since it was formed about a year ago. The fund also has promised Dr. Frosch about another $40,000 before spring.

Small change. The $100,000 is small change.

With paid vacations, sick leave with pay, insurance, cash for suggestions, paid holidays, premium pay for nightwork and other benefits, it wouldn't cover more than three Postal Service *distribution clerks, machine.*

It wouldn't buy many of the 30,761 tiles that cover the fuselage of the space shuttle Columbia, which NASA rolled to its launching pad as the New Year dawned.

The Columbia is NASA's showpiece. Its construction program was plagued with delay; it is two years behind schedule. But it is *built* and soon we will see if it will fly. What other government agency can claim as persuasively as NASA that it knows what it is doing?

The Columbia has cost $8 billion and change and it has been described as a penny-pinching program. The shuttle's liftoff, however, will put the United States back spectacularly into the manned spaceflight business. The two astronauts on the shuttle's first flight, possibly in March, will be the first men sent into space by NASA in almost six years.

The Columbia program, on the other hand, is almost irrelevant to NASA's present financial straits.

What it needs, and what the supporters of the Viking Fund were concerned about, is money enough to insure that the data collected on the unmanned flights to Mars and beyond can be processed.

Administrator Frosch remarked that the gift to the government shows that the public may be ahead of the nation's leaders in recognizing the importance of space exploration.

"As important as the money," he said, ". . . is the expression of a deep and abiding commitment by many people to a strong and ongoing space effort."

Amen to that.

The Pittsburgh Press

Pittsburgh, Pa., October 25, 1981

Inasmuch as almost every other federal agency is being asked to take its budget lumps, there's no reason the National Aeronautics and Space Administration should be an exception.

NASA is under orders to lop $367 million from its fiscal 1982 spending, $1 billion from its budget for fiscal year 1983, and another $1 billion from fiscal 1984.

The question is where the cuts can or should be made.

One option being considered by the space agency is to turn off the Voyager II craft, which sent back those amazing closeups of Saturn in August and is now heading toward a rendezvous with Uranus in 1986 and Neptune in 1989.

NASA figures it would save $222 million over the next eight years in salaries for Voyager scientists and engineers and in the expense of maintaining the worldwide network of deep-space antennas used to communicate with Voyager II.

There aren't many other options. One would be to scrap the Galileo mission to Jupiter planned for 1987, which would save about $300 million.

Another would be to forget the Large Space Telescope scheduled to be put in Earth orbit in 1985.

Still another: Delay indefinitely the planned 1985 delivery of the fourth space shuttle, for a saving of around $1.2 billion.

Although Voyager is the easiest choice, to abandon it now would represent a dismaying waste of effort and opportunity.

The spacecraft is already on its way to Uranus and Neptune. It will be generations before the outer planets are lined up in such a way as to make another such mission feasible. Tracking, guidance and communications facilities on Earth are in place.

And when you stretch out the savings over eight years, scrapping Voyager would net less than $28 million a year.

Whatever NASA decides, however, it is clear that the glory days of space exploration are over, at least for this decade. 'Tis a pity.

The Ann Arbor News

Ann Arbor, Mich., September 13, 1981

HEAVEN KNOWS the country's space program has seen better days and more secure funding. Even the race to the moon seems like ancient history.

The tight NASA budgets of recent years and the need to concentrate money on development of the problem-plagued space shuttle have left big gaps in our space outreach.

Even so, the stunning successes of the Voyager probes and the launch and re-entry of Columbia earlier this year show how technology continues to enhance our lives and expand knowledge.

The consumer benefits of the space age are now schoolboy stuff. We already take for granted the spinoffs from space research and exploration such as Teflon pans and satellite communications.

The question before the house is whether we proceed to close the space window having just opened it a crack.

TV ASTRONOMER Carl Sagan personifies Americans' revived interest in space and space exploration.

Sagan was a member of the team of scientists that analyzed the most recent Saturn data. He hailed the turnout of reporters that later crowded into Pasadena's Jet Propulsion Laboratory as evidence the news media "are catching up with public interest in scientific exploration...."

Then there are the words of Esker Davis, Voyager 2's mission director:

"What Voyager demonstrates is that we can build the automated robots to do anything we want to in this solar system. It's not a hard thing to do. We can fly to these planets, we can land on them, we can orbit them. We know how to do that and do it well."

WILL THE PUBLIC support space research, admittedly a costly undertaking?

Will Americans seeking tax cuts, coping with inflation and hoping for better times fund the likes of space probes with their tax dollars and demand that their government alter its spending priorities?

The fact is we're into space already...with our growing solar power industry. All we have to do is make solar work on a massive scale.

Space exploration is merely one more expression of man's questing nature. Clearly, we'll never know what space offers in energy resources or colonization possibilities or other benefits unless the space program goes ahead.

Billions of miles across space, man's tiny machines talk to us. Just as airplanes and jets shrank the globe, these machines shrink the solar system.

It's important to try to understand that system, man's place in it and its potential for good.

The Miami Herald

Miami, Fla., October 17, 1981

ANYONE who hasn't felt a sense of pride, excitement, or simple fascination at America's exploits in space just hasn't been paying attention. Space is a barely charted, vast frontier. It holds clues to man's existence.

Yet planetary missions of the National Aeronautics and Space Administration (NASA) could be scrapped. Reagan Administration budget cuts dictate taking $367 million from NASA, about 15 per cent of its overall budget. For fiscal 1983, the cuts will be $1 billion, and for 1984, another $1 billion will be sliced.

The White House has pledged full funding for the costly space shuttle, a big-ticket item, and NASA agrees with the commitment. NASA's other large programs where cuts of this magnitude can be made are the Galileo mission to Jupiter in 1987 and the Space Telescope program, which will fix a huge telescope in earth orbit in 1985. And, of course, the Voyager mission.

NASA officials have said that Voyager could be the program they recommend for cutting. Already, the mission to Halley's Comet has been axed.

A few weeks ago, photos from Saturn taken by Voyager added exponentially to knowledge of the solar system. Voyager is now on its way to Uranus and could go on to Neptune by 1989. Pulling the plug on Voyager now would save $222 million over the next eight years. A similar savings could be made by forgetting the Galileo program.

It would be disturbing if the space program is forced to eliminate any of these programs. Yet there may be no alternative. NASA's next "exploration" must examine the space between its scientific aspirations and the funds it will have to finance them.

Fort Worth Star-Telegram

Fort Worth, Texas, January 14, 1981

At a time when most citizens are complaining about how much money the government takes out of their pockets, it is refreshing to learn that about 10,000 people are voluntarily sending money to the government to assist a program that is dear to their hearts.

The people, ranging from school-aged children to *Star Trek* fans to aerospace engineers, have raised $100,000 to help the National Aeronautics and Space Administration, which is operating under a tight budget that is likely to grow tighter in the near future.

The money dribbled in in varying amounts—dimes from school kids and large checks from concerned older citizens — and the money will be used this summer to analyze data from NASA's two Viking missions to Mars.

Viking 1 and its twin, Viking 2, orbited Mars in the summer of 1976 and returned more than 52,000 pictures and much other data, some of which has never been analyzed because of a shortage of funds.

So the Viking Fund, a private group made up of volunteers and based in San Francisco, began soliciting donations from private citizens to provide funds so that NASA could continue the analysis of the Viking voyages.

The money, of course, is important, but perhaps the symbolism is even more-so. Contributors to the Viking Fund are to be applauded for digging deep into their pockets for a cause in which they obviously believe deeply.

To resurrect some terminology that was overused during the dawn of the space age, such conduct by concerned citizens is A-OK.

San Francisco Chronicle

San Francisco, Calif., January 9, 1981

A CHECK FOR $60,000 was turned over to the federal government the other day to keep an aspect of the Viking exploration project on Mars going. The action was something of a historic switch, and it carried a message: let's not chop the space program any more. The amount of money wasn't much, of course. Federal grants running the other way tend to total a great deal more. But it represented donations from 10,000 individuals. And the spirit of the giving tells us a good deal about the zeal that still supports this country's ventures into space.

These dollars — many of them provided by space workers — will pay for at least two more months of study and analysis of reports being sent out from the surface of Mars by the Viking 1 lander, the first spacecraft to undertake such studies on another planet. Viking's famous soil scooper no longer prods the Martian crust, but every 37 days the robot dutifully dispatches photographic and other information to Earth.

VIKING'S MESSAGES are no longer full of surprises. But they should not come sailing through the ether only to be ignored by a nation whose political establishment has turned stingy toward space exploration. As George E. Cranston, executive secretary of the American Astronautical Society, a professional group that helped to organize the fund, said: "... if you think about it, you have something almost alive out there — on Mars — and it keeps on handing out signals. It kind of appeals to the imagination."

It does indeed. The whole concept of learning more and more about what is "out there" appeals. This extraordinary banding together of space workers and other enthusiasts to prolong Viking's functional life with money from their own pockets makes the point well. As the scientist who presented the check said: "Our message is hands off the space program if you want to cut the budget." That message should be listened to on Capitol Hill.

THE INDIANAPOLIS NEWS

Indianapolis, Ind., October 30, 1981

While David Stockman and Co. figure and congressmen calculate and confer in order to come up with more budget cuts, many programs are left in limbo, including the space program.

The people at the National Aeronautics and Space Administration (NASA) have yet to learn what their budget will be. "OMB (Stockman's Office of Management and Budget) is being very quiet," said Fletcher Reel, a public information staff member. NASA "may be acting on a continuum until January," he said.

Reel couldn't say for certain what projects would survive in the new budget. It has been reported that Voyager 2 may be left unmonitored for the remainder of its fantastic voyage — to save some $220 million. But Reel couldn't confirm that cut, either.

He did say the chances for sending a probe to explore Halley's comet when it passes in 1986 are slim. "Everybody's kind of ruled that out," he said. That rewarding project is not high on NASA's list — not when the agency is struggling just to maintain its major projects. But it should be.

NASA should not have to scramble for survival.

True, these are times for fiscal austerity, particularly in government programs. But the space program should not be starved. There are plenty of other places to trim, as we have pointed out on many occasions.

The space program has been a great technological adventure, as well as an economic advantage for this country. But a story in Aviation Week & Space Technology points out another reason for maintaining a strong space effort: The Soviet Union reportedly has put into orbit an antisatellite battle station equipped with interceptors that could destroy U.S. spacecraft.

A chilling thought. But what's scarier is the idea that the U.S. may not be able to counteract such a weapon because of NASA cutbacks.

We don't know if that's the case. We do know the combined efforts of the Carter and Reagan administrations and Congress have applied the budget brakes to the space program. It is not a wise cutback.

Building on what's been accomplished in the last two decades of planetary exploration, there are significant economic and military frontiers yet to be conquered. This is not a time for retreat in space.

The Boston Globe

Boston, Mass., December 24, 1981

Among the various targets of the Reagan Administration in its attempts to trim deficits in the fiscal 1983 budget has been NASA's deep-space planetary program, including the next exploratory flight to Jupiter. The savings involved are partly illusory. The cuts should not be made.

Eliminating the flight will be wasteful in two ways. It would lead directly to the dismantling of the nation's deep-space scientific team of about 1200. The nation wants a space program and it needs a team like that working at Jet Propulsion Labs. Reassembling such a group will be difficult and expensive.

Furthermore, the Administration's proposals would simply throw away $300 million already invested in the flight, dubbed Galileo, to save $300 million for completing the effort.

The style in which the reductions may be made are instructive about the way the Administration is handling budget decisions in the area of science and technology. The office of Management and Budget claims that NASA is getting a fixed amount of money and has the right to set its priorities within that total. NASA reportedly blames OMB for the proposed cuts, in effect denying that it enjoys autonomy in setting its own program priorities. NASA needs no badgering.

The Galileo decision is of a piece with another Administration cutback, the two solar-polar flights scheduled for 1985 or 1986 in co-operation with an 11-nation European consortium.

In eliminating the US spacecraft, the Administration stranded European experiments that were to be carried on the American shot. The Europeans have honored their commitment to carry American experiments rather than bump them for their own stranded experiments. There is some danger that the Administration will fail to approve funds for the launch of the European craft, as originally agreed, and that the whole program will be lost – at a loss to the Europeans in excess of $100 million for their expenditures so far. Such a snub would be gratuitously insulting and should certainly be avoided.

More fundamentally, the Administration is building a reputation for hostility to basic scientific research. In other budget-cutting situations, the Administration has claimed other ways were available to handle the problem: states to develop social programs if they really want them; private charity to replace public assistance.

Those alternatives are feeble reeds, but were at least alternatives. In space science, there is no substitute for federal funding. Without funding, there will be no deep-space program and the United States, which has built up an impressive leadership role over the past 20 years, under Democratic and Republican Administrations alike, will relinquish that leadership role.

THE SAGINAW NEWS

Saginaw, Mich., August 27, 1981

A billion miles away, Saturn and its storied rings were as close as our living rooms Tuesday night as the Voyager 2 spacecraft sent back pictures from a mere 63,000 miles. It was enough to make you feel downright small, a speck in the cosmos — which, in fact, is precisely what we are.

But the vision of the mysteries around us should not diminish the human race in its own eyes. The farther we range, the larger will be our understanding of who we are, and what our true place is.

Space has been called our last great frontier, but the journey of Voyager 2 is more than exploration for its own sake. Nor are its glimpses of our planetary neighbors a mere transitory thrill, to be experienced, perhaps marveled at, and forgotten.

It is our future, and it is our past. From Voyager's wanderings, we are learning more about where we might have come from — and we may be seeing our eventual destiny. Who can believe that someday, humans will join the voyage? Who, knowing what we have achieved in just this incomplete century, can believe that they will not?

"Star Trek" and "Star Wars" were pleasant diversions. But this week we have witnessed truth, not fiction. It was not always clear what we were seeing. It will take long analysis to determine the meaning of what we did see. The fact that it was real, coming to us from such a distance that the signals took an hour and a half, is what stretches even the widest imagination.

We wish that Channel 19 at Delta College had found it possible to join many other public TV stations in expanding on the coverage provided by the commercial networks. If it can show us the finals of a softball tournament in Saginaw, there ought to be room on the schedule for the universe. But then, who's perfect? Certainly not Voyager 2, as its camera problem reminded us.

Our consolation is that Voyager 2 will continue. Our hope is that the human voyage it represents will not be curtailed.

It does not come free, a point some down-to-earth politicians are making much of. But if we wait until all our needs on earth are solved, we will never know how great our opportunities may be out beyond the stars.

The Wichita Eagle-Beacon

Wichita, Kans., October 19, 1981

Reports recently emanating from the National Aeronautics and Space Administration that the deep-space probe Voyager II might have to be shut down because of budget strains would stand as a monument to shortsightedness if they were to prove to be true.

Voyager II completed a spectacular fly-by of Saturn less than two months ago, and now is headed for a 1986 rendezvous with the planet Uranus and a 1989 pass by Neptune. It will take years for scientists to collate and decipher the vast body of new information the spaceship and its sister ship, Voyager I, radioed back about Saturn.

To let the opportunity pass to collect even more data on the Earth's more distant planetary neighbors, even if it would mean short-term savings in expenses, would be to throw away a major investment already made in the mission.

A NASA space sciences official has said no final decisions have been made, and that it is highly unlikely Voyager II will be abandoned. For the sake not only of scientific knowledge, but of common-sense fiscal responsibility, it must be hoped he's right. The possibility of quitting before the job is done is one that never should have come up.

TULSA WORLD

Tulsa, Okla., October 19, 1981

ELSEWHERE on this page, Columnist Art Buchwald raises a humorous argument against President Reagan's plan to cut back on space exploration. But the prospect of a second-rate aerospace research program isn't altogether funny.

Through this admittedly expensive effort, American scientists have raised — ever so slightly — the curtain of ignorance that has always blocked humanity's vision not only of the universe but of our our own planet. To save a comparatively small sum of money, the Reagan Administration proposes to lower that curtain. It is shortsighted economy.

Oddly enough, most of the opposition to space exploration, until Reagan came on the scene, came from liberals. Their argument: Why spend money on rocket ships and radiotelescopes when people are hungry?

The same argument could have been made against much of the important scientific research throughout history. Hunger was common in the world when France's Madame Curie began her costly, government-supported experiments with radium. And there was even more poverty abroad in the land when Leonardo was living on the subsidy of Italian princes and fooling around with drawings of flying machines and other far-fetched scientific subjects.

In the long run, the poor would have been much worse off if the liberal argument against scienctific exploration had prevailed.

Buchwald is right. It would be a shame to pull the plug on Voyager. It would, in fact, cheat future generations out of something much more valuable than a few million dollars

Los Angeles Times

Los Angeles, Calif., August 27, 1981

Voyager 2 has hurtled past the rings of Saturn and filed another dazzling report and plunged back into inky space, racing toward the outer edges of the solar system.

And, once more, the people that the spacecraft left behind on Earth are bombarded with images and impressions of the vastness and mystery of the universe that make the senses ache and make mankind seem rather noble after all.

It took Voyager four years to cover the billion miles between Earth and the great bubble of gas that the astronomer Galileo could barely make out through a crude telescope only a dozen or so generations ago.

Now, people on Earth can see Saturn and the collection of cosmic debris that makes up its rings far more clearly on television than they can see their own moon from the backyard.

As Times science writer George Alexander put it, the spacecraft crammed a lifetime of science into a few hours of photography and scientific measurement when it grazed Saturn Tuesday night at 33,000 m.p.h.

The mission was a scientific marvel that surely must thrill the men and women who built and launched and nursed the spacecraft to its target.

But what is it about these ventures that touches the spirit of people who do not even know enough about space to be baffled by what they see, as some of the scientists are themselves?

Surely it is not the spacecraft itself—an awkward lump with booms and antenna all akimbo, looking like something that the kid next door did with old coat hangers.

Most people could live happily not knowing that it is so cold around Saturn that helium gas in the planet's atmosphere condenses and falls like rain.

They now know that there is a tiny moon called Hyperion that is shaped like a can of tuna and circles Saturn at a cockeyed angle and has been beaten to a pulp by asteroids over billions of years. But people always knew that it was rough out there.

The best answer that we have seen comes from Carl Sagan, the poet laureate of space. There are many things on Earth that divide mankind, and many pleasures and sorrows that cannot be shared. But the one thing that all people have in common is the great unknown that lies all around Earth. They can all share the lift and the sense of pride that come from rummaging around in the unknown, one step at a time.

The U.S. space program has been winding down in recent years, with bursts of enthusiasm for projects like the Voyager mission to Saturn coming less and less often.

The nation must reverse that trend in this peaceful use of its highest technology. It must find the resources to reach farther into space, and more often.

When Voyager 2, booms akimbo, makes its final pass at a planet five years from now, it will sail free of the solar system, carrying a record of greetings into the unknown.

The record was placed aboard the spacecraft in hopes that a life form beyond our stars might find the message and puzzle over it. Whether that happens is not the important thing. The important thing is that mankind sent the message

ST. LOUIS POST-DISPATCH

St. Louis, Mo., August 30, 1981

College freshmen often learn first how much they have to learn, which is the way scientists view results of Voyager II's space trip past Saturn. Or, as one said, "Once again we underestimated everything."

There, from the mission's cameras, were pictures of some of the planet's 17 moons, one the size of the planet Mercury, another shaded half-white and half-black, another shaped like a hamburger patty. There were those mysterious rings, some braided, and one of them most visible from Earth seems to have a vertical thickness at its edge of only 500 feet. And, of course, there was Saturn itself, its clouds swirling at 1,100 miles an hour, with ammonia rain possibly falling over much of the gaseous planet. So the telemetry brought back both information and puzzles.

From the layman's point of view, though, much of the drama of Voyager II lay simply in its 950 million-mile trip — so distant that its radio signals reached earth nearly an hour and a half later. And now it is heading on a 1.5 billion-mile voyage toward a date with Uranus in 1986, and what then for space exploration? Well, in 1986 Halley's Comet again will approach Earth, but budgetary restraints make it appear unlikely that the United States will join West Europeans, Russians and Japanese in a space rendezvous with the comet. America's great space effort is slowing down.

In a way that is odd, because since man walked on the moon, polls have shown that an increasing number of Americans support the space program. Meanwhile the budget for the National Aeronautics and Space Administration has shrunk by 42 percent when inflation is taken into account. But as Voyager II indicates, it is not necessary for man to walk in space to persuade the public, as well as scientists, that the nation should have a continuing role in space pioneering.

The State

Columbia, S. C., September 3, 1981

THE AWESOME achievement of Voyager II's billion-mile mission to Saturn must not be the "last hurrah" of America's exploration of outer space, but that is a possibility.

Our aerospace scientists and managers expressed apprehensions about the future of the space program even as Voyager flashed past the thousand rings of the second largest planet, shooting back to Earth magnificent pictures of things never before seen by man.

Where there should have been euphoria at the Jet Propulsion Laboratory in Pasadena, Calif., for our technological triumph and harvest of knowledge, there was bridled excitement. They knew the National Aeronautics and Space Administration will not escape the President's fiscal ax.

The Administration has pledged, of course, that our space program will not dry up. The Administration is "very receptive" to the space program — within budgetary restraints. But who is to say now how deep fiscal cuts will impair this nation's most successful peacetime undertaking?

It is a time for setting priorities. But in the admirable struggle by the Reagan people to seize control of the economy and national budget, we fear nearsightedness. The explorations of space, the development of technologies, the extraordinary planning must be started years in advance of the launches. Voyager II has been four years on its way to Saturn, and now heads to Uranus, arriving by 1986, and Neptune, with a 1989 flyby schedule.

The United States is the pioneer in exploring our planetary system. Ours is a position of prestige and leadership in the world, where we have slipped otherwise. The benefits of our space program are shared by people the world over, in terms of knowledge of our universe and in the development of new and useful Space Age devices.

Our satellites help to forecast weather all around the world, identify the mineral wealth of the lands and the seas, and provide instant communication among the nations. There is a prodigious new medical technology springing from our manned space flights.

Literally thousands of new consumer products are spun off from our space program.

We do not argue that NASA should be immune from budgetary cuts — nor should we allow the Defense Department to go scot-free, but that is another matter. But any trimming in the NASA budget should not impair this nation's progressive, active space program.

The budget cutters should know exactly what they are doing, and what the short- and long-range results and ramifications will be. And those who know best, the scientists of NASA, should not hesitate to speak out if they disagree. The United States has never mounted an undertaking that has provided greater benefits to mankind than the space program. We must not cease the efforts.

The Houston Post

Houston, Texas, October 23, 1981

The U.S. unmanned space program accounts for only 3 percent of NASA's budget, which is itself less than 1 percent of the federal budget. But the latest administration spending cut proposals could ground our exploration of the planets. The White House Office of Management and Budget wants to cut $367 million from the NASA budget in the current fiscal year and another $1.1 billion in fiscal 1983 without reducing funding for the space shuttle, which gets the lion's share of the money.

The proposed new fiscal '82 reduction in the NASA budget would be on top of a 10 percent cut voted earlier. But funding for NASA's Office of Space Science, which operates the unmanned space exploration program, was slashed 20 percent. The second round of cuts would take another 15 percent. David Morrison, chairman of the American Astronomical Society's Division of Planetary Science, warns that further cuts may end two decades of unusually successful interplanetary missions.

Ending the program now would mean shutting down the Voyager II spacecraft that recently sent back such spectacular pictures of Saturn and is headed for a rendezvous with Uranus in 1986. The only new space mission now being planned — an orbiter and probe of Jupiter, on which $300 million has already been spent — would be scrubbed. Plans for any future interplanetary missions would be ended, along with further analysis of data already gathered by our interplanetary probes. And the team of scientists that has operated the program would be disbanded, costing us invaluable expertise that would be extremely difficult to recoup.

Before the program is killed for economy reasons, those who must make the decision on its life or death should consider the contributions those ingeniously designed spacecraft — the Mariners, Vikings, Pioneers and others — have made to our knowledge of our solar system and the universe. The probes, packed with sophisticated instruments, have raised as many if not more questions than they have answered. Yet the data they have gleaned from the other worlds of our solar system have already begun to change the way scientists look at our world in the context of the universe.

The results of NASA's interplanetary exploration have been made available to the public in an intelligible way that has caught the imagination of the man and woman on the street. The program has wide popular support. And, as with the manned space program, the unmanned flights have also produced technological spinoffs with down-to-earth applications. The unmanned space program, in short, has been a remarkably sound investment of our tax money. Seldom has a federal project produced so much for so little.

Apart from its contribution to basic scientific knowledge and its technological fallout, however, the program should be kept alive for another compelling reason. The Soviet Union has its own ambitious unmanned space program, which includes a planned mission to inspect Halley's Comet when it swings by Earth in 1986. Western Europe and Japan plan similar missions, but the United States may be conspicuously absent because of a lack of funds. One mission more or less will not make or break our program of interplanetary research. But to sacrifice our lead in exploring this new frontier would be a false economy that could ultimately cost us far more than we saved.

America Urged to Get Ready for Approach of Halley's Comet

Halley's Comet, subject of legend and superstition since earliest times, passes the earth every 76 years. As the 1986 rendezvous date drew near, astronomers around the world embarked on preparations to send up satellites for a closer look. The National Aeronautics and Space Administration, however, disqualified the U.S. from the start, saying it would not have enough money to launch a probe. President Reagan's attack on federal spending was greeted with enthusiasm, but as usual, complaints arose the moment a favorite program was threatened.

Lincoln Journal

Lincoln, Neb., July 3, 1981

Halley's comet is THE comet of history, romance and legend. Humans have been wondering about it at least since 240 B.C., when it supposedly was first seen. It sails through our solar system every 76 years, exciting wonder and fear. In 1456 Europeans were so frightened by it that Christian churches added a prayer to be saved from "the devil, the Turk and the comet."

Today our civilization has the capacity, both in understanding and in equipment, to learn more from the comet's visit than ever before. So it is unthinkable that the United States should look the other way when Halley's shows up on schedule in 1986.

The National Aeronautics and Space Administration would like to launch an unmanned space probe that would intercept the comet, snap some photographs, sample the gases that make up its tail and maybe even fly alongside it for several months.

If NASA is to do this, however, it must start planning now. And neither the Carter administration nor the Reagan administration has been very receptive to new space ventures. The U.S. House, to its credit, is a little more enlightened. The other day it voted to spend $5 million to keep the possibility of a probe alive.

That's not very much, considering that even a modest project of this nature would cost perhaps $350 million. But that figure, in turn, if spread over five years would represent less than 1 percent of NASA's budget.

The Senate should go along with the House action. And President Reagan should indicate a willingness to approve funds for this purpose in future years.

Halley's Comet is coming to call. European space shots will be on hand to greet it. So will one from Japan. And undoubtedly the Soviet Union will put in an appearance. The United States, the world's space pioneer, would be conspicuous by its absence — and most neglectful of a rare educational opportunity.

Detroit Free Press

Detroit, Mich., September 2, 1981

THE WORLD is expecting a distinguished visitor from outer space in 1985. The appearance of Halley's Comet is a once-in-a-lifetime event — it comes around only every 76 years. But it appears that the opportunity to find out more about Halley's is not going to be seized by the U.S.

Oh, others will take it. Japan, the Soviet Union and the European Space Agency are preparing to send spacecraft to meet Halley's. But the administration argues that the nation blew so much money on the space shuttle program that it can't afford the Halley's project.

A pity, although the $300 million the project would cost is a lot of beans. What an opportunity it would be to learn more about a phenomenon that has awed and frightened humans since prehistoric times.

There are millions of comets, and at least 80 of them are regular visitors of the solar system. But Halley's is imbedded in folklore the most deeply. It may have given rise to the story of the Star of Bethlehem, and was believed to portend the destruction of Jerusalem in 70 A.D. and the Battle of Hastings in 1066. It created panic in the U.S. when it made an unusually close visit in 1910.

Yet not much is known about the nucleus and the surrounding emanations of comets. Some scientists believe life on earth resulted from the deposit of organic molecules from impacting comet material.

Astronomers say that if the U.S. wants part of the comet action, preparation would have to begin immediately. Halley's has already entered the solar system and is midway between Uranus and Saturn.

It is not to be unless somebody relents quickly. Halley's will not be around again until 2061. Quite a few of us will not be here then.

SAN JOSE NEWS

San Jose, Calif., September 8, 1981

FROM devices that could save the life of an accident victim to data that could unlock the secrets of life itself, America's space program has immensely enriched the store of human knowledge. Two examples — one millions of miles away, the other within the borders of our own county — illustrate how much we can gain from the program, and how much we stand to lose by its disintegration.

Take the close-to-home example first: the Stanford Medical Center Biomedical Applications Team (BATeam).

The Stanford BATeam has been in operation for 10 years. Its job is to identify medical problems and needs and then to hunt for solutions through the vast mounds of information NASA has accumulated as the result of its space programs. The search has often paid off. Among the successes are a wheelchair which allows its occupant to "talk" by way of a keyboard, a device for safely and continuously monitoring pressure inside the skulls of brain-injury patients, and magnetic measurement devices which allow doctors to monitor heart problems without dangerous and intrusive surgical or chemical procedures.

Last year, the Stanford team was budgeted at $200,000. Next year's appropriation is zero. The program will cease to exist on Dec. 31, 1981.

The far-out example is Halley's comet, which will be making a trip through Earth's neighborhood in 1986. The comet flies by only once every 76 years. The Soviets, the French, the Japanese, and the European Space Agency are all planning to take advantage of this once-in-a-lifetime research opportunity by sending unmanned probes, a recent edition of Discover magazine reports.

The United States has had more experience with deep-space exploration than any other country and could do this job better than anybody else. According to Discover, NASA Jet Propulsion Laboratory scientists believe they could get a probe within 600 miles of the comet's surface. But at this point, an acutely budget-conscious NASA is planning nothing. While the Soviet and other spacecraft are sending back the first close-up views of a comet, America will be sitting at home as a spectator.

However, as Discover points out, "Much more than injured pride is at stake. Comets are believed to be chunks of ice, primordial dust, and rock preserved virtually unchanged in the refrigerator of deep space since the birth of the solar system some 4.6 billion years ago. Some scientists have even suggested that organic molecules from comets that crashed into the Earth were the precursors of terrestrial life . . ."

NASA estimates the cost of an unmanned mission to Halley's comet at $300 million over the next five years — about 25 cents a year for each American. Unless some $25 million to $30 million is allocated in next year's budget, NASA will miss the launch date. Of course, we can always wait until the comet comes back again — in 2062.

The relentlessly businesslike Reagan administration probably considers unlocking the secrets of the universe a bad buy compared to, say, a few new tanks or a B-1 bomber. After all, where's the payoff, what's the bottom line?

There's something to be said for such hard-eyed realism, we suppose. Still, one wonders where President Reagan and the rest of us would be now if Ferdinand and Isabella had run a cost-benefit analysis on Columbus' crazy idea.

THE PLAIN DEALER

Cleveland, Ohio, August 28, 1981

The remarkable performance of Voyager 2, following two decades of U.S. technological triumphs in space exploration, is as good an argument as can be made for commitment by this country to a close examination of Halley's Comet when it returns in 1986.

The bottom-line purpose of the U.S. space program has been to explore matter outside Earth's confines to try to learn more about the origins of the universe and of life itself. The visit by Halley's Comet to the inner solar system five years from now offers a unique opportunity to add significantly to mankind's knowledge.

The comet drags along with it through space a 50-million-mile-long tail of cosmic dust and gases that scientists believe to be as primitive as any matter in the universe. To capture some of that matter for first-hand inspection and analysis not only would be a stunning accomplishment but also would help further human understanding of the beginnings of things.

The almost complete operational successes ot spacecraft such as Voyager 1, which startled the world with the precision of its flyby of Saturn and the information it sent back to Earth last November, and Voyager 2, which now has taken a separate but no less spectacular look at the planet, proves that U.S. science is capable of doing the job. The question is whether the $300 million or so needed for the job can be made available in this troublesome economic time. A decision must be made in the next several months.

Cutbacks in the space program have been the rule as the United States has begun to concentrate on more mundane concerns. In addition, overruns in the project cost of the space shuttle damaged NASA's credibility — but, it should be noted, neither muffled the cheers nor dulled the pride felt when the Columbia finally flew flawlessly.

There can be little doubt that priorities point to use of federal resources for the basic needs of the American people. But there also should be room for funding programs that hold as much promise for adding to human knowledge as does a Halley's Comet mission. It is, in truth, a once-in-a-lifetime opportunity. The comet only comes as close as it will in the winter of 1985-86 once in 75 years. The next such visit will be in 2061.

There are compelling fiscal reasons for lowered priorities for a number of proposed space projects. But we believe the Reagan administration and Congress should take advantage of available expertise and provide the means for examining Halley's Comet up close. The Voyager flights and other achievements show that it can be done. All that is lacking is the will.

THE  SUN

Baltimore, Md., August 26, 1981

The major purpose of planetary probes, such as the U S Voyager 2 spacecraft that is having a close encounter with Saturn this week, is to uncover the history of the solar system, its various features and ultimately the cosmos This history exists on innumerable time scales, from the billions of years of the life of the universe to the milliseconds it takes for solar radiation to ionize an atom broken loose from a comet. Planetary probes can throw light on all of these processes. They represent one of man's most exciting scientific ventures. Yet the United States, the undisputed pioneer, may soon play a second-class role.

Measurements of chemical, electrical and physical aspects of the planets, their rings, satellites and other features lead to conclusions about their history. Venus, the second planet from the sun, for instance, may have developed a "tectonic plate" system of geology like earth's, in which giant continental plates gradually have floated apart on a semi-fluid sub-surface medium But Venus's surface is obscured by dense clouds, and only sophisticated radar "photographs" from a planetary probe will reveal its surface in detail enough to verify the tectonic hypothesis.

Chemical analyses of planetary atmospheres can be tremendously revealing about the origins of the solar system, because certain rare gases that are chemically inert have changed little since the solar system was first formed. But measurements of these gases are difficult from remote sensors such as the Voyager ones, and more intimate approaches are necessary. The U.S. Galileo mission to Jupiter in 1985 will probe the Jovian atmosphere directly.

Yet apart from Galileo and a 1988 mission to Venus to do radar imaging and other things, U.S. planetary scientists don't have much left on their plates. After a decade and a half of highly active U.S. planetary exploration beginning with the Mariner 2 flyby of Venus in 1962—and climaxing with the Viking landings on Mars in 1976—the U.S. almost seems to have lost interest. Meanwhile, Western Europe and the Soviet Union are stepping up their activities.

NASA should not be immune from budget restrictions that hurt others, such as the poor, even more. It would be a tragedy, though, if the U.S. ceased to be the world leader in solar system studies. The least the Reagan administration might do is give NASA funds for a probe of Halley's Comet in 1986, about as much money as it takes to build a relatively small dam out West. This probe could be immensely rewarding, because comets probably have not changed much since the early days of the solar system and therefore are a chemical record of what it used to be like. The U.S. needs to launch a Halley's Comet probe, in part to keep its preeminence in solar system science.

The Evening Gazette

Worcester, Mass., August 14, 1981

The Reagan budget slashes are welcomed by most Americans. But on a grander, cosmic scale, they are timed just wrong.

Halley's comet, thought by some to be the Star of Bethlehem, is due to cross the earth's orbit and swish its vast tail around the sun in February, 1986. The National Aeronautics and Space Administration (NASA) had hoped to televise and study it with instruments aboard a special spaceship powered by a revolutionary ion engine that could fly beside the comet at 36 miles per second.

But it appears now that the $5 million needed in fiscal 1982 to get the $300 million project started won't be forthcoming. The prevailing view in Washington is that you can't give people pictures of a dirty snowball from outer space when what they want is more money in their paychecks.

That estimate may be right politically, but scientifically deplorable. Study of this awesome comet isn't something you can put off for a year or so, like most space probes or the construction of a dam or post office. Halley's comet zooms past the planets and around the sun only once every 76 years. Its last visit was in 1910. It won't be back again until 2062.

Other nations are going ahead with space probes to study the comet, but they are much less ambitious than the NASA fly-beside plan. The Japanese, for their first venture into deep space, will fire two spacecraft toward it. The European Space Agency, a consortium of 12 nations, hopes to get a probe within 5,000 miles of it. And the Soviet Union plans two fly-bys within 1,000 miles of it bearing highly advanced cameras.

NASA apparently will have to be content with observations from space shuttles in earth orbit. Some American space scientists say this constitutes taking a back seat at the celestial event of the decade and portends the end of American supremacy in space exploration.

Maybe so. And Americans who have heard parents or grandparents talk about the comet's 1910 appearance, when it was four times as big and bright as the full moon, are likely to be disappointed with the 1986 visit. This time the earth will be positioned differently in its orbit, and the comet won't appear as large as the moon. If it turns out that no television close-ups are available, the long-awaited space show won't be much of a spectacular.

We have applauded nearly all the Reagan budget cuts. But this particular cut ought to be reconsidered. Failure to lead the world effort to glean scientific knowledge from this once-in-a-lifetime celestial event could prove short-sighted. It is an historic opportunity that won't recur for 76 years.

"Columbia" Makes Perfect Landing After Historic Flight

Three years late and costing $10 billion, the first reusable spacecraft, the shuttle *Columbia* made a perfect landing on a dry California lake bed April 14 after orbiting the Earth for 54 hours. The craft had lifted off Cape Canaveral launching center April 12, the anniversary of the first manned space flight, made by Soviet cosmonaut Yuri Gagarin. The flight of the *Columbia* was the first U.S. manned space venture in six years. Guiding the craft were pilots John W. Young, who had made four space flights, including two moon missions, and Robert L. Crippen, on his first space voyage after 15 years of training.

"We consider it a 100% successful flight," declared Donald Slayton, orbital test manager for the National Aeronautics and Space Administration. The *Columbia* and its booster rockets were designed to be reused, and both systems passed their tests. "I see no reason why we can't have 100 missions with this machine," Slayton added. During the 169-mile-high orbit, the shuttle opened and shut its cargo doors, demonstrating that it could be used to transport equipment into space. The critical problem was reentry into the earth's atmosphere, since friction would cause ordinary materials to burn up. The shuttle was protected with 31,000 ceramic tiles designed to withstand the intense heat, but several had fallen off during the launch. However, the ship landed safely without them.

President Ronald Reagan hailed the *Columbia's* flight as a "brave adventure" that "has opened a new era in space travel." Astronaut Young remarked: "We're really not too far . . . from going to the stars."

Los Angeles Times

Los Angeles, Calif., April 15, 1981

It is an awkward-looking machine, its wings too stubby for its high fuselage, its big and blunt rear end looking for all the world like a jetliner that had backed into a pile of booster rockets. But the space shuttle Columbia is a thing of beauty nonetheless, beautiful in conception and performance alike. If its takeoff from the Florida coast last Sunday morning was a spectacular sight, its landing in the California desert 54 hours 20 minutes 52 seconds later was a moment of grandeur. With that landing, a new era in the exploration and use of space began.

The Columbia will fly again—several score times or more, according to plan. That is its purpose and its value: to be, along with the rockets that help thrust it into orbit, a recyclable transporter of people and instruments, a laboratory for study and experiments. Its huge cargo bay, larger than any freight car, can haul up to 65,000 pounds of equipment. In time, it will not be necessary to launch space satellites with rockets pushing off from the earth. The shuttle will be able to carry them into their chosen positions. There will be other missions for the shuttle as well, in aid of science, of medicine, of national defense.

Columbia is the pathfinder for other shuttles that will be completed in the next few years. Some of the knowledge gained from its test flights will be the basis for modification designs in subsequent models. Years behind schedule in its development, billions of dollars more expensive than initially envisaged, the shuttle is now at last a reality of demonstrated capability and high future promise.

In time, shuttle flights will likely be looked on as routine, even as the later manned space flights of the 1960s and '70s came to be, even as the journeys to the moon ultimately were almost taken for granted. But there was nothing commonplace about the return of the Columbia from its first test mission on Tuesday morning. No one who saw John W. Young and Robert L. Crippen glide their great and complex craft into a perfect landing wil soon forget the sight.

It was a majestic end to a flawless mission, a stirring beginning to a bold new age.

The Boston Globe

Boston, Mass., April 15, 1981

A mingled sense of relief and accomplishment marks the historic landing of mankind's first reusable spaceship, exactly on time and in exactly the spot intended following a 36-orbit trip around earth. Columbia has landed and a new era has begun.

The flight itself was a technological triumph because it demonstrated beyond doubt that the highly complex spacecraft can indeed survive the rigors of launch, orbit and reentry into the earth's atmosphere. Astronauts Robert Crippen and John Young did some chores during their 54-hour odyssey but their outstanding contribution was a willingness to fly a pioneering trip. They flew in the face of frightening doubts about the integrity of the thermal tiles that were to protect them from the intense heat of reentry.

It remains to be seen whether the tile problem has truly been laid to rest. The spacecraft did survive the loss of about 17 tiles from the less-vulnerable tail section of the aircraft; the crucial tiles sheathing the belly of the craft have now shown that they can survive at least the first of the 100 flights planned for each of Columbia's sister ships.

Columbia has been a long time in coming. It left the launching pad at Cape Canaveral two years behind the original target date and two days after Crippen and Young first entered the craft expecting to soar into space. The waits were worthwhile and the care taken to provide maximum likelihood of success was a tribute to the concern for the safety of those who are to use this remarkable space system.

Young, Crippen and Columbia have done great service by providing the country with a fresh focus on space and the opportunities that may lie there. It is six years since the last Americans returned from the Skylab; America had in effect ceded manned space flight to the Soviet Union. Now, perhaps we can get moving again.

Space programs are, much to their credit, forward-looking rather than retrospective. The return of the shuttle will have done its most important job if it reignites support for the nation's future in space. The shuttle itself, which has already absorbed $9 billion in its development, is only on the threshold of regular operations. Its programs will be of most value if they help restore momentum to the rest of the nation's space program – notably the unmanned exploration of the solar system and such near-term projects as interception of Halley's Comet in May, 1986.

For the moment, though, it is proper to celebrate. Welcome home, John Young and Robert Crippen. It is good to have you safely back.

Oregon Journal

Portland, Ore., April 15, 1981

A new era in man's relationship with space has begun, with the successful landing of the "Columbia."

We have progressed from exploration pure and simple to extended use of the vastness that surrounds our planet.

It remains for us to keep the peace of space inviolate. And by keeping that peace, it may well enhance the possibility of assuring peace on Earth for all men.

We already use some of the space in the heavens above us for purposes of sending information. We already have experienced misgivings about the possibility of evil uses of this incomparable part of our environment. The Cain in mankind is not dead.

But perhaps the flight of Columbia — and subsequent journeys already planned — may prove that sinister uses of space will be unprofitable for those who may contemplate them.

The whole world has seen what Columbia performed. It brings dreams of being able to do wonders within the lifetimes of people now suffering deprivation and hardship. The possibility of harnessing directly the energy of the sun comes closer to probability.

We salute astronauts John W. Young and Robert L. Crippen and the NASA team for the completion of a task in the face of almost insurmountable obstacles, and for the flawless performance of that task.

In those few heart-stopping minutes of Tuesday morning, a new role for America was born, to which we must dedicate ourselves completely.

We have become the friend of all the world. And we can extend — in some long year from now — our oath, "One nation, under God, with liberty and justice for all," to all the people of the earth.

Not by military domination, but by good will. Not by fear, but by fairness. Not by threat, but by love. The time will come.

Chicago Defender

Chicago, Ill., April 1, 1981

There is a segment of the public, perhaps all around the world, that is interested in space travel and colonization of other planets. One of the unstated reasons for this hope or expectation is that there is widespread fear that our own sphere is doomed to some kind of nuclear oblivion. So, the idea runs, let's get out while we can.

It is farfetched thinking and its primary value is that it illustrates the serious human plight in the international political situation. It is for the human species which evolved here to try while there is still time to make the earth peaceful, habitable and workable. Dreaming of places in outer space won't quite do it. There may be no nearby planets that can be colonized.

Getting to know as much as we can about the universe is fine, but as for looking at it as "a way out" seems to us unrealistic.

The Providence Journal

Providence, R.I., April 15, 1981

Long years ago, when it was first brewing, the idea seemed wildly improbable: send men into space aboard a rocket (strapped to extra fuel tanks, yet), and design the rocket like a glider to return them and their craft unharmed to dry land. Yesterday, on a scorching desert lakebed in California, it *still* seemed improbable. Here, after all, was this boxy-looking flying machine of 160,000 pounds, skimming down the hot sky at fearsome speeds after 36 orbits of the earth, touching down on the desert with tingling precision and rolling to a proud halt almost exactly on schedule. Galileo and Newton would have been dazzled. The Wright brothers would have been dazzled. And today's Americans, even after experiencing two decades of Project Mercury and Project Gemini and Project Apollo and all those unmanned missions to the planets, could not but feel a bit dazzled themselves as the space shuttle Columbia brilliantly completed its maiden voyage.

Astronauts Crippen and Young are some pilots, and the vast team of engineers, designers and builders behind the shuttle project has seen prodigious effort well fulfilled. The event is being applauded in part as sheer, raw achievement — even by the millions who are not sure what the shuttle program is designed to accomplish. As the TV commentators kept reminding us yesterday, Columbia is the most complex machine ever built for any purpose. We Americans are skilled at such things — skilled at developing the hardware and skilled at conceiving the grand scope of thought that puts it to work. That our scientists can dream up such an idea, and forge it into a titanic reality of aluminum and computers and thousands of heat-absorbing tiles, and make it work as planned — this gives a satisfaction all its own, of a sort rare enough anymore.

And so there are to be more cheers and speeches and medals, all of them deserved, all of them mirrors of our need as a people to feel success and a share in what we are pleased to regard as a national accomplishment. We already had "done" the moon, in a way not so different from the eager tourist who "does" Paris or London or Vienna; now we are poised to "do" this enormous task of ferrying men and supplies (and who knows what else?) to and from earth orbits as a matter of routine.

The Columbia mission, we are told, marks the start of a new era in man's ability to employ the wondrous environment of space for his own ends, and this is probably true. Technology has a way of leapfrogging itself, of driving change at an increasing pace (and driving us along before it). Having once succeeded in a shuttle orbiting test, we are not awfully likely to put the feat away, a bit of history, in some dusty museum. There will be telescopes and other orbiting laboratory instruments galore, and another frontier will have been boldly breached.

Yet after the celebratory champagne, and as technicians groom Columbia for another flight, some more detached evaluation seems in order. The shuttle's costs are huge (a $10 billion total so far, by one calculation), and it is worth seeing where these might be shaved. The military implications of the shuttle program are vast and deserving of careful scrutiny. But today is not one for being grouchy about such things. Today belongs to the Columbia team who built and flew a most remarkable machine around our globe and into the annals of America's heroes. It was a flight to remember.

The Washington Post

Washington, D.C., April 15, 1981

A FEW minutes after 1 o'clock yesterday afternoon, space travel suddenly became a part of real life. There was nothing exotic about the landing of the Columbia on that desert in California—no splashdown, no frogmen, no rubber boats. As the space shuttle rolled to a stop and the trucks gathered around, it looked little different from any other airplane landing at any one of hundreds of airports. Even the steps that were brought out to provide access to the crew compartment would have been at home at National Airport. For the first time in the history of the space program, the machine and the men seemed life-sized, something most of us have seen before and can count on seeing again.

As he watched the perfect landing, Astronaut Joseph P. Kerwin remarked that a new airline had just been born. While it will be a while before the shuttle becomes the airline of space, the idea that ordinary people—not just super-trained astronauts—can orbit the Earth is no longer a dream. It is only a matter of time, if the government properly develops this great new tool, until the shuttle opens to travel the near reaches of space in the same way the airplane has opened the air immediately above the Earth.

The flight of the Columbia was a remarkable testimonial to American technology and to a government agency. Despite the long delays and the huge cost overruns, NASA and the scientific community produced a space vehicle that, once launched, performed precisely as they said it would. The theories and the engineering that went into this untested craft were without a major flaw. Not often have science and technology been able to produce a product that performed so well on its first trip out of the workshop. Indeed, historians will be hard pressed to find any major project in mankind's efforts to explore and exploit the environment in which the error rate has been so low.

There is still much to do before the new space transportation system, as NASA has formally designated the shuttles, becomes operational. More test flights will be conducted this year and next. Modifications in design are almost inevitable; the jetliners of today barely resemble the first commercial airplanes. But Columbia has demonstrated that the potential of space can be exploited on a regular basis at a price this nation can afford.

WINSTON-SALEM JOURNAL

Winston-Salem, N.C., April 15, 1981

When astronauts John Young and Robert Crippen guided the space shuttle Columbia to its Mojave Desert landing yesterday, they brought to a close one of the United States' most ambitious and successful space missions. They also opened what should be one of the most adventurous phases in the space program's 20-year history of manned flight. Successful beyond expectations, the mission of Columbia reaffirms American technological abilities and provides a new sense of optimism for the future of space exploration.

Optimism has not always been one of the shuttle program's more pervasive emotions. The National Aeronautics and Space Administration started quite expansively in the early 1970s with its plans to fashion a reusable space vehicle. The original proposal called for a $10 billion budget, but the Nixon administration would approve only a scaled-down program costing $5.2 billion. But that plan ran into unforseeable delays and problems which have run the figure back up to $10 billion — a 30 percent cost overrun when adjusted for inflation.

Budgetary problems gave way to even thornier engineering problems. One particularly difficult problem threatened to ground Columbia: The tiles or protective shields designed to keep the shuttle and its crew from being incinerated upon re-entry into the atmosphere refused to stick to the craft. Even after a year of additional work on the tiles, a few nonessential ones fell off during Sunday's launch.

NASA engineers and technicians remained undaunted, and it's clear that their perseverance has brought them to the big payoff. But the singular success of this mission leaves room for further debate and speculation. It's a given fact that a reusable spacecraft exists. The nation now has a proven system of carrying scientists and equipment into space for experimental and commercial purposes. The question remains as to how the shuttle will be used, and how successful those uses will be.

At the moment, cargo space is booked solid on some 60 future shuttle missions. One-third of those bookings are for the Air Force for deployment of communications and surveillance satellites. The rest is reserved for corporations such as RCA, which will use the shuttle for everything from launching communications satellites to testing technology in weightless conditions — a luxury never before available. Those uses exploit but do not explore outer space. With the installation of an orbiting space telescope, scheduled for 1985, further exploration of space can begin in earnest.

Beyond its commercial potential, the space shuttle's military applications cannot be overlooked. Debate has never been more intense over military use of space than it is now. Should the United States press forward with "killer" satellites, in keeping with its Soviet competition? Should the United States extend and develop sophisticated spy satellites and death-ray devices in anticipation of the possibility of future "star wars"? Those are but two of the questions NASA, the Pentagon and the nation must decide as the American space program accelerates into the 1980s.

For now, the people of the United States rejoice alongside NASA. Americans are back in space with one of the most spectacular demonstrations of daring since the space program began. Americans are once again proven technological leaders, challenging the doubts of experts and critics alike. The test of technology has been passed. Of course, questions and doubts about the space shuttle's future uses remain. But where there are doubters, there are also dreamers — and today, space-watchers have a lot to dream about.

The Des Moines Register

Des Moines, Iowa, April 16, 1981

Newspaper readers realize that not all front pages are equal. Some report news that will be forgotten within weeks. Others feature changes that will be felt for months or years. A few chronicle events so important that the front pages on which they appear become collector's items.

The Wednesday Register front page is in this rare third category. The successful completion of the space shuttle Columbia's maiden voyage seems likely to stand as one of the supreme technical accomplishments of the human race, an accomplishment that could change forever its relationship to space.

Previous space missions have been voyages of exploration, but the Columbia was designed to turn space trips into almost ordinary occurrences. The space shuttle has been designed to go back into space on mission after mission, for as many as 100 trips before being retired.

The shuttle should greatly expand humankind's ability to explore the universe, monitor and control events on Earth and make productive use of space. It might serve as a stepping stone toward the establishment of human settlements in space.

Few of the millions who watched the shuttle's flight on television fully appreciate the complexity of the machine. The astronauts and the National Aeronautics and Space Administration made it look too easy. The voyage was so flawless that perhaps only scientists could fully appreciate the risks and challenges that had been overcome.

The success of the shuttle should silence critics who claim that America cannot do anything right anymore. No other nation has the ability now to carry off such a complex mission.

Not all the news about the shuttle is good. One of its chief missions will be to gain a military advantage over other nations. The Pentagon already has booked more than one-third of the shuttle's scheduled flights.

While this incredible example of humankind's ability to master nature was circling the globe, there were race riots in London, diplomats were worrying about a possible Soviet invasion of Poland, and millions of people in the Third World were living in subhuman poverty.

The flight of the Columbia offers an ironic comment on the condition of the human race. On the one hand, as astronaut John Young said shortly after Columbia landed, "We're really not too far, the human race isn't, from going to the stars." Yet think how little progress people have made in learning how to deal with each other.

It is a sad, yet all-too-typical commentary on the human race that the main use of this marvelous piece of space machinery may be to speed up the arms race.

Chicago Tribune

Chicago, Ill., April 16, 1981

It was with a slight twinge that we read of the "Star Wars" aspect of the space shuttle program. Despite the breathless NASA publicity about space telescopes and high school science projects, it appears that the basic thrust of the shuttle program will be to beat the Soviets in a new phase of the arms race. As the people at the Pentagon like to describe it, space is the "high ground" of future battlefields and we must get control of it before the Soviets do. So the shuttle Columbia and its future sister ships will be devoting the most sizable part of their carrying capacity to military gadgetry ranging from spy satellites to laser ray guns.

It would have been far better had the two superpowers agreed from the start that space should be free of weapons, saving us the useless expense of a dangerous military competition. But already the Soviets have fired off a dozen killer satellites in tests that began in the mid-1970s, and both sides are well advanced in the development of laser guns capable of knocking out satellites, ballistic missiles, and bombers.

It is not too late to halt or at least limit the race. If the two powers could begin negotiating soon perhaps they could at least limit the military uses of space to surveillance and communications. A ban on tests of killer satellites and laser weapons would be fairly easy to verify and therefore fairly easy to enforce.

It cannot be stressed too often that we negotiate arms treaties with the Soviet Union precisely *because* it is a dangerous, warlike nation whose behavior in many areas is hostile to the United States. That is why arms control is needed: to prevent the latent hostility from turning into devastating war. Arms limitation treaties are agreements between two hostile powers. If they were anything else we would be negotiating with Britain or France, not with the Soviets.

Nevertheless, the political climate is not favorable for arms control agreements—new or old. It appears we will get into a space arms race whether we like it or not. And if we get into it, we had better plan on winning it.

We must win it because we cannot afford to lose it. Technology is advancing at such a fast rate that the Soviets could catch us by surprise with an array of space weapons that would render our strategic forces helpless. The Kremlin would be in a position then to dictate the terms of surrender in what would amount to a non-shooting World War III. That might be better than losing a shooting war, but better still if we avoid such nasty surprises and best of all if we do the surprising.

So fly aloft, Columbia; deliver your laser guns and satellite busters and spy eyes. Build your battlestars. May the force be with us and with all of us fragile earthlings together.

DESERET NEWS

Salt Lake City, Utah, April 4, 1981

Sometime within the next few days — perhaps as early as Tuesday — the space shuttle Columbia is supposed to lift off its launching pad at the Kennedy Space Center to inaugurate a new era in America's space program.

Initially, the Columbia is to circle the Earth 36 ½ times and spend 54 ½ hours in space before gliding to a landing at Edwards Air Force Base in California. Should anything go amiss, three alternate landing sites are available.

The Columbia might be described as the commercial application of space travel. Not that it will carry persons into space, but it is to serve as a vehicle for all sorts of engineering tests and scientific experiments for private industry.

After the excitement of the early manned flights and the moon landings, the Columbia's mission sounds rather boring. It isn't. It still involves the risks of an entirely new space vehicle, one that can be used over and over. And the possibility for significant new additions to man's knowledge is limited only by man's imagination.

Roanoke Times & World-News

Roanoke, Va., April 13, 1981

It's been a long time between shots for the U.S. manned space program, but Sunday's launch makes up for (so to speak) a lot of lost ground. With the shuttle, NASA has gone from throwaways to recycling. Ability to reuse a spacecraft, especially one the size of Columbia, opens up a range of new possibilities in space, from expanded scientific experimentation to manufacturing to — unfortunately — war.

Superhuman patience and persistence were needed to nurse the Columbia project this far. The ship went into orbit more than two years behind schedule; a book could be written about the technical problems that had to be overcome.

For his part, pilot John W. Young is pleased that his and Capt. Robert L. Crippen's training stretched out to three years; if it hadn't, he says, they would not have been fully prepared. As it is, they had an active part in the changing design of the craft, including trouble-shooting. Part of the first cargo were 22 books, each three inches thick, detailing safety procedures.

As that indicates, the bigger and more complex the project, the more that can go wrong. That's part of the price of expanding the space program's capabilities. Columbia is as big as a small airliner and affords a more roomy, relaxed, Earth-like environment than the cramped Apollo capsules. Those belonged to "the age of space exploration," wrote Rick Gore in the March *National Geographic*. "The shuttle begins the age of space exploitation." If all goes well, such flights eventually will become routine, maybe bi-weekly.

The military has a big stake in Columbia. President Nixon gave the project the go-ahead nine years ago, but it limped along until late 1979, when the Defense Department declared it vital to national defense. There is an argument now that American technical superiority in space should be used to counter the Soviet Union's superiority on the ground with conventional forces.

There is a counter-argument that the Soviet Union has an active anti-satellite program; that the United States and the Soviet Union should not become in the 20th Century the rivals for control of space that England and Spain (and later Britain and Germany) were in earlier centuries for control over the waters. There is a third belief that there seem to be too many shortages on the planet Earth to warrant diversion of human activity and materials into space.

The launching of Columbia was a great triumph and a salute is due to all who planned the vehicle, those who built it and those who flew it. The dangerous tragedy is that mankind's ability to govern in peace does not equal his scientific achievements in space. There is another gap — a man and missile gap.

Pittsburgh Post-Gazette

Pittsburgh, Pa., April 1, 1981

The great day approaches. After long delays and huge cost overruns, the space shuttle Columbia is being prepared for its scheduled April 10 blastoff. As excitement grows and expectations rise, one sad thought intrudes: Despite the vaunted scientific and commercial applications of the craft, there are those who would put it to more sinister purposes.

According to The New York Times, military planners at the Pentagon are rubbing their hands together and salivating at the prospect of this flying boon to space warfare. Some civilian scientists believe that military uses of the craft will exceed its other roles — a notion that hitherto has tended to be glossed over in the general euphoria surrounding America's move back into the manned spacecraft business.

There is, of course, no cause for surprise. The Cold War, particularly in the early days, was the real *raison d'etre* of the space program, for all the pious talk of new frontiers. And it was really crass superpower rivalry that enabled Neal Armstrong to take a noble small step for man and a giant leap for mankind.

In an age of rearmament, especially, it is unsurprising that space be considered as a kind of Star Wars arena for surveillance satellites and those designed to counter them. As awful as the prospect is — and out of keeping with the spirit of the UN treaty that proscribes weapons of mass destruction in space — there is a sad inevitability about the way strategic demands have conspired to make this so.

Therefore, when the space shuttle blasts off with its courageous crew, the ring of cheers will have one disturbing echo — that this world cannot aim at the heavens without contemplating hell.

ST. LOUIS POST-DISPATCH

St. Louis, Mo., April 15, 1981

The flight of the space shuttle Columbia was risky, daring and successful, and with the drama of it settled by the landing in the Mojave Desert, Americans can look to a new phase of the space age. In this phase mankind passes from pioneering to the *use* of space, and that has untold implications.

As to risk, Columbia was far behind its development schedule when it finally took off from Cape Canaveral, and its 54-hour space journey was the first to be made without a preliminary test in unmanned flight. Hence the daring belongs to its crew, astronauts John W. Young and Robert L. Crippen, and the success to them and more broadly to the host of scientists and technologists of the National Aeronautics and Space Administration who planned and prepared the Columbia and had enough confidence in their achievement to send it on its way.

For this was NASA's day. The landings on the moon in 1969 proved a difficult act to follow, and NASA's budget did not follow it. (Even now the 1982 Reagan budget for NASA trims some $600 million from that suggested by President Carter.) To be sure, the Viking and Voyager trips to Mars, Jupiter and Saturn returned scientific dividends and captured widespread interest, but they lacked the intense theater of a space odyssey. Now the nation has one, and NASA and the space program may well find more encouragement in Congress.

Indeed, Columbia has become an instant source of national pride in more ways than one. For a nation becoming somewhat doubtful of its traditional know-how, productivity and ability to compete, Columbia is proof that these qualities remain if properly marshalled. More than that, the shuttle is a demonstration, albeit nearly a $10 billion one, of what investment in research and technology can do. For the U.S. has been giving less to R&D in recent years, and the Columbia project is bound to call new attention to the essential framework for technological and economic advance.

Columbia's initial journey was, however, only the first of four test flights planned before the shuttle becomes operational in 1982. And it will be operational with a vast difference from the early space probes. It represents a space transportation system — a kind of space truck. Not only is the shuttle reusable, but it is usable in many different ways. It will introduce new types of astronauts, for example; NASA already is training 33 specialists, including women, for future ventures. These crews will supply orbiting vehicles, conduct scientific experiments in space, identify Earth's resources, launch satellites and orbit a large telescope that can explore the universe from beyond Earth's haze.

There is, of course, a further potential. Nearly one-third of the shuttle's first 60 or so missions have been reserved by the Air Force. Soviet Russia already has accused the U.S. of seeking to turn space into a future battleground, although the Russians themselves wasted little time in experimenting with space weaponry. The possibility remains, though, that space could become an extension of the arms race, if that has not already begun.

In the great explorations of the 15th and 16th centuries, the adventuresome caravels and galleons carried flags, cannon and hopes of empire. Is mankind in any better position today to limit such nationalism? The fact is that space does not invite political boundaries. The Columbia has opened a highway toward the stars, and they will remain however man approaches them.

The Burlington Free Press

Burlington, Vt., April 10, 1981

Men have been fascinated by the extraterrestrial since the first time they looked up at the infinity of the universe and wondered what mysteries lay in the regions beyond the Earth's limits.

But it was not until this century that technology provided the tools for the exploration of space. Through space flights, scientists hoped to learn more about the origins of the universe and our planet's place in that scheme of things.

Those missions already have yielded precious data for scientific analysis and have spawned more curiosity about the phenomena that exist outside Earth's boundaries. Some items of equipment designed for use by astronauts have been put to other uses.

Since putting men into orbit and sending them to the moon, America's manned space program has languished. Scientists instead have concentrated their attention on the exploration of distant planets with unmanned spacecraft.

In the meantime, the Soviet Union has enjoyed a monopoly on manned space flights, sending 43 astronauts into orbit since 1975 while the United States failed to launch any missions.

Today's flight of the space shuttle Columbia should be a step toward further exploration of space by this country and might well signal the beginning of an era when travel outside the Earth's confines becomes more commonplace. That the shuttle is a reusable vehicle represents a drastic departure from the concept that only conventional spacecraft should be used for such missions. It was created with the idea of making space travel as routine as overseas airline flights.

The eyes of the world will be on the Kennedy Space Center this morning when the shuttle is launched and success can boost the country's prestige at a time when it is sorely needed.

Many Americans no doubt will be hoping that the Columbia does what it is supposed to do when the signal for lift-off is given.

THE ARIZONA REPUBLIC

Phoenix, Ariz., April 8, 1981

MORE then just an expensive new gadget for space exploration is sitting on the launch pad at Cape Canaveral waiting for blastoff.

The shuttle spaceship *Columbia* — as big as a Boeing 747 jetliner — is America's hope for resuscitating a gasping space program, and the key to competing in space with the Soviet Union.

Except for the launching of unmanned satellites, the U.S. space program has been dormant for nearly six years.

However, the Russians have been busy making repeated manned space flights, and preparing to build an orbiting manned space station by 1985. They also have been developing a killer satellite that can blind, or destroy, U.S. military satellites.

If the *Columbia's* 54-hour Friday-through-Sunday test flight and landing on a California dry lake bed are successful, three other shuttle spaceships will join it in a new space program.

The four — part of a $9 billion program — will be the first U.S. vehicles designed to use, rather than merely explore, space.

Nearly 100 scientific, military and commercial missions are expected to be flown by each of the four reusable spaceships, including construction of a space station.

In time, as former astronaut Alan Shepherd has forecast, the shuttle will be the DC-3 of future space programs — provided the U.S. realizes the importance of maintaining a space program.

Friday's scheduled launch is a dangerous one. In previous manned space flight programs, prototype space capsules were launched before astronauts were allowed to ride them into space.

But there has been no test space flight of a crewless shuttle. The two shuttle astronauts literally are risking their lives in an unproven vehicle plagued by design and operational problems.

Blase as Americans have become about space flight, they should cross their fingers that the shuttle flight is flawless.

The News Journal

Wilmington, Del., April 5, 1981

Americans who look skeptically on the Space Shuttle and ask, "What's in it for me?" echo a question as old as exploration of the unknown. Ferdinand and Isabella may not have put it quite so crassly, but they weren't functioning solely as disinterested patrons of science when they underwrote Christopher Columbus' three-ship voyage out of the known world.

The visions of those who challenge the unexplored are almost invariably crackpot schemes or deranged hallucinations to their contemporaries.

What good could possibly come of getting from one part of town to another in a horseless carriage that moved at speeds the human body was not designed to endure?

How could anyone take seriously two bicycle mechanics who persisted in risking their fool necks trying to fly above the ground on which God had firmly planted them?

After they succeeded, some young fellow with more nerve than brains wanted to fly by himself all the way to Europe. What would that prove?

There were other idiots who risked blowing themselves to kingdom come by setting off toy rockets. If there was anything on God's Earth that was a greater waste of time and money, not many among the scoffers could imagine it.

These and many other commonly acknowledged follies of their day are the foundations of a late 20th century world almost unimaginable 100 years ago. None was an unmixed blessing. Few advances are.

The internal combustion engine and the airplane made possible mobility undreamed of a few years earlier. They also gave humanity new potential for self-destruction.

Nonstop transoceanic flight opened the doors to rapid international travel and an increased sense of world community. It also opened a horrible new dimsension of warfare, perfected with devastating consequences in the aerial bombing of World War II.

The rocket enabled a new type of explorer to break the gravitational bounds of Earth and float weightless around it as he worked out details for a successful round-trip visit to the moon. Combined with atomic energy, the rocket made possible a terrifying intercontinental ballistic weapon so formidable that superpowers compete in the proliferation of missiles that they shudder at the thought of using.

Even the Space Shuttle has potential applications not yet anticipated. The generals in the Pentagon are eager to exploit its military opportunities. The space scientists in the National Aeronautics and Space Administration are impatient to adapt its cargo-hauling capabilities for satellite maintenance and construction projects in space. No doubt someone will soon seriously propose its use as an interstellar garbage truck for dumping Earth's toxic wastes in the ocean of space.

"But still," say those skeptical of a program costing billions of dollars, "what's in it for me?" The shuttle's obvious and intended uses hardly justify the financial price and it is not yet possible to determine its ultimate value.

Mankind's history of risking the unknown has produced a harvest of unanticipated technological advantages, from transportation to communications to medicine. But manned exploration of space also has made us much more aware of the fragile interdependence of our world.

Columbia's first trip challenges two brave Americans to prove that it is possible to survive the extremes of space and glide safely back to Earth on the aerodynamic equivalent of a falling rock. In doing so, they will not only overcome the uncertainties of a trouble-plagued project but also give us a new tool with which to extend our reach into space.

It is impossible to guess all that awaits us out there. It surely includes material benefits and further technological gains of value, greater knowledge of our universe and, therefore, ourselves, and an increased sense of our obligation to use space in the best interests of everyone on this tiny Earth.

That's in it for all of us.

The San Diego Union

San Diego, Calif., April 9, 1981

Once again, excitement and anticipation are building up for a space launch from Cape Canaveral — the first time since 1975 that U.S. astronauts have suited up for a trip into space. Barring last-minute delays, the space shuttle Columbia will make its maiden flight into orbit tomorrow morning.

The scene at the launch site and the chatter of the countdown are reminiscent of the past — of Mercury and Gemini, of Apollo and the missions to the moon, of Skylab, and finally of the Apollo-Soyuz link-up six years ago that ended an era of manned space flight on the American side.

Among space enthusiasts there has been much lamenting over the fact that public interest in the space program went downhill so rapidly after the triumphs of Apollo. The Manned Orbiting Laboratory program that was supposed to follow Skylab was canceled. The Russians, meanwhile, pressed ahead with refining their technology for sustained operations in space, and have now amassed more than twice as many hours of space experience as the Americans.

The first launch of our space shuttle is coming two and a half years behind schedule, partly due to technical problems but also because of cutbacks in space budgets during the years of design and development. Man-in-space has had a low priority in Washington.

But some of the old sense of challenge can now be felt. Astronauts John Young and Robert Crippen tomorrow will be at the helm of a spacecraft never before flown into orbit — the largest, most versatile space vehicle ever built. They must see it through launch and re-entry maneuvers more complicated than any undertaken before and then bring it to a landing like an airplane at Edwards Air Force Base.

The flight plan, including 36 orbits of the earth if all goes well, promises familiar moments of suspense for us in the earthbound cheering section. There is irony in this. The aim of the space shuttle program is to take the suspense and excitement out of the launch and landing of spacecraft — to make these events the most routine part of space activity in the future.

Once the shuttle proves itself — and we can pray for a safe and successful test flight for the Columbia and its crew — the drama will come from another quarter. It lies in the vast possibilities for working and building in space which a fleet of space shuttles will provide.

The Columbia and its sister ships can carry four passengers in addition to the two pilots. Their huge cargo bays can accommodate 65,000 pounds of payload. As a workhorse designed for repeated trips into orbit, and with booster engines that can be recovered and used again, the shuttle will cut the cost of space launches in half.

The shuttle will become the delivery van and workshop for the approaching "industrialization" of space, whether the assignment involves orbiting and servicing new families of satellites or assembling large, permanent space stations. Not incidentally, it will provide the West with a means of answering whatever military advantages the Soviet Union sees for itself in its aggressive man-in-space program.

Americans took their "giant step for mankind" on the surface of the moon 12 years ago. There are more to be taken on the space frontier. The first launch of the space shuttle brings us to the threshhold of an era when the destination of those steps will be limited only by our willingness to take them.

The Houston Post

Houston, Texas, April 7, 1981

We are poised on the threshold of a new era in space with the setting of Friday as the tentative launch date for the space shuttle orbiter Columbia. If all goes as planned, the 54½-hour maiden flight of the Columbia will mark the first full-scale test of a reusable space vehicle — the first space plane. But more important, it will vastly expand our ability to use space for scientific, commercial and military purposes. This last is essential to national security in the face of the Soviet Union's ambitious manned space flight program with its various military applications.

It has been more than five years since American astronauts were in space — in a joint venture with the Russians that linked an Apollo spacecraft with a Soviet Soyuz in 1975. In the interim, 51 cosmonauts have blasted into Earth orbit, the last two on March 22. The cosmonaut corps has set records for duration of a single manned flight — 185 days — and for total man-hours in space. The Kremlin also has used its manned space-flight program for propaganda purposes, sending up several cosmonauts of other communist nations, including Cuba and Vietnam.

But this showboating is secondary to the experience the Soviets are accumulating about living and working in space. Their Salyut 6 space station has been in orbit for more than three years and has been manned much of that time by cosmonauts performing a wide range of experiments. It is the military implications of these experiments, however, that make U.S. defense planners increasingly anxious to see the shuttle fly.

A great deal rides on this first shuttle launch. Its success could revive public interest in our manned space program, interest that has lagged visibly since our spectacular moon missions of the late 1960s and early 70s. It could open unlimited opportunities to do scientific research and develop new products and processes in the weightless environment of the Spacelab that future shuttle flights are scheduled to carry into orbit. It would give the "go" signal to a plan for several Western European nations to participate in using the Spacelab's unique facilities. Just as the development of airplanes for commercial use revolutionized our world, so development of the shuttle could lead to profound change. Ironically, the success of the shuttle program will be established only when its flights become so routine they no longer stir much public interest. By then, of course, it would have proven its indispensability, just as the communications, navigation and weather satellites have.

Unfortunately, the shuttle program has been marred by repeated launch delays and large cost overruns that eroded much-needed support for it in government and from the public. The vehicle's engines have been plagued by malfunctions, and the silica tiles that form the protective shield against overheating on reentry into the Earth's atmosphere have peeled off in tests. But the beginning of the countdown for the flight is a hopeful sign that these and other technical problems have finally been solved. When Astronauts John W. Young and Robert L. Crippen man the controls of the Columbia and blast off, they will put the United States back into the manned space flight field from which we have been absent too long. Bon voyage, gentlemen.

Detroit Free Press

Detroit, Mich., April 10, 1981

IT IS FANTASY come to life, and the greatest question it poses for us is apocalyptic. The first true space ship will ascend from Cape Canaveral Friday with a fiery crescendo, if all goes well: a manned orbiter that may fly another hundred times, to carry cargo, launch satellites, study the earth and its galaxy, and perhaps to repair other spacecraft, go on rescue missions or build a space station. The Columbia also may open an arms race in space.

From the time Sputnik I went aloft in 1957, speculation about the military uses of space has intensified. Somehow it seemed inevitable that a planet that could not control arms on its surface, in the air and in the depths of its seas would not control them in space, though a 1961 United Nations agreement formally banned spatial weapons of destruction.

The word "shuttle" seems pale to describe this vessel, which is far different from anything that has risen from the earth before. The Columbia, the gem of our galactic ocean, can be used for peace or war. If it proves worthy, if the risk in this first flight does not reveal serious flaws, we will be entering a new era.

The future of the U.S. space program is fuzzy. Critical policy decisions have not been made about additional shuttles, unmanned space probes or levels of funding. The country's space policy is being guided more by the momentum of rapidly developing technology than by conscious, broadly conceptual decisions and vision.

The Reagan administration is not the first to fail to answer the larger questions. We have had no clear policy on space since President Kennedy determined that we would go to the moon. Construction of the Columbia was authorized by President Nixon, at a time when Washington was reconsidering whether to spend huge sums for great leaps forward in space. As it turns out, this ship has cost almost twice the $5.2 billion originally envisioned. While it was being developed, the country turned away from plans for more manned flights and toward less costly voyages of unmanned explorer satellites. The Columbia brings us back to man in flight far from the blue planet.

It is imperative now that we give more attention to the costs of future space programs, measured against their results. No one should miss the contradiction inherent in our launching of a $10 billion space ship at a time when budgets for human services face big cuts because of a belief that the government cannot afford them.

The Columbia project has gone forward in large part because of its potential military uses. One day these shuttles may be equipped with lasers to destroy missiles fired below or satellies orbiting above. They may be used to construct orbiting, manned space stations that can warn of missile attacks — or abet them. One reason the Columbia was built was that the Soviet Union is known to be developing "killer satellites" that can destroy other satellites, and another is a belief that the Soviets are developing a shuttle of their own.

This first space ship, then, carries with it the great war-and-peace questions that have not been resolved by civilizations capable of giant technological strides but seemingly incapable of restraining destructive military forces. No one can stop the march of technology. But again we must ponder the deeper questions that threaten our humanity, and humankind itself.

The Charlotte Observer

Charlotte, N.C., April 10, 1981

Over the last three years, NASA officials have set several dates for the takeoff of the space shuttle, only to postpone them because of engineering problems or accidents. Americans might have given up long ago on the first re-usable spacecraft ever getting off the ground, but they haven't.

In fact, Kennedy Space Center officials were predicting a million people would gather in Cape Canaveral this morning to watch the anticipated launching, and all the major TV networks planned to cover it live because they, too, expect millions of viewers.

There are probably several reasons for this show of public interest, but none so compelling as one given by John Noble Wilford, New York Times science writer, in an April issue of the Times' Sunday magazine. Americans are counting on the shuttle to prove to themselves, and to the world, he writes, that American technology is still inferior to none, and that America hasn't lost its frontier enthusiasm and know-how for exploring the unknown.

Americans have grown up believing they are the world's mechanical geniuses, Mr. Wilford writes, but recent events have tended to dispute that. Recalled automobiles, collapsed arena roofs, a nuclear plant breakdown, helicopters that crash in the desert on a hostage rescue mission — all force Americans to ask "if we've lost the technological touch."

Recognizing that we embellish the past, we still ask ourselves if we are today the equals of the Western pioneers, the Wright brothers or Thomas Edison. If only the space shuttle could make its 36 Earth orbits successfully, we say, paving the way for the first space station in the universe, then the world would see once again the values of American entrepreneurship.

That's a simple and understandable wish, but it's based on too narrow a view of American entrepreneurship. America hasn't lost its technological know-how or its inventiveness; U.S hardware and computer software are in constant demand by other countries, including Western Europe and the Soviet Union.

Also, let's not forget how American farmers and scientists have figured out ways to grow food in places in the world where no food grew before. And there are probably more U.S. inventors than ever before — they're just not as well known as their forebears because most of them work for big corporations or the federal government.

Our good wishes and prayers should ride with the space shuttle when it takes off. But let's not expect it to prove things about ourselves we already should know.

The Birmingham News

Birmingham, Ala., November 2, 1981

Even as the space shuttle Columbia prepares for a Wednesday liftoff which offers once again to astound the world with America's space technology, the National Aeronautics and Space Administration is facing the prospect of debilitating budget cuts which well could blunt further U.S. advancements in space.

NASA already has seen its budget reduced 7.5 percent for this year, while the White House is asking for another $367 million cut to reduce total space agency spending to $5.7 billion. The administration, further, is asking for an additional $1 billion slash in each of the next two fiscal years.

The effect of such deep reductions would be to shut down practically all NASA work, excepting the shuttle program, and even it would suffer some elimination in delay of flights. Lost altogether would be space exploration activity — such as the Voyager flights which this year added so greatly to man's understanding — and other such planned missions such as that to monitor Halley's comet as it swings near Earth.

The proposed cuts would, in short, constitute "a full retreat across the board from the challenges of the future," as U.S. Rep. Ronnie Flippo of Alabama so correctly has noted.

To do so would be a mistake, regardless of the administration's well-intentioned ambitions in budget-cutting. The space program is at least as important to the world standing — and contribution to the world — of the United States as are some proposals for defense.

Herald News

Fall River, Mass., April 9, 1981

Barring some last minute hitch, this country will return to space tomorrow by sending the first reusable space ship into orbit. The Columbia will take off from Cape Canaveral on a test flight intended to last 54½ hours. It is scheduled to end Sunday with the Columbia returning to Edwards Air Force Base in California.

In an era when the United States is plagued by chronic economic problems and international tension, it is easy to view the exploration of space as a kind of luxury which we can ill afford.

It is worth noting, however, that the Soviet Union does not view its own space program in that light, and that it has continued to send astronauts into space in spite of increasing its expenditures on military equipment.

There is no reason to think the Kremlin is idealistic in this respect. Rather, it views the exploration of space as essential to its own future security. The United States, it seems, should in this respect adopt the viewpoint of the Kremlin.

Space exploration is equally essential to this country's security, and the security of its allies. It is also the single most important thrust of scientific research in our lifetime, with implications just as vital in terms of the future as there were in the discovery of America by Columbus 500 years ago.

In this instance, the Columbia is intended to initiate shuttle service into space with all that will mean for the lifting of heavy objects that will become the first space stations.

This is, indeed, the introduction of shuttle service into space and back. In a sense, this may amount to a pathway to the future for large numbers of people, however strange that may seem to us right now.

By any standard, therefore the launching of the Columbia into space tomorrow, and then its safe return on Sunday, are major events. The country will follow the Columbia's journey with pride as well as with genuine concern for the astronauts who are manning the mammoth space ship.

A safe voyage to the Columbia and to the astronauts is the heartfelt wish of us all!

The Dallas Morning News

Dallas, Texas, April 10, 1981

BY the time these words are read, the space shuttle will be high above the earth — or still on the ground, the victim of still more mechanical foul-ups.

This is in some sense beside the point. If not today, then tomorrow the shuttle will soar — pursuing its appointed mission of helping acclimate man to outer space. What an invigorating prospect that makes.

America's space program reached its high point of public support in July, 1969, when, upon the moon's surface, Neil Armstrong took one giant leap for mankind — an event that President Nixon somewhat hyperbolically called the most important since the Creation.

Manned space missions shortly afterward fell victim to the Vietnam malaise. Even at the time of the moon voyages, the likes of Edward Kennedy — whose brother is forever linked in the public mind with space exploration — were urging that we "reorder our priorities." These voices meant we should quit scattering amid the sterile planets billions of dollars we could be spending to combat poverty, hunger and disease right here at home.

Vietnam destroyed the buoyant, patriotic mood of the "soaring '60s," as they were first called. The nation lost interest in the spectacle of brave, skilled Americans, catapulted to regions more distant than mortal minds could comprehend.

The space program, though never stopped, wound slowly down. No manned missions have been undertaken since 1975. The space shuttle itself has been delayed three years by serious mechanical problems. Space shuttle planners have been described as "white-knuckled" lest something else go wrong with the $8.6 billion program.

The space program as a whole, however, does not depend on one craft's performance, however precedent-setting the role of that craft. (Columbia is the first reusable space vehicle.) Had Columbus' whole fleet sunk on leaving Palos, Spain, America would in due course have been discovered — more precisely, rediscovered — by Europe.

The necessity, to be sure, is for the space program's lost momentum to resume as soon as possible. For which reason we hope the shuttle is a roaring success. As clearly demonstrated by experience, the program's benefits are economic, scientific, medical and military. So are they spiritual. A society perpetually reaching for the stars, figuratively or literally, is the kind of society we used to be but haven't been at least since the '60s. There are signs of recovery these days, indications of a renewed understanding that dynamic societies prosper but static ones fall. If Columbia can't raise our sights, it's hard to know what can.

Shuttle Boosts U.S. Morale, but NASA's Future Still in Doubt

Columbia's third successful flight in March 1982 continued the rebuilding of public confidence in American know-how that the first shuttle flight had started. However, the shuttle program seemed fated to be the only accomplishment of the limping U.S. space program. A spending-conscious administration kept the National Aeronautics and Space Administration on a tight rein, although President Ronald Reagan proposed increasing NASA's fiscal 1983 budget to $6.6 billion. That figure, which former President Jimmy Carter had wanted to spend in the previous fiscal year, was only $750 million more than NASA eventually had received in fiscal 1982. However, the amount would cover the once-threatened Galileo mission to Jupiter and keep alive Voyager 2's transmissions once it reached Uranus. No money was earmarked for a Halley's Comet probe or another Venus satellite. The emphasis on the space shuttle program, which remained the largest item in NASA's budget, indicated that the administration's priorities were close to home rather than in the far-flung reaches of the solar system.

Houston Chronicle

Houston, Texas, March 31, 1982

For those who lament the often reported decline in American know-how and who wonder what ever happened to the "can do" attitude that has characterized American ingenuity in the modern age, the landing of the U.S. space shuttle is always a bolstering sight.

Many of those who witnessed the desert landing of the shuttle Columbia at White Sands, N.M., called the experience a beautiful "thrill of a lifetime." President Reagan, watching television coverage of the landing from the White House, exclaimed, "That's marvelous!"

The numbers racked up by the shuttle's third test mission are impressive in their own right. After traveling 3.9 million miles and circling the Earth 129 times in eight days, the 105-ton craft glided without power to a pinpoint landing. Braking from a speed of over 17,000 mph, Columbia streaked across the West Coast at 13 times the speed of sound. An automatic landing system controlled the shuttle for part of the descent, and pilot Jack Lousma took over just above the runway.

On the way down, Lousma confirmed what was obvious to those watching from the ground: "This is a beautiful flying machine." The nose rose precariously when the craft touched down but quickly settled before Columbia came to a stop just short of the runway's 15,000-foot mark.

Although it was plagued by minor technical difficulties, the mission amassed a mountain of data from an exhaustive series of tests, which indicated that the shuttle program should be well able to fulfill the United States' future objectives in space. When Columbia touched down, an official at the Johnson Space Center's Mission Control in Houston conveyed the sentiments of Americans when he told astronauts Lousma and Gordon Fullerton, "Welcome home. That was a beautiful job."

St. Louis Globe-Democrat

St. Louis, Mo., May 3, 1982

Many Americans may look upon the Space Shuttle Columbia as just another interesting experiment of the National Aeronautics and Space Administration.

They will soon discover that Columbia is actually a "railroad" into space which can open space to industry as surely as the railroads opened the West to industrialization in the latter half of the 19th Century.

St. Louis' own McDonnell Douglas Corp., along with Johnson & Johnson, will launch the new era of space industry when the Columbia takes off on its fourth space mission in July. McDonnell Douglas and Johnson & Johnson will place aboard the Columbia its first commercial cargo: an experimental pharmaceutical processing machine (called an electrophoresis device) — a 2-foot by 7-foot box weighing 475 pounds that will electrically separate compounds to determine the feasibility of producing pharmaceuticals in space.

Earth gravity prevents the electrophoresis process from producing compounds of the purity and quantity believed possible in space. Astronauts will send six samples of natural body materials through the unit to study its ability to separate components of the material under weightless conditions.

If successful, this experiment could open up a vast new space pharmaceutical processing industry.

Right behind McDonnell Douglas and Johnson & Johnson are a number of other companies preparing space industry projects. Barron's reports that GTI, Inc. of San Diego — aware of the potential for combining alloys in gravityless space — is planning to build a sophisticated furnace, with room for many different crucibles in which metal-combining processes can be tested. Already more than a dozen corporate customers of GTI have expressed interest in renting the crucibles when the furnace is flown into weightless space by the Space Shuttle in the not too distant future.

Microgravity Research Associates of Coral Gables, Fla., also expects to sign a contract with NASA in two weeks to investigate a new process for growing semiconductor crystals in space. And these are only a few of the myriads of possibilities for new industries in space.

By 1986, NASA will have four Space Shuttles. Interest in the spaceplanes is so intense that all the room available in their cargo bays already has been booked solid even though the cargo bay of each shuttle has a diameter of 15 feet, is 60 feet long and can carry up to 35 tons.

A private firm, Space Transportation of Princeton, N.J., is so intrigued with the commercial possibilities of the Space Shuttle that it says it intends to raise about $1 billion to pay NASA to build a fifth shuttle. Space Transportation would then pay NASA to operate the Space Shuttle and, in return, would be allowed to rent out the commercial cargo space on the shuttle.

Patently astronauts in the Columbia are opening the door to fantastic new ventures in space and opportunities which are limited only by man's imagination.

The Seattle Times

Seattle, Wash., March 31, 1982

AMERICAN ingenuity, more often identified in recent times with the past than the present, was magnificently on display yesterday, when the space shuttle Columbia rolled to a stop dead-center on a substitute runway.

The Columbia's steady-nerved astronauts and her small army of flight controllers mastered with aplomb a number of minor breakdowns and weather disturbances that dictated changes in the time and place of the third space-shuttle touchdown.

That demonstration of versatility and durability by man and machine made clear that Uncle Sam is winning his multibillion-dollar gamble on the shuttle as a reliable, long-term workhorse in space.

Glynn Lunney, flight-operations director, says nothing now stands in the way of "flying payloads for paying customers" beginning with the fifth flight in November.

That will open the door to numerous scientific advances, including those with practical commercial applications and some of vital military import.

No other nation, including the Soviet Union, has anything remotely matching the space shuttle — a rocket on launching, a satellite in orbit, and an airplane on descent.

Its success demonstrates the foresight of space-agency planners who, shortly after the moon-landing program of the 1960s, shifted their emphasis to developing a reusable vehicle for accomplishing numberless diverse missions in earth-orbit space.

The success also justifies the decisions of successive administrations in keeping space-agency funding on an even keel.

Hail Columbia!

Herald News

Fall River, Mass., March 31, 1982

The successful, although delayed, return of the space shuttle Columbia ends seven days of complicated maneuvers and experiments in space. Many of these maneuvers and experiments were carried out under less than ideal conditions because of minor but irritating malfunctions of equipment on the spaceship.

The two astronauts, Colonel Jack R. Lousma of the Marine Corps and Colonel C. Gordon Fullerton of the Air Force, were also handicapped in the earlier stages of the space flight by attacks of nausea.

It was not, in other words, an untroubled flight, and the fact that in spite of minor setbacks and hindrances, it worked out so well is a great credit to the astronauts and to their back-up crew here on earth.

It also displays what the Columbia's flights were basically intended to demonstrate, the durability of these vast structures that are made with a view to putting space travel on a regular basis.

The success of the Columbia's flights proves that a space ship can be used more than once after relatively minor overhauls and repairs.

It is obvious that since this is so, the enormous expense of constructing a space ship at least will not have to be duplicated for each voyage.

This is the immediate significance of the Columbia's space flights, that they were a first step toward regular shuttles between earth and space as well as the space stations that will sooner or later be placed there.

In retrospect, the successive steps of man to move into space will come to seem the most significant human activities of the latter half of this century. They will be seen to have the same kind of significance that the voyages of Columbus and the other European explorers had in the 16th century. They have been the indispensable introduction to what comes after.

The fact that the United States has been in the forefront of this expansion of the range of human knowledge and activity will demonstrate to generations in the future that this country was the most important factor in enlarging human consciousness in our time.

This, rightly, should be a source of pride to every American.

The Columbia's spectacular return from space yesterday had more than dramatic impact. It heralded a time when space ships will make regularly scheduled trips to way stations between earth and the moon or to other planets.

In spite of the constant setbacks to human progress, mostly because of man's own warring nature, there is progress, even though at times the setbacks obscure our awareness of it.

That progress takes different forms. Sometimes these forms are scientific, sometimes political or social. In our time they have been mainly scientific, and the most important way in which science is progressing in our time is in opening man's pathway to the stars.

Sometimes the expense, the effort and the risk involved in space travel seem out of proportion to the benefits derived.

But the benefits should be thought of in terms of permanent extension of human knowledge and capacity, rather than immediate, short-range returns.

In these terms the Columbia's latest voyage is one more triumph of American scienitifc and engineering ingenuity and skill, and of sheer human grit.

The Columbia's latest flight is another demonstration of the capacity of this country's scientific and military establishments to work together for the advancement not merely of the United States, but the whole human species.

It's no wonder that the country is enormously proud of its achievements in space. It has a right to be.

DESERET NEWS

Salt Lake City, Utah, March 3, 1982

An earnest debate is taking place in government and science circles about the future of the U.S. space program.

The issue is whether to continue exotic scientific probes of the planets or to concentrate on earth orbit missions that have possible military or commercial applications, such as the space shuttle.

In a time of budget reductions, planners feel the emphasis ought to be in one place or the other, but say the nation can't afford to pursue and improve both programs at the same time.

If the military emphasis wins at the expense of purely scientific programs, then NASA, the nation's space agency, is likely to dwindle into a more obscure department with limited funding.

But two reports this week have put the spotlight back on planetary probes.

The first was not a NASA triumph, but a Russian one. The Soviets soft-landed a spacecraft on the blisteringly hot, cloud-covered surface of Venus, something the U.S. has never done. For 127 minutes the craft sent back pictures before the enormous pressure and heat on Venus knocked it out. Soviet scientists said it lasted twice as long as expected.

The second item was the 10-year anniversary of Pioneer 10, a scarred veteran of nearly 3.3 billion miles of space flight. After 10 years it is halfway between Uranus and Neptune and still functioning. So far, Pioneer 10 has transmitted more than 125 billion bits of scientific data. Scientists are devising more experiments for the craft before it leaves the solar system about 1989 — a representative of Earth hurtling into the galaxy.

A good argument can be made for exploiting the many uses of the space shuttle after the heavy expense of its development. But the NASA budget is such a small fraction of the total U.S. budget that it seems a shame to let future planetary probes die on the drawing board.

One of the things that made America a great country was its willingness to strike into the unknown, to push back frontiers, to take a risk, to indulge a sense of adventure.

If we give up that in a search for more security, we will be giving up a vital part of ourselves.

The Philadelphia Inquirer

Philadelphia, Pa., March 31, 1982

It came in, clear to the eye and the television lens, its main landing wheels touching down perfectly, with a barely visible jolt, on the white gypsum sands of New Mexico's Tularosa Basin. It leaned back a bit, holding its stall attitude for a graceful, sure moment, easing its speed and solidifying its marriage with earth, exactly as the tiniest, simplest of airplanes would in the hands of a confident pilot. The nose was gentled down by practiced hands, the forward landing gear touching the ground without impact.

It was a perfect landing. It was like thousands of other perfect landings every day on an earth swarming with flying machines. But it was different. It was the perfect ending of a flight that, for all its precedents in space travel and exploration, was a brilliant, moving, convincing testament to mankind's and America's capacity to excel.

The simplest details would fill pages. The technical data will fill hundreds of volumes. There were failures during the eight-day flight. One was a one-day, weather-caused delay in the landing. The landing site itself was in doubt, and crews were shifted and reshifted, never losing step. Communications systems broke down.

None of those failures interfered consequentially with the totality of the operation. In the end, all were welcomed parts of what may be the most painstaking and elaborate testing program of a machine, and man's capacity to manage it, in history.

Of course, it was an accomplishment to earn glory for the commander, Jack R. Lousma, and for the pilot, C. Gordon Fullerton. But thousands share that accomplishment, all members of an intricate and superbly coordinated team.

The expense of the space program is immense, and there are legitimate arguments about the practicality of the ultimate objectives of the shuttle experiment series, of which Columbia is the heart and soul. The fact that the same might be said of every one of the human race's most daring accomplishments will not still that controversy, nor should it. But taken away from the proper practicalities of mundane economics, there is — and yesterday dramatically was once again — something in the program that has no parallel, no competition in contemporary human endeavors.

That drama was a strong, positive one. It spoke confidently and resolutely of the ability of humans — and particularly of Americans — to bring intelligence, technology, courage and will to a challenge that was imaginable only as the wildest fiction a generation ago. It spoke of the capacity to prevail.

The Christian Science Monitor

Boston, Mass., March 23, 1982

Congratulations, Columbia! Making it back into orbit a third time after a near flawless countdown is another big step toward the goal of an operational Space Transportation System.

Astronauts Jack Lousma and Gordon Fullerton are to be commended for their skillful performance. They will understand, however, if the nation's admiration includes one Todd Nelson. As the first high-school science student to fly an experiment on the shuttle, he is opening a new era of space exploration — an era of wider opportunity for interested people to share in one of the great adventures of the age. The so-called "getaway specials" — shuttle canisters in which small experiments can be carried cheaply — extend this opportunity beyond the subsidized student participation program.

Certainly the shuttle concept brings important new dimensions to space flight. President Reagan is wise to maintain strong support for the shuttle program at a time of severe budget cutbacks. It is reasonable to trim other parts of the space budget to hold down overall costs. It is not reasonable to shut down operating spacecraft which still are returning valuable data just to save a few million dollars.

Provision in the fiscal 1983 budget to turn off the spacecraft now orbiting Venus and Pioneers 10 and 11, which are entering unexplored regions of the outer solar system, seems especially misguided. Many hundreds of millions of tax dollars have been spent putting those spacecraft in position. To throw away an important part of the return on that investment is questionable fiscal management, to say nothing of depriving the country and humanity of new scientific knowledge which otherwise would be gained.

Thus it is that enthusiasm for Columbia's third mission has to be tempered with misgivings about the overall thrust of the US administration's space policy.

The Bismarck Tribune

Bismarck, N. D., April 14, 1982

Among the many areas being considered for budget cuts by the Reagan administration is space exploration.

Certainly, if the federal budget is to be cut, the cuts have to come somewhere.

But some of the proposed cuts in the space exploration budget look as though they would be almost counterproductive, given the amount of money already invested in the programs and given that they still have the potential of yielding valuable information for relatively small cost.

For instance:

• A U.S. spacecraft is orbiting the planet Venus. The orbiter has already made radar maps of Venus and is capable of providing other data.

But because of proposed budget cuts, it might be shut off.

That seems a waste, because we've already spent the money to get the spacecraft there.

• Pioneer 10 and Pioneer 11 have passed some of the larger planets and are now in the outer reaches of the solar system. They have provided the first data ever from so far out in the solar system.

But because of proposed budget cuts, they might be shut down.

This also seems like a waste, given the amount of money already invested in the program.

• The Lunar Curatorial Facility, where moon rock samples are kept, is available to scientists.

But, because of proposed budget cuts, it might be closed.

This also seems like a waste.

Those are just some of the proposed space exploration cuts.

All told, the planetary and lunar research budgets may be cut by about half, and mission operations and data-handling may be cut by about two-thirds.

In the exploration programs, the amount proposed for cutting is about $40 million, after hundreds of millions have already been spent to get the craft to their current locations in space.

Expanding man's knowledge of his universe isn't something one can readily place a cost-benefit figure on. But if we could come up with such a figure, it would seem that the programs being eyed for cuts should stand to be continued, because most of the costs have already been met.

It would be a shame to see these efforts go by the wayside, especially when they've come this far.

The Times-Picayune
The States-Item

New Orleans, La., February 6, 1982

Much of the scientific muscle will be put back into the space program by President Reagan's proposed 1983 budget. It would be a relatively modest $6.6 billion well and wisely allocated.

The National Aeronautics and Space Administration's budget has been pared down to the point where private donations were offered last year to keep NASA from having to switch off the Viking I lander that has been transmitting data from the surface of Mars since 1976. The space shuttle has taken up most of NASA's recent budgets, and that program will continue to account for the major portion of spending. But the '83 budget will also fund planetary and deep space projects the Office of Management and Budget had marked for elimination.

The proposed budget includes $92.6 million to continue development of the Galileo satellite to orbit Jupiter ($300 million has already been spent on it), $21 million to keep alive U.S. participation in a sun probe with several European nations, money to maintain the deep space tracking network and to keep Voyager 2 alive for the first closeup look at Uranus and Neptune later in this decade.

Two projects did not make the cut: an orbiter for Venus and a probe to intercept and study Halley's Comet at its next pass in 1986. Choices are cruel, but since there is not enough money for everything, the decisions on what to fund and what to leave till later seem justified.

The immediate effect of this funding would be to save 1,200 jobs at the Jet Propulsion Laboratory in Pasadena, Calif., which directs deep space programs. The larger effect would be to preserve American leadership in space exploration and scientific and technological advances. Though it may be derided as blue-sky stuff by some who are rightly concerned about finding money to solve immediate problems on Earth, space science and engineering is a growth field in military and industrial applications that the nation will neglect at its real peril

Detroit Free Press

Detroit, Mich., February 8, 1982

THE U.S. SPACE program is suffering the torture of a thousand little cuts. The big budget decisions have already shut down planetary exploration: We've given up a once-in-a-lifetime chance to look at Halley's comet close up, abandoned plans to probe the atmosphere of Jupiter and to map the surface of Venus. Now the data from a decade or more of space probes is doomed to sit in storage untapped, for lack of the relatively modest funds to analyze it.

The space program has always had to defend itself against charges that the astronomical sums required for space exploration would be better spent on earth. But if we really can no longer afford interplanetary flight — a point the scientific community would argue fiercely — this would seem to be a good time to consolidate the knowledge gained from past missions.

Only 10 percent of the data from the last Viking flights has ever been analyzed, according to a recent issue of Science News; thousands of photos have never been touched by researchers. The data sent back by the early Mariner missions to Venus and Mercury in the 1970s has not been adequately studied. The funding for analysis of the Mars fly-bys has not kept up with inflation.

We're not getting the knowledge out of the numbers, as a staffer in the National Aeronautics and Space Administration puts it. Some NASA researchers, however, are being diverted to work on defense projects, where the money is.

Data analysis programs are relatively inexpensive compared to the cost of launching a space probe. But having paid for the fireworks, the government is uninterested in what they mean. Yet who knows what it would do for the economy or the creative spirit of a nation to decode those sounds and pictures and numbers from outer space?

There *are* things that it is good and wise for government to spend money on. There *are* investments that only government can make, because the cost is too great or the payback too distant for anyone else to undertake them. The space program is one such investment, though there are others. It is a curious irony that the planets are crowding our quadrant of the sky this month — just as we've decided we're no longer looking.

THE PLAIN DEALER

Cleveland, Ohio, January 29, 1982

Congress has much to ponder in President Reagan's challenging State of the Union address but in the meantime we urge it not to lose sight of the nation's need to preserve and finance NASA. Of course we have a selfish interest. We want to save the space agency's Lewis Center here. But we also see Lewis as an indispensible part of NASA and NASA, in turn, as a golden investment by the nation.

To recapitulate, NASA is threatened by Reagan fiscal economies and nothing we have heard recently suggests that the president has been sold on the agency's importance to this country. We are not only talking about flights to the stars. We are talking about solid work in the area of defense on the one hand and civilian projects on the other. It is good to keep the space shuttle. But what about research into fuel-efficient aircraft and automobile engines, wind turbine technology, reduction in airplane noise, communications by satellite and so on?

NASA's supporters in Congress, such as Sen. John H. Glenn of Ohio and Rep. Dan Glickman of Kansas, are fearful of what Reagan has in store for NASA and, by extension, Lewis in his 1983 budget, due to be unveiled Feb. 8. They assert that NASA does much important work with ultimate commercial benefits that private industry is either unwilling or financially unable to do. They are afraid that America is about to engage in "technological surrender" to competing nations, among them Japan and West Germany.

We know that in times of austerity a space-oriented agency is vulnerable when down on earth people are jobless and hungry and frightened. But we are not talking about a NASA that goes off only on poetic and expensive missions to the planets. We are talking about the one that ploughs its benefits into the earth and yields many treasures to the American people and to Mother Earth. Reagan wants to give many federal responsibilities back to the states but you can't turn the space shuttle over to Idaho. We think Reagan is aware of this but we expect Congress will have to take up the fight to remind him of NASA's other good qualities.

The Orlando Sentinel

Orlando, Fla., March 20, 1982

LIKE AN automobile without flashy trim, Columbia will have a plainer look Monday when Jack Lousma and Gordon Fullerton drive her off the Cape's used shuttle lot. But under the hood she's the same machine. For this third time out, the huge fuel tank to which the shuttle is attached is a dingy brown rather than the bright white of earlier appearances, saving NASA a $1,700 paint job and opening up another 600 pounds for the payload.

Taking the plain look for this trip is just one example of cost cutting by the space agency. It also has called it quits on some of its research programs, including deep-space probes, and delayed the start of others. It's only right that the space agency participate in economizing, but the truly beneficial parts of the program should not be thrown out.

There are those, of course, who question the repeated throwing of this 94-ton ship into an orbit 150 miles away. The answer is that it is just good business. Without the continuous honing of technology, the world's economy couldn't support increasing populations and there could be no enhancement of lifestyle.

That the space program has already given us communications satellites, non-stick frying pans and advances in medicine is pretty well known. But during this orbit, astronauts will conduct an experiment that could mean a new heart treatment drug, affordable and on the market three years from now. Another medical experiment holds hope of advances in fighting cancer.

These and dozens of other experiments are being paid for by companies that hope to capitalize on the improved technology. Indeed, shuttle management expects that the repeated trips into space will be paying for themselves within four years.

Parochially, of course, there are additional benefits. In Central Florida the space program has been one of the magnets that brought such job makers as Martin Marietta to Orlando, Harris to Melbourne and a small communications equipment builder to St. Cloud.

Even without a paint job and even though space shots are already becoming old hat, Columbia's liftoff next week will be awe inspiring and dramatic. But there's a lot more to that sudden burst of energy than show business. Once again, civilization will be moving out.

Technology Office Warns U.S. Space Lead in Danger

A report released June 14 by the Office of Technology Assessment warned that the U.S. was in danger of losing its lead in space exploration. It said the civilian space program needed clearer goals and more money or it would lose ground to Western Europe and Japan. According to OTA, the U.S. faced increasing competition from those countries in rocket-launching services, communications satellites and probes of the earth's surface from space. The congressional agency blamed Congress and the presidency for failing to set long-term policies for the U.S. space program. Space programs were usually designed during yearly budget negotiations and were inconsistent as a result, the agency noted. "In order to focus the U.S. civilian space program and to introduce more consistency into all U.S. space activities, the president or the Congress must set forth new goals. In the absence of such direction, the current drift will continue and worsen." The report suggested that the government adopt a policy of encouraging private industry to participate in space research. "A great part of the success of the European and Japanese programs," OTA observed, "results from their institutional arrangements within which private and public sectors can work well together." The report urged better coordination of military and civilian space projects and stressed the importance of deciding future space policy before the shuttle program ended.

The Times-Picayune The States-Item

New Orleans, La., June 17, 1982

In a major critique of the U.S. space program, the Congressional Office of Technology Assessment has warned that the United States must establish long-range goals, rally a coordinated effort and provide stable funding for its space effort or lose both technological and operational leadership to the aggressive Japanese and European space programs.

The OTA is not talking about exotic deep-space exploration or of U.S.-Soviet military and propaganda competition, but about commercial market dominance in Earth orbit, where the most immediate practical uses of space are to be made. The study considered four programs and finds the United States is being paced by the Japanese and the 10-nation European Space Agency in all of them.

In satellite communications, the most profitable of current space operations, the National Aeronautics and Space Administration has done much research on advanced technologies, but has not funded anything for development. Japan and the ESA are working on the same track, and the OTA says our relinquishing leadership would lead to a major loss in revenues for U.S. firms.

In earth resources sensing, the U.S. Landsat program does not look beyond the mid-'80s, which could leave another lucrative field to the French SPOT satellite and even to the Soviets.

In space transportation, the problems with space shuttle timetables and launch costs (a doubling of launch costs has just been announced) have led several customers — including U.S. firms — to switch their bookings to the European Ariadne rocket.

In materials processing — producing in space such high-value items as pharmaceuticals, electronics, chemicals and metal alloys — all space programs are conducting research, and no leader has emerged from the pack.

The OTA makes several recommendations, but its fundamental suggestion is to improve the coordination of the public program, civilian and military, with private participation. It is public-private cooperation, typically, that has given the Japanese space program its powerful momentum. Allied to that is the need for stable, adequate funding instead of the annual-appropriation cliffhanger.

"Given the likely constraints on the federal budget," the report says, "it will be important to decide in what areas the U.S. wishes to compete, because attempts to maintain a comprehensive program without additional capital and manpower may lead to second-best technology." That is something the United States cannot settle for.

The Orlando Sentinel

Orlando, Fla., June 26, 1982

When Columbia blasts off from the marshes at Cape Canaveral on its fourth and final test flight, the question will be, where to?

The shuttle is, of course, headed for California's Edwards Air Force Base — the long way: 112 Earth orbits in seven days. But the direction of the space program is not as clear: On one side is a winking military suitor and on the other an uncertain civilian future. Clearly, it is time for some decisions.

First, the question of NASA's relationship with the military must be resolved and that should be to turn aside the well-heeled soldier. NASA was created as a civilian, and it should remain independently so. Civilian uses of space science are far too important to our technological leadership and economy to let them lag.

And as far as the military in space goes, here again we have world leadership, and we should maintain that position. Such things as surveillance systems should be refined, but we seem to be headed for an age of Buck Rogers, and that bodes well for no one.

That aside, what NASA and the nation need is a defined space policy. The lack of such a policy, faltering financial support and rising foreign competition are threatening our leadership in civilian space technology. And to lose that would mean we lose the edge on an important part of tomorrow's commerce — jobs.

That is the gist of a study released by the Congressional Office of Technology Assessment. Advances by the 11-nation European Space Agency and the Japanese have been phenomenal. Both threaten to overtake us in advanced technologies.

The space program has only scratched the surface in what it can contribute to our lifestyle and economy. Unless we keep scratching, our foreign competitors will bury us. We'll not only be driving Toyotas but when we reach out and touch someone it might be via Ma Tokyo.

President Reagan will soon outline a new long-term space policy, perhaps at the Columbia landing. Whenever, the policy must look far into the future, to an inclusion of private enterprise as a full partner, assuming part of the risks and making quicker applications of technology.

Serious foreign competition in civilian space programs is new to America. But, as Rep. George E. Brown Jr. of California, a member of the House Space Committee, puts it, this is a reminder that American leadership in technology is "not a God-given right." That being true, there's only one new direction for NASA that's for the good: a new and even higher orbit.

AKRON BEACON JOURNAL

Akron, Ohio, June 21, 1982

A LACK OF direction and a poor sense of purpose has the U. S. space effort, well, lost in space.

That is the appraisal of the Congressional Office of Technological Assessment, which released a gloomy report last week on the state of American leadership in space technology. The report found the American effort drifting aimlessly just as it is on the verge of reaping huge dividends.

The matter goes deeper than mere prestige, which is important enough for those who remember the heady days when Americans walked the moonscape and returned with extraterrestrial souvenirs. In that one segment of our culture, the United States was second to none.

But there was more to the space effort than pride. Advances in medicine, computer guidance systems and other sophisticated technology made the space race a sound investment. Those advances laid the groundwork for important non-military uses of space, just now being launched.

The question that concerns the Office of Technology Assessment is: Will the United States take advantage of its head start? Unless policies change, it won't.

"Policies," according to the report, can be defined loosely. Policy becomes simply what happens year-to-year during federal budget negotiations; there is no "long-term perspective" of what our space goals should be.

For a while, that didn't matter. Now competition is arising from — you guessed it — Japan and Western Europe. Advanced nations with new technology soon will be cutting the United States out of such potentially important future markets as missile launching services, communications satellites and vehicles for map-making, weather study and other functions.

To get the U. S. space effort back on track, the study recommends closer cooperation between government and private industry in technological financing and development, and long-range planning to decide where funding should go after the space shuttle project. A high-level agency is needed to oversee the planning and coordinate the efforts.

The report found that Japanese and European space technology is moving forward because of "institutional arrangements within which private and public sectors can work well together." America progress, however, is hampered by a "lack of foresight and, especially, lack of coordination."

Those statements may apply to other industry-government relations. If so, perhaps a renewed space race could be a rallying point for greater unity, as it was in the 1960s.

One thing is certain — if America's industrial future is to be based on high technology, we cannot let an opportunity like commercial space markets slip through our fingers. Through lack of direction, we might be blowing our lead in the space race. But it is not too late to regroup and push forward.

Los Angeles Times

Los Angeles, Calif., June 18, 1982

The United States has learned to navigate in space with such precision that it can skip a satellite through Saturn's rings. Ironically, the agency that learned that trick has no sense of direction of its own at a time when Japan and Europe are challenging U.S. space supremacy.

That is the general conclusion of a new study by the Congressional Office of Technology Assessment in a report that is detailed, disturbing and persuasive.

"Unless the United States is prepared to commit more of its public and private resources to space than it now does," the report warns, "it will lose its pre-eminence in space applications during the 1980s."

The United States still has technical superiority, the report says. What it lacks is a long-range policy on which government and commercial interests agree for directing its engineering talents and a decision about which agency should manage and implement the policy.

The study concludes that the first step must be a broad and prolonged national debate on two questions: Does the United States really want to commit itself to staying ahead in space? Does it want to stay ahead in everything, or should it settle for carving out specialties and letting other space-oriented countries do the same?

The report notes that the National Aeronautics and Space Administration, which was created in 1961 to provide the research-and-development base for space exploration, was never intended to manage all of the systems it developed.

But neither was the job of managing systems assigned to any other agency. As a result, managing space operations has been assigned rather haphazardly among several public and private agencies, with no long-range policy.

One example of a need for a broader space policy involves communications satellites, the most profitable sector of space services today.

Europe and Japan are taking a growing share of the space-communications business. They also are deeply into research on the next generation of space-communications technology.

But the United States, without a comprehensive space policy, is still debating the size of its commitment to the new communications technology.

The report is another timely early warning from the future-oriented congressional study group. Congress and the White House should take the early warning seriously.

THE BLADE

Toledo, Ohio, June 19, 1982

DESPITE the pride that Americans have shared as a result f three successful space shuttle ights, there is no reason to assume hat our country will invariably reiain a leader in communications-atellite technology and other uses f space.

That is the gloomy message to be ound in a study conducted by the ederal Office of Technology Assessment. It reports that the Japaese and Europeans, in their recent pace efforts, have been gradually utting into our lead in technology. he OTA report states, "Their ineased activities threaten the loss f significant revenue opportunities or the U.S. as well as potential loss f prestige and influence."

Our first problem may be adjusting to the idea that the competition omes from Europe and Japan, not om our long-time foe, the Soviet nion. In general, the increasing hallenge from other nations that ie National Aeronautics and Space Administration must contend with ; tied in part to some of the problems that the shuttle's popularity may produce.

For example, the report states that a shortage of launch vehicles to accommodate some of the more elaborate projects may limit what can be achieved in the shuttle program. The United States may have to consider developing a new and costly launch vehicle in order to orbit communications satellites at affordable prices and thus stay in step with our competitors.

The most frustrating problem as always is the lack of money. At a time when reduced federal spending is trimming almost all domestic programs except defense, the space progam is no exception. One solution might be to make policy decisions that permit only a few key scientific and space efforts, as the OTA report suggests.

But perhaps too the United States must accept the economic realities and be thankful that at least our friends are going to be among the ones who will join us as deriving key benefits from development of space-related technology.

Appendix

America's Technology Lead: In Danger or Still Holding?

There was major concern over U.S. technology by 1978. In a National Academy of Sciences panel discussion that year, there was general agreement that the U.S. had abandoned the drive for dominance that it had begun in the 1950s with the "space race" against the U.S.S.R. As one participant declared: "America is backing away from technology." In the view of the scientists and researchers on the panel, there was a dangerous national tendency to neglect new technologies. This gloomy assessment, on the eve of the U.S. explosion into computer technology that gave California's "Silicon Valley" its name, seemed premature, but the concern has grown, not diminished, over time. (Even "Silicon Valley" is threatened by foreign competition.) According to a 1980 National Science Board report, the proportion of U.S. patents granted to American inventors dropped by 26% from 1971-78, while the proportion of foreign inventors receiving patents rose by 11%. U.S. inventors still obtained the largest share of American patents, but their share declined from 71.5% in 1971 to 62.3% in 1978. Of the foreign recipients, West Germany received the most, about 25%, while Japan ran a close second, with 20%. It was clear that the U.S. was weakening, but what were the implications?

DESERET NEWS

Salt Lake City, Utah, July 14, 1979

Is the decade of the '60s — when the United States first put a man on the moon — destined to be the most spectacular in U.S. scientific and engineering achievement?

Ten years ago this month, on July 21, 1969, Neil Armstrong took his first step on the moon with these words: "One small step for man, one giant leap for mankind."

That "giant leap" has not been sustained, at least in the critical area of scientific research and development. Spending for research and development, in fact, reached a high in 1965, in preparation for the moon landings. It has been declining ever since, and in fiscal 1980 will account for less than half the 12.6 percent of the federal budget that went for R&D in 1965.

At the same time, U.S. competitors like West Germany, Japan, and Russia have doubled their R&D spending and manpower. What that means in military leverage is displayed by the Russian buildup and the increasing accuracy of its ICBM missiles. One has only to point to Japanese and West German industrial and scientific advances to see the advantage of more R&D spending.

The American Chemical Society, for one, is becoming deeply concerned. It notes that private industry is increasingly unwilling or unable to invest in new and innovative ventures — partly as a result of increasing government regulation.

The federal government can do much to alleviate that situation. One way is to give tax breaks to those industries willing to venture capital on new ideas. Another is by cooperative arrangements with industry to finance projects like energy research that are too prohibitive for the private sector.

One thing is certain: Without adequate research and development, the U.S. position in the industrial world will steadily deteriorate. It's time to reverse that course.

DAYTON DAILY NEWS

Dayton, Ohio, July 23, 1979

The United State is losing its lead in science and technology, but the news is not altogether bleak. The United States also is vigorous in many areas and can improve its position.

Scientists worry that U.S. innovation has lost vigor, and there are some disturbing signs: Private sector spending for basic research has fallen 20 percent since the 1960s, foreigners are registering more U.S. patents than they used to and Americans are winning a smaller proportion of patents abroad. The U.S. space program, launched by President Kennedy, paid off handsomely but has receded as a priority.

But, as historian Daniel Kevles points out in the August issue of *Harper's*, part of America's "decline" is natural and only relative. The glory years of American science and technology, from the end of World War II through the 1960s, occurred in part because the European scientific community, so strong formerly, was crawling out of the rubble, as was Japan's.

It is natural that those industrialized societies, restored, would become competitive with the United States again. And the United States still leads in many scientific developments — microprocessing, for example. President Carter has kept as a priority federal funding for basic research and remained sensitive to its importance.

But that does not mean this country should become complacent. It must remain highly competitive, and can.

For one thing, the United States ought to quit giving away highly sophisticated innovations to countries that compete with us. The National Technical Information Service (NITS) opens to the public elaborate files on all innovative research developed with the help of federal grants.

The intent is good: to share among taxpayers the bounty of what taxpayers paid for. But when Japan, Russia and the Soviet Union snatch the information and exploit it, American taxpayers lose. A way should be sought to serve the nation's economic-development interests without lapsing to secrecy.

The United States, which has antitrust laws to keep small domestic enterprises from being trampled, also has to temper its antitrust provisions to allow American companies to compete worldwide — especially with nations such as Japan which uses its government to support and coordinate private conglomerates so they can undersell the world markets. To their credit, American officials are now thinking of developing government-industry cooperative technology centers.

One new and immediate key, then, is more internal cooperation — including, say, university sharing of new, elaborate and expensive research equipment — for external competition, while continuing to support diverse basic research. Certainly a country as innovative as this one can pull it off.

FORT WORTH STAR-TELEGRAM

Fort Worth, Texas, January 11, 1978

"If the United States can put a man on the moon..."

What a powerfully persuasive way to open a conversation. "If the United States can put a man on the moon, why can't it solve the energy problem...why can't it provide for Social Security... why can't it supply jobs for all who want them?

Part of the answer may be that the nation, after the considerable adventure into space, has been coasting downhill technologically.

Sen. Adlai Stevenson, D-Ill., said the nation's slipping commitment to research and development is costing it technological pre-eminence.

The results can be seen in foreign-made steel, television sets, automobiles and other high-technology products which pour into this country.

"Our balance-of-trade problems, declining productivity rates, unemployment and sluggish economic growth are some of the more obvious by-products of the slippage in U.S. industrial technology," Stevenson said.

The senator noted that the U.S. economy is based largely upon abundance of raw materials and on technology.

The era of cheap raw materials is gone, he said, so the key to development lies in improved technology. In building a better mousetrap, so to speak.

Foreign competition once came mainly from countries where labor was cheap, but this no longer is true, Stevenson said. Technologically-advanced nations now are beating the United States in its world markets.

One suggestion for dealing with the Organization of Petroleum Exporting Countries was that the United States should "maximize our non-energy leverage with these (OPEC) nations..."

To which Sheik Ahmed Zaki Yamani of Saudi Arabia scoffed, "They are out of their minds. Someone else will take their markets." The greater danger than a lessened commitment to technological superiority is the possibility that we may have lost our stomach for competition.

Like pigs at the trough, nations that are hungry will root for business. One way of doing this is to produce a better performing product, or learn to make it cheaper, or to improve its quality.

If the United States can put a man on the moon, we can do this, too.

If we want to.

The Philadelphia Inquirer

Philadelphia, Pa., November 7, 1980

Not so long ago, a report commissioned by the White House observed that the nation had lost its post-Sputnik commitment to science and that most Americans are headed "toward virtual scientific and technological illiteracy." That came on the heels of a report prepared for the National Academy of Sciences that claims the United States is in danger of losing its place as the leader in technology development.

If all that is true, then don't tell the people who frequent the U.S. Patent and Trademark Office. American ingenuity is in full flower, or so it may be argued by Vitale Catalano's recently patented "ketchup-rapping apparatus," guaranteed to extract that last little bit from the bottle, and by the inventor of a plastic container lid that can be folded into a spoon.

In the current edition of *Smithsonian*, editor Paul Trachtman reports on one recent month's business at the Patent Office and proclaims that all this hand-wringing about the end of American creativity is worthy of a place in the portable spittoon an unidentified inventor has perfected. Who but a bunch of clever Americans would come up with an "anti-tip-crossing" device for skis, or replacement tines for rakes?

Would a Frenchman or a Russian perfect a fast-food French fry suitable for microwave reheating? Certainly not one that fits this criteria: "Each potato being in the form of an elongated body having a generally rectangular cross section and first and second pairs of generally opposing side surfaces, each side surface having an alternative sequence of laterally extending, rounded hills and valleys of substantially uniform dimensions, each of said hills having a top and each of said valleys having a bottom, each hill of one side surface of said first pair opposing a corresponding valley of the other side surface of said first pair."

A New Jersey psychologist patented a device to cure those unfortunates who snore. An alarm awakens the offender each time a snore escapes and in the morning the snorer is provided with a night-time tally of his noisome affronts on the principle that the numbers — if not the interrupted dreams — will encourage an end of the practice.

A San Francisco inventor designed a "magnetically unlocked pet door" which operates on the same principle as automatic garage-door openers. A radio transmitting device is attached to Fido's collar and he has complete freedom to come and go as he pleases, according to Mr. Trachtman's report.

Is there a doubt that Americans are unceasingly motivated toward new and better technology? Consider the two Michigan men who obtained U.S. patent number 4,216,606. It is for a mousetrap.

The Seattle Times

Seattle, Wash., May 7, 1979

TO VALIDATE the familiar warning that America is in danger of losing its world technological leadership, the American Chemical Society recently pointed to these disturbing trends:

— The slight drop in the proportion of scientists and engineers in the U.S. population from 1965 to 1975, while the proportion roughly doubled in Russia and West Germany.

— The steady drops in research and development (R&D) as a percentage of the U.S. gross national product since 1964, while the R & D percentage has steadily increased since 1961 among this country's chief competitors.

— The decline in R & D expenditures in the federal budget from 12.6 per cent in 1965 to only 6 per cent in 1980.

— A drop in the number of industries willing to invest in new-venture R & D.

In 1977, U.S. industry spent only $17.5 billion on R & D, with 85 per cent of that concentrated in just six industries. It so happens that the U.S. balance of trade is favorable in those industries, such as aircraft, that are R & D intensive, while it is unfavorable in industries that are not.

Substantial and well-tested incentives for research are available in Canada, Japan and many West European countries. As a result, those countries' share of the international high-technology market is increasing at the expense of the U.S.

The need for this country to stir itself to change patent, tax and other laws to offer equal incentives ought to be obvious. The notion of the U.S. as a "technologically backward" nation is by no means so ludicrous as many Americans might suppose.

Seattle, Wash., August 24, 1981

THE public prints and airwaves seem crowded these days with strictures on how American industry must set about imitating the Japanese — the sooner the better — as the proven road to efficiency and productivity.

No doubt this is good advice as applied to certain sectors of the economy. Yet it might be instructive to view the international marketplace through the eyes of a Japanese expert.

Writing in Nikkan Kogyo Shimbun, Japan's equivalent of The Wall Street Journal, Ichio Takenaka, director of Tokyo's Research Institute on the National Economy, notes the loss of competitive advantage by some American industries. But he observes that "it would be a mistake to leap to the conclusion that U.S. economic strength is on the decline or that American industry is no longer competitive." He adds:

"The United States remains second to none in several fields. Space and aircraft companies and the related high-technology electronics firms are flagship industries. American agriculture is in the same class, and the education industry — the services provided by universities — is also outstanding."

"U.S. industry," Takenaka writes, "is restructuring itself toward specialization in complex technology, and the less sophisticated manufacturing industries are being left to Japan and other countries. This reshuffling explains the apparent decline of the U.S. auto and steel industries. But the United States still has a commanding lead in those manufacturing industries that require highly specialized know-how."

It is not always a discouraging thing to see ourselves as others see us.

Is Technology Worth the Fuss?

At the heart of the debate over America's technological superiority or inferiority is the assumption that technology is a positive force. Despite worry over technology's capacity for evil as well as good, Americans are optimistic over the ability of technology to improve human life. Americans respect inventiveness, and American ingenuity has done much to make daily life more comfortable than in the past. Not very long ago, however, there was a period when many people questioned the value of technological progress. In books such as Charles Reich's *The Greening of America,* authors argued that a simple life, uncomplicated by gadgets, was healthier. Rebels raged against technology for fostering an "unnatural" lifestyle. They denounced scientists for engaging in a headlong rush to invent for the sake of inventing, without considering the human cost. They blamed technology for polluting the earth and atmosphere, for speeding up the pace of life and finally, for threatening the world with extinction by means of the atomic bomb. Young people streamed into the country in search of a "natural" life. Despite the passions, technology was not banished. On the contrary, once the "rebellion" was over, computers proliferated in private homes. Technology undoubtedly has both bad and good aspects, but in the end, few of us really want to do without it.

Chicago Tribune

Chicago, Ill., June 25, 1980

Several recent incidents have raised questions about how well modern technology is serving us.

First, a computer in the North American Air Defense Command blipped out an erroneous warning of a Soviet missile attack. Announcement of the mistake led to some panicky speculation about what might have happened if the United States had sent its own missiles off in reply. The propagandists in the Kremlin made the most of this alleged picture of a trigger-happy America poised with a finger on The Button and an eye on a misbehaving computer.

Then, just the other day, the Federal Aviation Administration shut down its traffic control computer at O'Hare field because of a malfunction—and the announcement brought instant concern about the safety of flights arriving and departing from O'Hare.

Lastly, on Monday, a spectacular fire broke out on one floor of a modern "fireproof" office building in New York; and in reply to questions a spokesman for the New York Fire Department acknowledged that "there's no such thing as a fireproof building."

The tendency to question the infallibility of modern technology probably goes back to the nuclear incident at Three Mile Island. It seems to be based on the assumption that if one valve or one silicon chip misbehaves, our lives and our entire society are in danger.

Yet what impressed us about these recent incidents is not that things went wrong, but that the ill effects were so limited. The missile warning was recognized within seconds as a probable error, because other sophisticated devices failed to confirm any attack. Even in our automobiles, we don't assume that machinery is infallible; we have warning lights on the dashboard to monitor the behavior of unseen — and to many of us mysterious — gadgets under the hood.

The computer failure at the airport was handled by a readjustment of landing patterns so that safety could be assured by human controlers working with radar, as they had for years.

Certainly anybody who saw photographs of the fire in New York, with flames leaping up the exterior of the building, must be impressed that there was no "Towering Inferno," that the fire was confined to one floor. Can we expect much more when tenants in "fireproof" buildings furnish them with things that burn?

In short, modern technology is not so very different from people. The secret of dealing with it lies in knowing what its weaknesses are, in recognizing when it is misbehaving, and in having other systems to rely on when it does misbehave.

SAN JOSE NEWS

San Jose, Calif., October 29, 1980

FROM Galileo to Darwin and on down to the present day, the relationship between religion and science has been uneasy at best and destructive at worst. The traditional "enlightened" attitude has been that religion should leave science alone and confine itself to moral issues.

Yet science — especially medical science — has, inescapably, a moral dimension. This never has been more true than today, when technologies like organ transplants and genetic engineering seem about to give humanity the power to reshape and even create life.

Pope John Paul II forcefully underlined this important truth Monday in addressing a group of physicians at the Vatican.

Technological progress, the pope told his listeners, "suffers from a deep ambivalence: While on one side it allows man to take control of his own destiny, on the other hand it exposes him to temptations to go beyond the limits of a reasonable domination of nature, threatening the survival and integrity of the human being.

"We must consider . . . the implicit danger to the rights of man from discoveries in the field of artificial insemination, birth and fertility controls . . . of genetic engineering, of psychic drugs, of organ transplants, etc."

It's temptingly easy to dismiss this as the raving of a medieval zealot who'd like to burn science at the stake and command the sun to revolve around the earth. But that would be grossly unfair to both the man and his message.

The theme of the pope's statement is that science and morality cannot be conveniently put into separate compartments. The scientist cannot and must not ignore the moral dimensions of his research; the physician cannot and must not ignore the moral dimensions of the treatments he applies. Both must defend the right of every person not just to live, but "to live in a way worthy of a human being."

Implicit in this is the idea that a human being is more than an arrangement of cells, more than a collection of symptoms to be treated, more than a subject for experimentation; he or she is a complete entity with spiritual needs and rights.

"Science," the pope said, "is not the highest value to which all others are to be subordinated." Man does not exist to serve science; science exists to serve man — in all his manifold dimensions.

That is a truth that scientists, and the rest of us, sometimes forget. The pope's message was an eloquent and timely reminder.

DAYTON DAILY NEWS

Dayton, Ohio, September 2, 1980

It's 1900. Lots of jobs are tied to the manufacture of wagons and buggies. The car is coming on the scene. Workers and businesses want their future secured.

Strategy: Have the government keep cars out, and underwrite the buggy business.

That would have been foolish, of course. But that is the approach some would to take toward new trends now. One attitude is that technology only jeopardizes jobs; another that some of the old, hard-pressed industries — such as U.S. shoe and textile manufacturing — ought to be protected against imports.

Some technology does kill jobs, unfortunately, as Dayton knows from its experience when NCR shifted from mechanical cash registers to electronics. But new technology creates jobs, too — at NCR (where there are now more technical and professional jobs) and elsewhere. Within the last decade, hand-held electronic toys alone have boomed into a $400 million-a-year industry in this country.

High technology is providing jobs at a much faster rate than low-technology enterprises. Technology has improved the job conditions and rewards in some industries. One example: Coal mining, where workers once picked, shoveled and dynamited in the grimmest of conditions but now drive modern seam-eating machines.

Instead of trying to protect dying industries, this nation ought to be working hard to get out of jobs that South Koreans and Taiwanese can take more cheaply and stress jobs more likely to grow with technology and investment.

Many analysts warn that the United States is losing its technological edge at a time it needs the improvements to increase productivity. Increased productivity remains vital. *The Economist* recently noted the example of the phone industry. "In 1910, America's Bell system handled 6 million telephone calls with a staff of 120,000, which meant 50 a year each. Last year, it handled 185 billion calls with one million employees, which meant 185,000 a year each. To handle today's telephone traffic at 1910 manning and technology levels would require 40 times the entire labor force of the United States."

HOUSTON CHRONICLE

Houston, Texas, April 26, 1982

There is a tendency, in periods of rapid change, to fear technology and to regard its advance as a malevolent force that can be neither controlled nor restrained, and hence must be rejected outright. But technology is a tool that can and must be made to serve mankind. The energy and material needs of the world's 4 billion-plus people require it; their very survival depends on it.

Concern about the harm that swiftly changing technology can do to society and the environment is not completely without basis, of course. If not properly employed and controlled, new technology can pollute, injure, even kill. But this potential for harm is more than matched by technology's potential for good. In the future, technology will provide us with more energy and help us to use it more efficiently; recent advancements in the biological sciences hold seemingly unlimited possibilities for manufacturing useful products, combating disease and increasing the food supply for a multiplying, and hungry, population.

Along with these benefits come added responsibilities. In recent speeches at Rice University, retired Adm. Hyman G. Rickover, the "father" of the nuclear Navy, and distinguished physicist Edward Teller pointed out that advanced technology demands an increasingly educated work force to use it, and an increasingly informed and rational electorate to decide when and how it will be used.

Viewed as a tool, technology holds the only real chance for a better quality of life for all. If properly controlled, it offers bounties beyond limit; unreasonably feared, its challenges cannot be met, its rewards cannot be gained.

The Charlotte Observer

Charlotte, N. C., October 21, 1980

Americans are showing more interest in science these days than at any time since the space program put a man on the moon. According to a recent Time magazine article, a half dozen new science magazines are on the market; the New York Times now includes a weekly science section; two television networks are considering weekly science shows; and public television is running seven separate science series concurrently.

Sagan

Scientists finally are realizing that communication skills are as important as significant discoveries in winning public support.

Take Carl Sagan, for example, a Cornell University scientist and the creator, chief writer and host of "Cosmos," a 13-week PBS look at the universe and the earth's place in it. (Sundays, 8 p.m., Ch. 42). PBS says "Cosmos" may be attracting more viewers than any other series in PBS history and Sagan, though he sometimes oversimplifies to the point of being hokey, is largely responsible for its popularity.

The Charlotte Nature Museum has its own, more down-to-earth version of Sagan — astronomer Ray Shubinski. Each Wednesday night, Shubinski leads Central Piedmont Community College students in discussion after they've watched "Cosmos" replayed on a large screen.

Shubinski has a double master's degree — in astronomy and in communication. By relating the workings of the universe to everyday experiences, he makes astronomy seem like something everyone should know at least a little about.

The CPCC "Cosmos" course is so popular — 160 people have registered — that college officials are thinking of offering a sequel. Nationwide, about 10 million people watched each of the first two PBS segments.

Such enthusiasm may be related, in part, to our desperate desire for order in a rapidly changing world. It also may stem from people's "hunger to talk about something other than baseball and babies," in the words of a CPCC spokesman. Maybe it has been there all along, waiting for scientists who were willing and able to translate science into language that non-scientists can understand.

THE INDIANAPOLIS STAR

Indianapolis, Ind., September 22, 1980

The ingenuity of man continues to startle even in this age of commonplace technical miracles. Two examples came under discussion recently, and they involve applications of older ideas.

One is a new recording disc patented after development in the RCA Laboratories. Twelve inches wide, it can store 100 billion bits of information. That's billion. A laser burns the tiny message into a coating of tellerium, and another laser reads it out.

That brilliant device, which can duplicate the data from big X-ray films so that their silver can be recycled, is a descendant of Edison's dogged phonograph work of a century ago.

Meanwhile, engineered wood is on the way, the product of genetic manipulation and new production techniques. The U.S. Forest Service's researchers are cloning especially fast-growing trees, and five-year farming cycles are becoming a practical possibility.

An entire mirror image of the petrochemical industry is claimed to be possible with wood as the source, though costs would be large. Much of the bright promise in wood technology elaborates on the concepts involved in plywood, an invention which improved on lumber's strength, and from the crop genetics miracles with corn and other edibles.

To be only pessimistic about the future today is to have your eyes shut.

Index